Ecological Studies, Vol. 152

Analysis and Synthesis

Edited by

M.M. Caldwell, Logan, USA
G. Heldmaier, Marburg, Germany
O.L. Lange, Würzburg, Germany
H.A. Mooney, Stanford, USA
E.-D. Schulze, Jena, Germany
U. Sommer, Kiel, Germany
I.T. Baldwin, Jena, Germany

Ecological Studies

Volumes published since 1995 are listed at the end of this book.

Springer Science+ Business Media, LLC

F. Stuart Chapin III   Osvaldo E. Sala
Elisabeth Huber-Sannwald

Editors

Global Biodiversity in a Changing Environment

Scenarios for the 21st Century

With 35 Figures, 7 in Full Color

 Springer

F. Stuart Chapin III
Institute of Arctic Biology
University of Alaska
Fairbanks, AK 99775
USA

Osvaldo E. Sala
Department of Ecology and IFEVA
Faculty of Agronomy
University of Buenos Aires and CONICET
Buenos Aisres, 1417
Argentina

Elisabeth Huber-Sannwald
Departmento de Ingeniería Ambiental
  y Manejo de Recursos Naturales
Instituto Potosino de Investigaciones
  Científicas y Tecnológicas
San Luis Potosí, SLP
México

Cover illustration: A map of a scenario of the expected change in biodiversity for the year 2100. See Figure 15.3. Reprinted with permission from Sala, O.E., et al. 2000, Global biodiversity scenarios for the year 2100, Science 287:1770-1774. © 2000 American Association for the Advancement of Science.

Library of Congress Cataloging-in-Publication Data
Global biodiversity in a changing environment: scenarios for the 21st century/editors,
  F. Stuart Chapin III, Osvaldo E. Sala, Elisabeth Huber-Sannwald.
    p.   cm.—(Ecological studies; v. 152)
    ISBN 978-0-387-95249-9   ISBN 978-1-4613-0157-8 (eBook)
    DOI 10.1007/978-1-4613-0157-8
  1. Biological diversity.   2. Climatic changes—Environmental aspects.   I. Chapin, F. Stuart
  (Francis Stuart), III.   II. Sala, Osvaldo E.   III. Huber-Sannwald, Elisabeth.   IV. Series.
  QH541.15.B56 G59 2001
  333.95—dc21   2001020200

Printed on acid-free paper.

Production coordinated by Chernow Editorial Services, Inc., and managed by Tim Taylor; manufacturing supervised by Erica Bresler.
Typeset by Best-set Typesetter Ltd., Hong Kong.

9 8 7 6 5 4 3 2 1

ISSN 0070-8356
ISBN 978-0-387-95249-9

Preface

The scientific community has voiced two general concerns about the future of the Earth. Climatologists and oceanographers have focused on the changes in our physical environment—changes in the climate, the oceans, and the chemistry of the air that we breathe. These concerns led to major environmental treaties such as the Convention on Climate Change and the Montreal Protocol on Substances that Deplete the Ozone Layer. A second general concern addresses issues of conservation and the extinction of species. This, too, has been addressed internationally through the Convention on Biological Diversity; however, there is increasing evidence that these two broad concerns are intertwined and mutually dependent. Past changes in the biodiversity of the Earth have both responded to and caused changes in the Earth's environment. An assessment by the United Nations Environment Programme has documented the current status and large recent changes in the Earth's biodiversity.

The Intergovernmental Panel on Climate Change (IPCC) has compiled scenarios of changes in climate, atmospheric composition, and land use. These scenarios have had considerable impact on national and international policies aimed at reducing these global changes or mitigating their impacts on society. Despite the large magnitude and societal importance of past changes in biodiversity, however, there have been no comparable scenarios of how the biodiversity of the Earth may change in response to global changes in environment.

The purpose of this book is to develop future scenarios of biodiversity for the next century in 10 terrestrial biomes and in freshwater ecosystems based on global scenarios of changes of the environment and the understanding by ecological experts of the sensitivity of biomes to these global changes. The scenarios presented in this book are not intended to be predictions; however, we hope that they will provide a starting point for careful assessments that must be made at national and regional scales. These regional assessments would provide a basis for planning future policy and research.

The book is based on a workshop supported by the U.S. National Center for Ecological Analysis and Synthesis (NCEAS) and the InterAmerican Institute for Global Change Research in which experts on biodiversity change were assembled for the major terrestrial biomes of the world. This exercise stems from an activity of the Global Change and Terrestrial Ecosystems (GCTE) core project of the International Geosphere-Biosphere Programme (IGBP).

F. Stuart Chapin III
Osvaldo E. Sala
Elisabeth Huber-Sannwald

Contents

Contributors

Gina Adams

Natural Resource Ecology Laboratory, Colorado State University, Fort Collins, CO 80523, USA

Paul L. Angermeier

U.S. Geological Survey, Virginia Cooperative Fish and Wildlife Research Unit and Department of Fisheries and Wildlife Sciences, Virginia Tech, Blacksburg, VA 24061-0321, USA

Juan J. Armesto

Laboratoria Sistematica and Ecología Vegetal, Facultad de Ciencias, Universidad de Chile and Instituto de Investigaciones Ecológicas Chile, Ancud, Chile

Mary T. Kalin Arroyo

Departamento de Biología, Facultad de Ciencias, Universidad de Chile, Santiago, Chile

William J. Bond

Department of Botany, University of Cape Town, Rondebosch 7700, South Africa

Josep Canadell   CSIRO, Division of Wildlife and Ecology,
                                 Canberra, ACT 2601, Australia

J. Caspersen   Department of Ecology and Evolution-
                                 ary Biology, University of Connecticut,
                                 Storrs, CT 06269, USA

F. Stuart Chapin III   Institute of Arctic Biology, University
                                 of Alaska, Fairbanks, AK 99775, USA

Scott D. Cooper   Department of Ecology, Evolution,
                                 and Marine Biology, University of
                                 California, Santa Barbara, Santa
                                 Barbara, CA 93106, USA

Kjell Danell   Department of Animal Ecology,
                                 Swedish University of Agricultural
                                 Sciences, S-901 83 Umea, Sweden

Rodolfo Dirzo   Instituto de Ecología, UNAM, Mexico
                                 City 04510 D.F., Mexico

Kurt D. Fausch   Department of Fishery and Wildlife
                                 Biology, Colorado State University,
                                 Fort Collins, CO 80523-1858, USA

William A. Gould   Institute of Arctic Biology, University
                                 of Alaska, Fairbanks, AK 99775, USA

Alex Haxeltine   Climate Impacts Group, Plant Ecology,
                                 Department of Ecology, Ekologihuset,
                                 Lund University, 223 62 Lund, Sweden

Richard J. Hobbs   CSIRO, Division of Wildlife and
                                 Ecology, LMB 4, P.O. Midland,
                                 Western Australia 6056, Australia

Elisabeth Huber-Sannwald   Departmento de Ingeniería Ambiental
                                 y Manejo de Recursos Naturales, Insti-
                                 tuto Potosino de Investigaciones Cientí-
                                 ficas y Tecnológicas, San Luis Potosí,
                                 SLP México

Laura Foster Huenneke   Department of Biology, New Mexico
                                 State University, Las Cruces, NM
                                 88003, USA

P.S. Lake   Department of Ecology and Evolu-
                                 tionary Biology, Monash University,
                                 Clayton, Victoria 3168, Australia

Sandra Lavorel

Centre Ecologie Fonctionnelle Evolutive, CNRS UPR 9056, 34293 Montpellier, Cedex 5, France

Rik Leemans

Department of Global Environmental Assessment, National Institute of Public Health and the Environment (RIVM), 3720 BA Bilthoven, The Netherlands

David M. Lodge

Department of Biological Sciences, University of Notre Dame, Notre Dame, IN 46556-0369, USA

Leal A.K. Mertes

Department of Geography, University of California, Santa Barbara, Santa Barbara, CA 93106-4060, USA

Harold A. Mooney

Department of Biological Sciences, Stanford University, Stanford, CA 94305, USA

Ronald P. Neilson

USDA Forest Service, Forest Sciences Lab, Corvallis, OR 97331, USA

Mark W. Oswood

Department of Biology and Wildlife and Institute of Arctic Biology, University of Alaska, Fairbanks, AK 99775, USA

Andrew N. Parsons

Natural Resource Ecology Laboratory, Colorado State University, Fort Collins, CO 80523-1858, USA

N. LeRoy Poff

Department of Biology, Colorado State University, Fort Collins, CO 80523-1858, USA

Frank J. Rahel

Department of Zoology and Physiology, University of Wyoming, Laramie, WY 82071, USA

James Reynolds

Alaska Cooperative Fish and Wildlife Research Unit, University of Alaska, Fairbanks, AK 99775, USA

R. Rozzi

Department of Ecology and Evolutionary Biology, University of Connecticut, Storrs, CT 06269, USA

Osvaldo E. Sala   Department of Ecology and IFEVA,
                                   Faculty of Agronomy, University of
                                   Buenos Aires and CONICET, Buenos
                                   Aires, 1417, Argentina

Martin T. Sykes   Climate Impacts Group, Plant Ecology,
                                   Department of Ecology, Ekologihuset,
                                   Lund University, 223 62 Lund, Sweden

Brian Walker   CSIRO, Division of Wildlife and Ecol-
                                   ogy, Lyneham, ACT 2602, Australia

Marilyn D. Walker   Boreal Ecology Cooperative Research
                                   Unit, USDA Forest Service, University
                                   of Alaska, Fairbanks, AK 99775, USA

Diana H. Wall   Natural Resource Ecology Laboratory,
                                   Colorado State University, Fort Collins,
                                   CO 80523-1858, USA

Kirk O. Winemiller   Department of Wildlife and Fisheries
                                   Sciences, Texas A&M University,
                                   College Station, TX 77843-2258, USA

1. The Future of Biodiversity in a Changing World

F. Stuart Chapin III, Osvaldo E. Sala,
Elisabeth Huber-Sannwald, and Rik Leemans

The expansion of human populations and their increasing consumption rate and access to technology have led to two general environmental concerns (NRC 1994; Vitousek et al. 1997): (1) the increasing human impact on the earth's environment and ecosystems through changes in the carbon pools of the biosphere, element cycling, and climate, and (2) changes in the earth's biota and communities, including species introductions and extinctions, and the fragmentation of natural communities through changes in land use. These changes are occurring more rapidly than at any time in the last several million years. These two areas have received largely separate research efforts to date, the first by ecosystem ecologists and the second by conservation biologists and community ecologists. Individual organisms, however, gain carbon and nutrients from the environment, transfer plant tissues to higher trophic levels, and decompose plant litter, and these organisms function within a complex landscape mosaic. It would be surprising, therefore, if the traits of individuals and their abundances and diversity did not determine ecosystem and landscape traits and processes (e.g., the pool sizes and rates of energy and material flux at local or regional scales). Species also have substantial indirect effects on ecosystem processes through shading, thermal insulation, tissue-quality effects on decomposition, and the like. We therefore expect that global environmental change could have substantial *indirect* effects on ecosystem processes through its effects on the species composition and diversity of communities and landscapes.

1

Changes in biota of ecosystems result from habitat conversion and land-use change, which reduces genetic and species diversity, from environmental changes, which causes changes in competitive balance, and from the introduction of exotic species, which leads to a gradual homogenization of the global biota. In addition to the ethical, aesthetic, and economic concerns raised by this situation, these biotic changes will likely influence ecosystem processes sufficiently to alter the future state of the world's ecosystems and the services that they provide to humanity (Chapin et al. 1997). The current global extinction rate, which is 100- to 1000-fold greater than prehuman levels (Lawton and May 1995; Pimm et al. 1995), and the loss of local diversity due to management practices have the potential to affect ecosystem processes strongly at both local and global scales.

It is becoming increasingly clear that biotic change is not simply a consequence of global environmental change; rather, it is also an important *driver* of global change through massive increases in both species invasions and extinctions. Humans depend strongly on biological resources for food, construction materials, medicine, and energy. Humans currently directly or indirectly use most of the available resources of the biosphere (Vitousek et al. 1986). These resources are in principle renewable and, with proper management, can be used sustainably. Human use unfortunately often exceeds the renewal capacity, after which the biological resource base becomes degraded. Resource management thus ultimately defines the fate of biodiversity.

The recognition of the functional importance of biodiversity resulted in the wide adoption of the Convention on Biological Diversity at the UNCED Earth Summit in Rio de Janeiro in 1992. The main objective of this convention (Article 1) is to conserve biodiversity, the sustainable use of its components, and the fair and equitable sharing of the benefits arising out of its utilization. The convention encourages parties to develop national strategies for conservation and sustainable use, which are linked with other environmental and societal issues (Article 6), and to develop monitoring systems to identify and quantify the processes and activities that threaten biodiversity (Article 7). The convention focuses on protected areas as a strategy to conserve biodiversity (Article 8). These are areas where natural habitats and viable populations can be maintained and the influence of humans and exotic species is reduced. The convention also promotes the development of research, training, educational, and awareness programs (Articles 12 and 13). Finally, the convention urges the parties to develop the appropriate assessment capacity to evaluate processes and activities that potentially can have an adverse impact (Article 14). This article led to the Global Biodiversity Assessment (Heywood and Watson 1995), which presents and discusses many aspects of biodiversity, changes therein, and some of the causes of change. Some central questions arising from the Convention on Biological Diversity are:

1. How do humans influence biodiversity?
2. What are the underlying causes for these influences?

3. What are the socioeconomic and ecological consequences of changes in biodiversity?

The ultimate question is probably:

4. How do human-induced changes in biodiversity (and the species and ecosystems responses to these changes) affect the societal goods and services provided by biodiversity?

These questions remain largely unresolved. Most of the literature on global change and diversity focuses at the genetic and population levels or describes regional variation in species and community diversity (Leemans 1996). There has been surprisingly little attention in the global change literature to the patterns or causes of change in diversity or its ecosystem and societal consequences. Arising from the growing concern about loss of biodiversity, an international program (Diversitas) was launched to (1) monitor changes in biodiversity, (2) determine the origin and maintenance of biodiversity, and (3) determine the causes and consequences of these changes. The Scientific Committee on Problems of the Environment (SCOPE) launched a program in 1991 that synthesized the current scientific understanding of the relationship between ecological complexity and ecosystem functioning (Schulze and Mooney 1993; Mooney et al. 1996). This international synthesis provided evidence suggesting that biodiversity might affect both the day-to-day functioning of ecosystems and the resilience with which ecosystems respond to environmental change. It also provided a framework for predicting how future changes in biodiversity might influence ecosystem processes that are relevant to society.

The Intergovernmental Panel on Climate Change (IPCC) developed scenarios of changes in climate, atmospheric composition, and land use. These scenarios have had considerable impact on national and international policy and in guiding research. As described earlier, however, biological diversity is changing simultaneously with these other global changes. Because of the rapid changes in biodiversity, it seems equally critical to develop plausible scenarios of future change in biodiversity to guide researchers in understanding its ecological consequences and to guide policy makers in carrying out the agreements reached in the Global Biodiversity Convention.

The purpose of this book is to develop future scenarios of biodiversity for the twenty-first century in 10 terrestrial biomes and in freshwater ecosystems based on global scenarios of changes of the environment and the understanding by ecological experts of the sensitivity of biomes to these global changes. Chapters 2 and 3 describe the general patterns at the global scale of the drivers of biodiversity change. Chapters 4–14 describe, for each biome, patterns of biodiversity, the major threats, and the expected patterns of change. Finally, Chapter 15 synthesizes all of the biome information into a common framework and develops global biodiversity scenarios.

Our definition of *biodiversity* includes diversity at levels from genetic diversity within species-to-species diversity to landscape diversity. The long-term goal of these biodiversity scenarios is to describe the ecosystem consequences of its change in terms that are useful to ecologists, managers, and policy makers so this information can be used to guide policies influencing land development, reserve design, and so on. The information here in is intended as the first step toward the development of that framework. Future efforts must provide sufficient regional detail to be useful for development of management policies at national and regional scales.

The book is based on a workshop supported by the U.S. National Center for Ecological Analysis and Synthesis (NCEAS) and the InterAmerican Institute for Global Change Research in which experts on biodiversity change were assembled for the major terrestrial biomes of the world. This exercise stems from an activity of the Global Change and Terrestrial Ecosystems (GCTE) core project of the International Geosphere-Biosphere Programme (IGBP).

References

Chapin FS III, Walker BH, Hobbs RJ, Hooper DU, Lawton JH, Sala OE, et al. (1997) Biotic control over the functioning of ecosystems. Science 277:500–504.

Heywood VH, Watson RT (1995) *Global Biodiversity Assessment*. Cambridge University Press, Cambridge.

Lawton JH, May RM (eds) (1995) *Extinction Rates*. Oxford University Press, Oxford.

Leemans R (1996) Biodiversity and global change. In: Gaston KJ (ed) *Biodiversity: A Biology of Numbers and Difference*. Blackwell Scientific, London, pp. 367–387.

Mooney HA, Cushman JH, Medina E, Sala OE, Schulze ED (eds) (1996) *Functional Roles of Biodiversity: A Global Perspective*. John Wiley and Sons, Chichester.

NRC (1994) *The Role of Terrestrial Ecosystems in Global Change: A Plan for Action*. National Academy Press, Washington.

Pimm SL, Russell GJ, Gittleman JL, Brooks TM (1995) The future of biodiversity. Science 269:347–350.

Schulze ED, Mooney HA (1993) *Biodiversity and Ecosystem Function*. Springer Verlag, Berlin, Heidelberg, New York.

Vitousek PM, Ehrlich PR, Ehrlich AH, Matson PA (1986) Human appropriation of the products of photosynthesis. BioScience 36:368–373.

Vitousek PM, Mooney HA, Lubchenco J, Melillo JM (1997) Human domination of earth's ecosystems. Science 277:494–499.

2. Modeling the Response of Vegetation Distribution and Biodiversity to Climate Change

Martin T. Sykes and Alex Haxeltine

The natural ecosystems of the world may be divided into a small set of biomes, each characterized by the dominance of one or more functional types of plants. At regional-to-global scales climate exerts a dominant influence over the distribution of these plant functional types (Woodward 1987). Smaller-scale variations in distribution may be controlled by smaller-scale features of the environment (e.g., soils and topography). Specific climatic controls on the distribution of these dominant plant functional types may be categorized as ecophysiological constraints, resource availability, and competition mediated by the effects of climate. Ecophysiological constraints on individual plant functional types account for the gross qualitative features of biome distribution, and such constraints have been incorporated into a number of rule-based vegetation models (Woodward 1987; Neilson et al. 1992; Prentice et al. 1992).

Woodward (1987) showed how absolute minimum temperatures are critical in controlling the poleward spread of different physiognomic types of plants. For example, a minimum temperature (T_{min}) of 0°C defines the poleward limit for tropical broad-leaved evergreen tree species. A T_{min} of −10°C controls the poleward spread of many broad-leaved evergreen species. A T_{min} of −60°C defines most continental climates in which boreal evergreen conifer species are found. The equatorward spread of more frost resistant species, however (e.g., cold-deciduous broad-leaved species), is not obviously explained in terms of extreme temperatures; rather, it must be understood in terms of the competitive relationships between species. Ecophysiological responses to extreme levels of drought stress exert similar controls on vegetation distribution (Larcher 1983), but it has proved far more difficult to generalize such controls to regional scales.

Mean growing-season temperatures and mean water-stress conditions also control vegetation distribution by affecting vegetation height, foliage cover, and net primary production (NPP), and by affecting the relative competitive performance of different plant-functional types in terms of these variables. Foliage cover and NPP are constrained by resource availability (i.e., water, nutrients, CO_2, and light). These constraints act differentially on different plant types, affecting the outcome of competition among these plant types. For example, in tropical climates the length of the dry season affects the competitive balance between evergreen and drought-deciduous woody plants. In savannas, rainfall seasonality and soil texture affect the competitive balance between woody plants and grasses. In temperate climates, the length of the summer growing season affects the competitive balance between evergreen and cold-deciduous trees. In grasslands, growing season temperatures and ambient CO_2 levels affect the competitive balance between C_4 and C_3 grasses. Such mechanisms, which describe the climate-mediated resource availability and competition constraints on the distribution of vegetation, have been incorporated to varying degrees into a number of regional and global scale vegetation models (Neilson 1995; Woodward et al. 1995; Haxeltine and Prentice 1996). These models have been used to study the potential impact of climate change on vegetation distribution at the level of plant functional types (Melillo et al. 1996).

Such models help to explain the broad-scale distributions of different functional types of plants; however, they do not directly provide information about how species richness might be related to climate. Correlations have been found between terrestrial biodiversity and measures of temperature and precipitation, but the exact relationships are often highly specific to the taxon and ecosystem concerned (Richerson and Lumm 1980; Turner et al. 1988; Wright et al. 1993; Hawksworth and Kalin-Arroyo 1995). In general, however, conditions that favor biological production (e.g., warm temperatures and abundant precipitation) are often associated with high diversity (Richerson and Lumm 1980; Currie and Paquin 1987; Turner et al. 1988; Currie 1991; Rosenzweig and Abramsky 1993; Wright et al. 1993). Furthermore, peaks in diversity have often been found at intermediate levels of biological productivity (Al Mufti et al. 1977; Rosenzweig and Abramsky 1993; Wright et al. 1993), leading to the idea that habitats with very low resource availability have lower diversity due to the effects of environmental stress, whereas habitats with high resource availability have lower diversity due to the effects of competitive exclusion or reduced environmental heterogeneity (Rosenzweig and Abramsky 1993; Tilman and Pacala 1993). In addition to the mean climate, the variability of the climate may also have a large effect on biodiversity. In conclusion, despite the many correlations found between climate and biodiversity, generalizations cannot be made in the same way that they can be for the relationship between climate and the distributions of different plant functional types. For this reason, there have been very few modeling studies of the direct impact of climate change on biodiversity.

There is every reason to expect, however, that climate change will have significant effects on biodiversity (Mooney et al. 1995). Areas favorable to the survival of individual species will move as climate changes. In combination with habitat fragmentation due to land-use changes (Ojima et al. 1994), this will potentially have a large impact on biodiversity. It is thus important to gain an understanding of how climate change and land-use change interactively affect biodiversity.

Current Vegetation Models

A number of modeling methodologies have been developed in recent years that simulate vegetation from the local to global scales. In local-scale models, such as gap models, vegetation is generally modeled at the level of individual species, whereas in the global-scale models, such as global biogeography models, vegetation is modeled at the level of plant functional types (PFTs). A number of different types of models exist to simulate different aspects of vegetation such as vegetation distribution or vegetation functioning (biogeochemistry); however, there is currently a trend toward constructing more process-based models that simulate both biogeography and biogeochemistry within a single modeling framework. Models have proven to be reasonably successful at simulating the distribution and functioning of vegetation, but the modeling of biodiversity is less well developed, particularly at the global scale.

Vegetation can be modeled at the individual species level, but only for a limited number of species, and usually at scales smaller than the global. At the regional to subcontinental scale, methodologies exist to predict distributions of common species. For example, STASH, which is a bioclimatic equilibrium model, uses the temperature of the coldest month, growing degree days, chilling response, and response to drought to predict species present and possible future distributions in Europe (Sykes, Prentice, and Cramer 1996; Sykes 1997) [e.g., for *Picea abies* (Fig. 2.1)]. Other approaches include Nix and Switzer (1991), who defined bioclimatic envelopes for vertebrates in Australia, and Huntley et al. (1995), who used climate response surfaces (Bartlein, Prentice, and Webb 1986) to predict present and future ranges of some European plant species.

At the local or landscape scale the dynamics of species growth and interactions can be simulated in detail. The growth and interactions of different species (usually tree species) are typically simulated over time. One of the more successful approaches to this has been through the use of forest gap models, many originating from JABOWA (Botkin, Janak, and Wallis 1972).

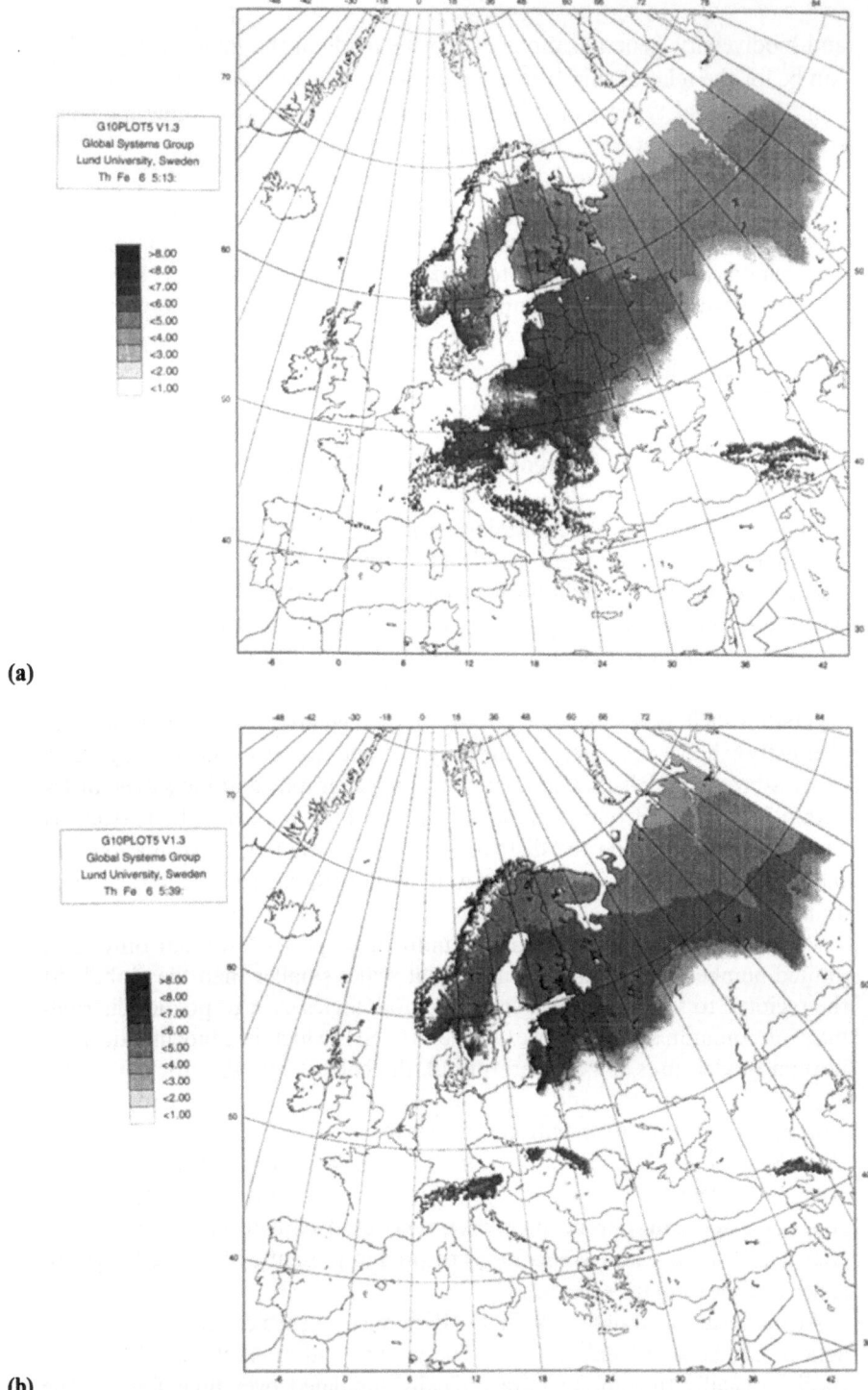

(a)

(b)

Figure 2.1. Simulated European distributions of *Picea abies* (L.) Karsten using the STASH bioclimatic model. Using (a) present climate (b) a future $2 \times CO_2$ climate (Hamburg ECHAM3 climate model).

Gap models have been developed to simulate gap-phase dynamics on small patches in a wide range of forests from tropical to boreal (e.g., Kienast 1987; Urban 1990; Prentice, Sykes, and Cramer 1993; Desanker and Prentice 1994; Bugmann and Solomon 1995). Within models of this type individual trees establish in forest gaps, grow, compete with neighbors, and eventually die. The idea is based on the theory of gap phases (Watt 1947) and associated autogenic successional dynamics. Varying degrees of detail are required depending on the model, although species-specific values such as height–diameter relationships, maximum growth rate, and climate responses are commonly needed (Bugmann et al. 1996). An example of this approach is FORSKA, an advanced, nonconventional (i.e., not originating from JABOWA) model (Prentice and Leemans 1990; Leemans 1991; Prentice, Sykes, and Cramer 1993). A later version, FORSKA 2 (Prentice, Sykes, and Cramer 1993) was developed to simulate forest landscape dynamics. FORSKA 2 is centred around a gap model in which competition between species for light and nutrients are handled within a patch typically of size 0.1 ha. Climate data (i.e., mean monthly temperature, precipitation, and sunshine hours) are used to calculate the environmental scalars that modulate establishment and growth. The differing effects of climate on different species are simulated through species-specific functions that convert the environmental scalars into multipliers in the process equations within the model. The model simulates a forested landscape using an array of replicate patches. Over the landscape a disturbance regime representing logging or fire operates. A Weibull probability distribution of disturbance (Clark 1989) is used that is usually set for an increasing likelihood of disturbance with the age of the patch. The model normally runs for hundreds of years and outputs species diversity, biomass, productivity, leaf area, and density either yearly or at a selected interval. FORSKA 2 has been used extensively to model the effects of climate change on forests (e.g., Prentice, Sykes, and Cramer 1991; Price et al. 1995; Sykes and Prentice 1995, 1996) (Fig. 2.2).

It is not possible at the global scale to model vegetation at the species level comprehensively; instead, potential natural vegetation is modeled using a small number of plant functional types (PFTs). One class of global model is the biogeography model (e.g., Woodward 1987; Neilson et al. 1992; Prentice et al. 1992; Neilson 1995) that simulates the natural distribution of PFTs as a function of climate and soils. One of the original and now well-distributed models of this type (BIOME 1; Prentice et al. 1992; Prentice et al. 1993; Prentice and Sykes 1995) uses mean temperature of the coldest month, number of growing degree days (as an indicator of growing season warmth), and a drought index to make global predictions of the equilibrium distribution of 17 biomes (formed from different combinations of a set of 14 PFTs). Another class of global model is the biogeochemistry model (e.g., Running and Coughlan 1988; Melillo et al. 1993; Parton et al. 1993; Woodward et al. 1995) that simulates carbon and water cycle fluxes through an ecosystem and predicts variables such as NPP, which is the net amount of carbon captured through photosynthesis; however, these models traditionally, use a prescribed

Boa Berg Halland Region

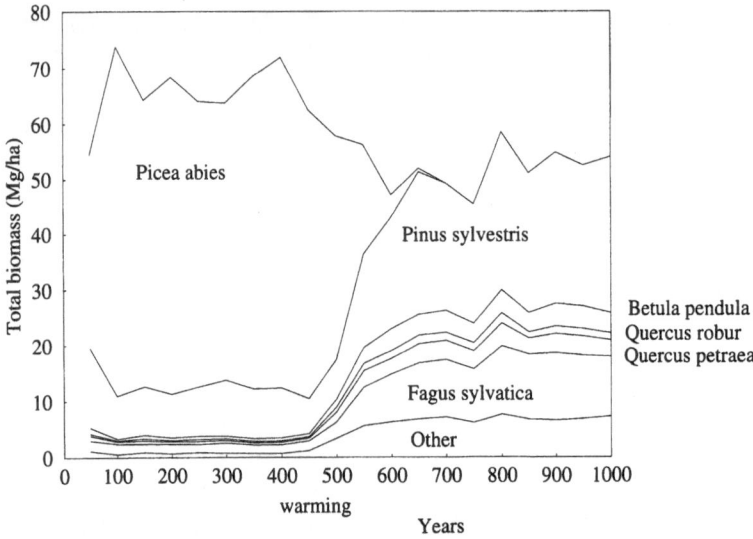

Figure 2.2. A 1000-year simulation of forest dynamics using the FORSKA 2 model at Boa Berg Halland, Sweden. The model was run for the first 400 years under the present climate, a climate warming was linearly imposed between year 400 and 500. The model was then run under the new $2 \times CO_2$ climate (Hamburg ECHAM3 climate model) for the remaining 500 years.

distribution of vegetation types obtained, for example, from data sets on the mapped distribution of natural vegetation.

A synthesis of the biogeographic and biogeochemistry modelling approaches has been achieved within the BIOME3 framework of Haxeltine and Prentice (1996), which couples vegetation distribution directly to vegetation biogeochemistry. BIOME3 uses a coupled carbon and water flux model to calculate the maximum sustainable leaf area index (LAI) and NPP for each potential PFT in a particular global grid cell. It includes seven PFTs: five woody and two grass. Potential woody PFTs are selected on their responses to absolute minimum temperature (Prentice et al. 1992; Haxeltine and Prentice 1996). Rooting depth and phenology affect a PFT's performance. Rooting depth and soil hydrology are modeled using a two-layer bucket approach (Neilson 1995; Haxeltine et al. 1996). Grasses only extract water from the upper layer, whereas trees extract from both, but mostly from the lower. Phenology is simply modeled using a ramp approach, where for summergreen PFTs leaf growth starts after a specified growing degree day threshold is past. Raingreen PFTs lose their leaves when soil water falls below 20% of available water-holding capacity in the rooting zone. Competition among PFTs is modeled using NPP as an index of competitiveness for each type. The dominant PFT is the one with the highest NPP. Output is usually classified into potential (not influenced by land use) biomes for each grid cell using dominant and secondary PFTs (Fig. 2.3; see color insert).

A number of global models of this type have been used in intercomparison projects. For example, a prototype version of BIOME 3 (Biome 2 Haxeltine, Prentice, and Creswell 1996), the Dynamic Global Phytogeography model (DOLY, Woodward and Smith 1994, Woodward, Smith, and Emanuel 1995), and the Mapped Atmosphere-Plant Soil System (MAPSS, Neilson 1995) in conjunction with three biogeochemistry models simulated the effect of climate change and increasing CO_2 on the vegetation of the conterminous United States (VEMAP members 1995). Results showed that all three models predicted potential natural vegetation reasonably well.

Global models of this type are equilibrium models (i.e., they describe vegetation or biome patterns that are static and not developed by dynamic processes). Vegetation clearly does change in response to a number of factors, particularly climate. Dynamic vegetation global models (DGVMs) have recently been developed (e.g., the Lund-Potsdam-Jena, LPJ-Devor, Sitch 2000, Sitch et al., in preparation). It uses carbon and water flux components from Biome 3 to predict distributions of seven woody PFTs and two grass PFTs (C_3 and C_4). It uses a yearly timestep, and at the end of each year an accounting is done of all flows in and out of various pools (e.g., heart wood, living wood, litter, etc.). Disturbances such as fire are simulated.

Developments include a first attempt at a global comparison of available DGVMs undertaken at Potsdam in the autumn of 1996 (Cramer et al. 2001). For this comparison, climatic input data fields were created using output from a transient simulation from the Hadley Centre HadCM2 coupled ocean–atmosphere general circulation model. Since this experiment started under pre-industrial conditions of radiative forcing, used realistic (i.e., observed) historical changes in forcing of both CO_2 and sulphur dioxide, and a standard scenario for the future developments until the end of the twenty-first century, the resulting data were considered an adequate place-holder for a realistic historical climatology, including the year-to-year variability that must be considered essential for assessments of vegetation change.

The comparison focused first on the biogeochemical fluxes which were found to be adequately simulated by most of the models. As a second step, the models' capacity to analyze changes in vegetation structure was compared by inspection of global time slice maps and the temporal developments at arbitrarily selected points. This comparison showed that the models, at least qualitatively, are now capable of yielding results similar to patch dynamics models, but with global coverage.

Foley et al. (1996) have further simplified the approach to ecosystem dynamics and biophysical processes through an integrated modeling framework, the Integrated BIosphere Simulator (IBIS). This has been incorporated directly into atmospheric general circulation models (agcms) (Foley et al. 1998).

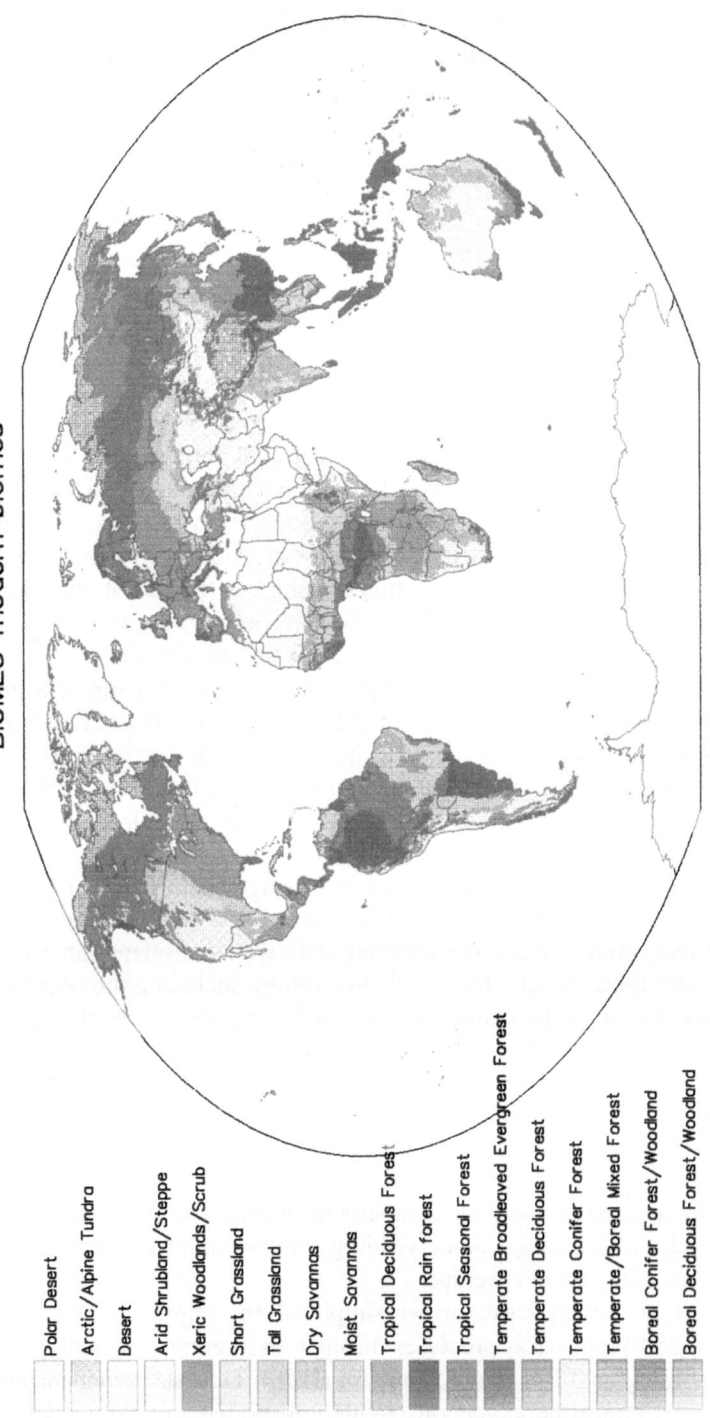

Figure 2.3. Global biome distribution under present day climate as simulated by BIOME3 (see color insert).

Attempts at modeling global biodiversity, however, have been much fewer. One of the few is the model of vegetation diversity described by Woodward and Rochefort (1991). They used Woodward's (1987) basic rules, based on experimentally described mechanisms rather than from correlations, for predicting vegetation distribution using climate, but modified to predict diversity. A value for diversity (maximum value = 100) was calculated using a multiple of three measures of diversity (i.e., diversity determined by minimum temperatures, heat sum, or growing degree days and water balance). For example diversity D(mt) influenced by changes in minimum temperature (mt) is calculated based on the observations that there is no temperature limit on survival if the minimum temperature exceeds 10°C and that approximately 9% of families can survive a minimum temperature of −90°C (Woodward and Rochefort 1991).
So

$$D(mt) = 90.9 + (0.909 * \text{minimum temperature}) \text{ where}$$
$$D(mt) = 9 \text{ when the minimum temperature is } -90° \text{ and}$$
$$D(mt) = 100 \text{ when the minimum temperature is } 10°C.$$

This combined diversity index was compared with global distributions of Angiosperms and Gymnosperms families [absolute number (313) normalized to 100] as defined by Heywood (1979) and Hora (1981). Both observed [Heywood (1979) and Hora (1981)] and model predictions had problems in some areas, so a further comparison was done comparing both with family diversity [absolute number (392) normalized to 100] as defined by 58 regional floras. Both were significantly correlated with the diversity as expressed by the regional floras; however, according to Woodward and Rochefort (1991) observed estimates from Heywood (1979) and Hora (1981) underestimated diversity in high-diversity regions, whereas predictions based on experimentally defined environmental limits overestimated diversity in these same regions. Woodward and Rochefort (1991) went on to use the model to predict the response of family diversity to environmental change. Increasing both temperature and precipitation led to increased diversity in moist cool climates (e.g., Andes, Europe, China, and Japan). Little change occurred in about half of the floristic regions of the world; in seven regions diversity decreased (e.g., Atlantic North America, Mediterranean, E. African steppe, and India). If precipitation was reduced, however, diversity decreased in a further 13 floristic regions, although the effect of doubled CO_2 in reducing canopy transpiration mitigated the effect of decreasing precipitation on diversity.

Climate and Models

Vegetation models of any scale or complexity need some form of climate data to drive them. These data are used to calculate environmental factors that directly influence vegetation at the PFT or species level. For example, a certain amount of warmth in the growing season is required by most plants to complete their life cycle. This is usually expressed as the number of growing degree days (GDD) above a certain base temperature (e.g., 5°C in temperate climates or 0°C in the arctic). These calculations are done directly from the monthly mean temperatures that are interpolated to pseudodaily values (Prentice et al. 1992). The level of moisture in the soil is likewise important in controling species composition in communities. Many models work with some variation of the Priestly-Taylor coefficient and the ratio between actual evopotranspiration (AET) and potential evapotranspiration (PET) (Prentice et al. 1993).

Present-day global or regional climate datasets of the form required by the models are interpolated from 30-year means of climate station data (Leemans and Cramer 1991, W. Cramer, personal communication 1996). Interpolation at the global scale is usually to a 0.5 latitude–longitude grid or to a 10′ or finer grid for more local scale simulations.

One of the main uses of vegetation models is to simulate past or future vegetation using modeled climate data as input to produce the required environmental factors. Such predictive climate data are obtained using general circulation models (gcms) (see Harrison 1990 for a background introduction to gcms). Vegetation models use present-day mean monthly temperature, precipitation, and sunshine values for each grid cell to which the anomalies (differences) between a present and a doubled greenhouse gas climate (for example) as modeled by a gcm are added.

An atmospheric gcm ECHAM 3 run at T106-L19 resolution (Bengtsson, Botzet, and Esch 1995, 1996) uses proscribed sea–surface temperatures from a coupled transient (10 years) ocean–atmosphere climate model to produce current and doubled greenhouse gas climate data. The temperature anomalies (differences between modeled current and future climates) for January as produced by this model shows a clear increase in temperature in the center of continents in northern latitudes and a cooling in South America (Fig. 2.4; see color insert). July anomalies again point to a general warming in North America and a southward shift in the main area of warming in central Asia. North western Europe is predicted to become colder in summer. Precipitation is also predicted to decline in northwest Europe and increase in South America and southern and western Africa.

Using these anomalies and the BIOME 3 model we can predict the global potential vegetation under a doubled greenhouse gas scenario including the direct effect of increased CO_2 on photosynthesis, stomatal conductance, and leaf area (Fig. 2.5; see color insert).

Methodologies for Predicting the Impact of Climate Change on Biodiversity

The process-based biogeography and biogeochemistry models certainly contain information relevant to assessing the impact of climate change on biodiversity; however, these models all predict vegetation distribution at the PFT level. Furthermore, global models are unlikely to be implemented at the species level at any time in the near future. Regional scale models such as STASH are able to make predictions at the species level, but only for a limited number of species. The challenge, therefore, is to relate simulations carried out at the PFT level to biodiversity. The relationship between PFTs and individual plant species is not trivial. A PFT represents a grouping of species that all have the same functional characteristics in terms of physiology, growth morphology, phenology, and so on. The group of species represented by a certain PFT will vary with floristic region. An individual species may also be represented by more than one PFT (e.g., for species that can adopt different growth morphologies). Despite these complications, the most straightforward approach is to describe the key species represented by each PFT in different floristic regions. Predicted changes in the distribution of certain PFTs under a changed climate may then be directly related to the distributions of individual species. The terrestrial biosphere models (e.g., BIOME3) make predictions about the structure and functioning of vegetation in addition to predicting which PFTs are present. For example, BIOME3 predicts NPP, foliage cover, vegetation height, and the spatial coverage of each PFT. Species richness might thus be empirically related to modeled variables such as NPP.

Wright et al. (1993) reviewed the current state of understanding of the relationship between energy supply and species richness. They showed that a large number of studies have found strong correlations between species richness and productivity or surrogates for productivity, and, furthermore, that such correlations appear to apply especially to vegetation in the major terrestrial ecosystems. The relationship between productivity and species richness is scale dependent (Wright et al. 1993). At the global scale species richness appears to increase monotonically with productivity (Adams and Woodward 1989; Currie 1991; Wright et al. 1993), whereas at smaller scales species richness is often a peaked function of energy (Tilman 1982; Rosenzweig and Abramsky 1993; Tilman and Pacala 1993; Wright et al. 1993). The relationship may also depend on the taxonomic scale being considered (Wright et al. 1993). In view of this scale dependence and the fact that the relationship is sometimes a peaked function of energy, it is of course vital that any productivity–richness relationship is only used for the set of species for which it has been developed. Experimentally derived relationships between productivity and species richness could be applied to modeled NPP values to produce estimates of species richness. On the other hand, spatially explicit databases on biodiversity (e.g., Sisk et al. 1994) might be used to relate species richness empirically to simulated vegetation parameters (e.g., NPP). In either case, the idea would be to derive a relationship between production and biodiversity for the current climate. Model simulations with possible future climates could then be used to gain a qualitative understanding of how climate change might impact upon species richness.

Such empirical methods can also provide a way of estimating how the diversity of birds and other animals might be affected by climate

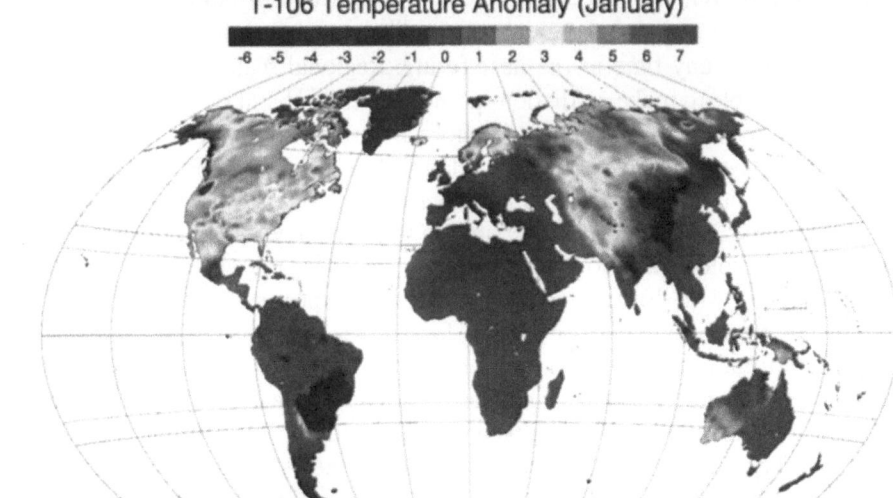

Figure 2.4. Global anomalies (differences between simulated current and $2 \times CO_2$ climates) using the Hamburg ECHAM3 atmospheric general circulation model: (A) Temperature anomaly (January). (B) Temperature anomaly (July). (C) Precipitation anomaly (cm/yr) (see color insert).

T-106 Precipitation Anomaly (cm/yr)

-70 -60 -50 -40 -30 -20 -10   0   10 20 30 40 50 60

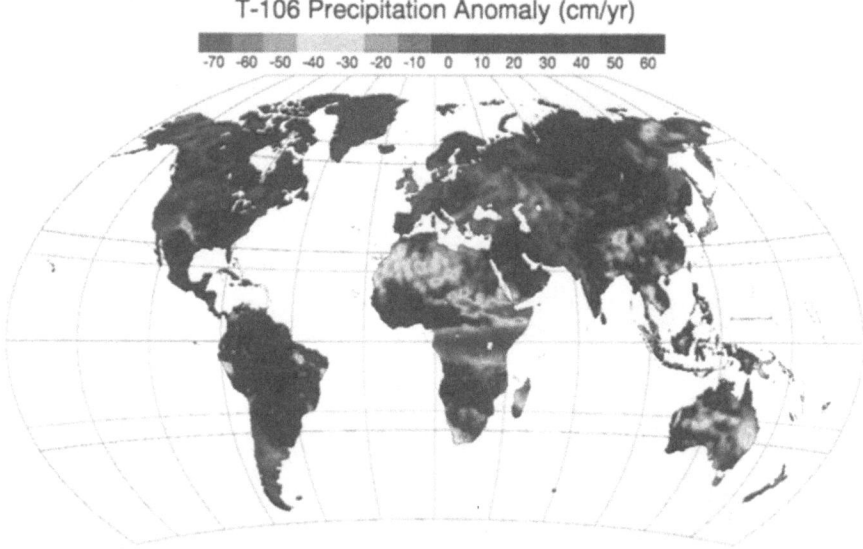

(C)

Figure 2.4. *Continued*

change–induced vegetation changes. For example, bird species richness has been related to vegetation height diversity (Cody 1975), which suggests that predicted changes in vegetation structure could be related to expected changes in bird populations. Working in North American deserts, Pianka (1967) also found relationships between lizard species richness and plant structural diversity.

An eventual aim of the global-scale dynamical vegetation models currently being developed will be to simulate accurately the rates of migration that might be expected under a changing climate (Pitelka et al. 1997). In some scenarios the equilibrium response to a changed climate involves the influx of new species with little or no loss in the species already present, leading to an increase in species richness. In other scenarios the equilibrium response involves both an influx of new species and a loss of the species already present. As climate changes, the delay associated with the time taken for species to migrate to new sites may lead to transient decreases in species richness if changes that act to reduce species richness (e.g., forest dieback) occur more rapidly than changes that act to increase species richness (e.g., migration of new species) . When combined with the effects of habitat fragmentation due to changing land use, such transient effects may lead to further losses in biodiversity, irrespective of whether the equilibrium response was an increase or decrease in species richness. Global-scale dynamical vegetation models may be used for assessing the importance of such effects. However these models will still make predictions at the PFT rather than species level.

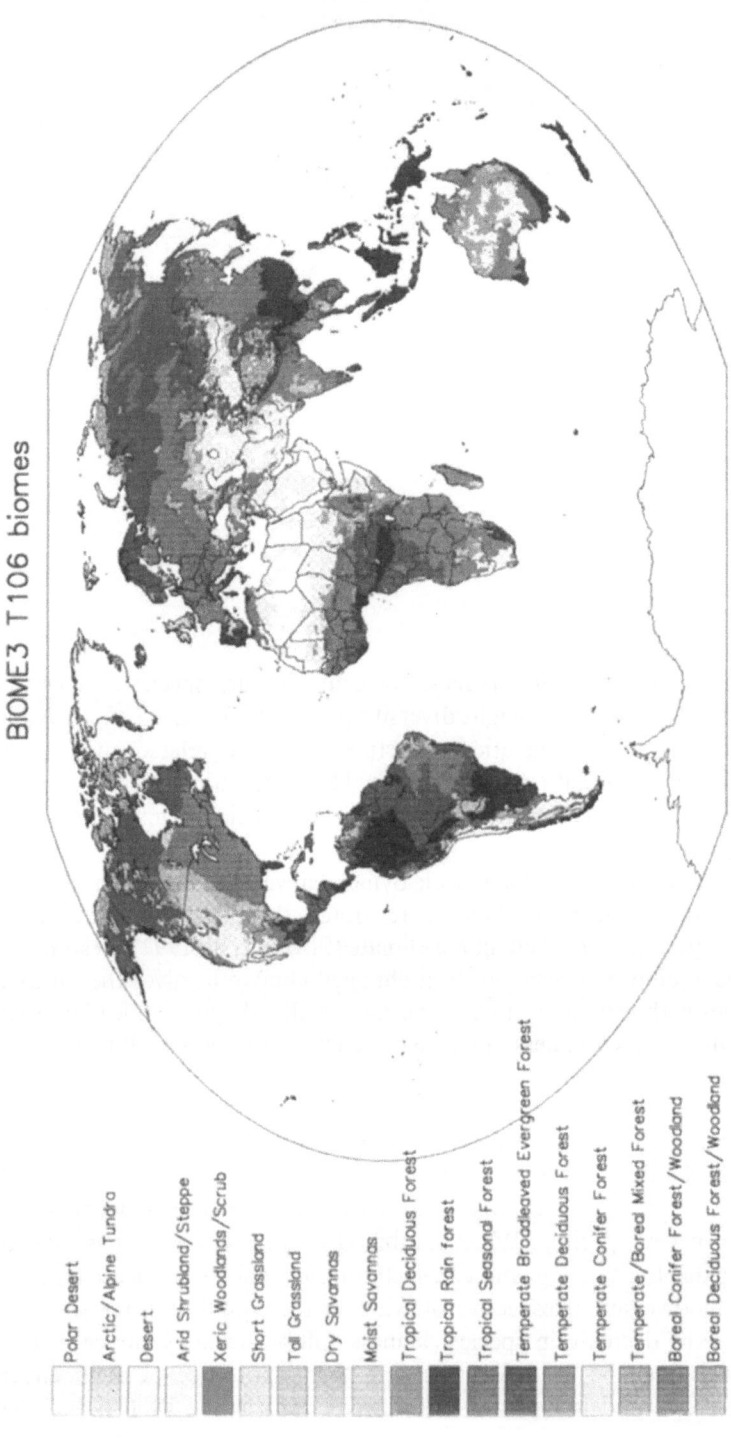

Figure 2.5. Global biome distribution under a $2 \times CO_2$ climate (Hamburg ECHAM3 climate model) as simulated by BIOME3 including direct effects of CO_2 (see color insert).

The impact of climate change on biodiversity will be greatly modulated by land use change. Where habitat has been severely reduced by land use changes, a reduction or shift in the area favorable for the survival of an individual species would result in the reduction or even total disappearance of the habitat available to that species. Changes in land use and the resulting habitat destruction are already the dominant drivers of biodiversity loss on land (Ojima et al. 1994). Thus, vegetation models must ultimately include land use and land use change if they are to provide meaningful indications of the impact of climate change on biodiversity. Dynamical vegetation models that couple land use with vegetation distribution could be used to study the spatial and temporal interactions between land use patterns and the migration of vegetation under a changing climate.

Acknowledgment. We acknowledge with thanks input from Wolfgang Cramer, Potsdam.

References

Adams JM, Woodward FI (1989) Patterns in tree species richness as a test of the glacial extinction hypothesis. Nature 339:699–701.

Al Mufti MM, Sydes CL, Furness SB, Grime JP, Band SR (1977) A quantitative analysis of shoot phenology and dominance in herbaceous vegetation. Journal of Ecology 65:759–791.

Bartlein PJ, Prentice IC, Webb T III (1986) Climatic response surfaces from pollen data for some eastern North American taxa. Journal of Biogeography 13:35–57.

Bengtsson L, Botzet M, Esch M (1995) Hurricane-type vortices in a general circulation model. Tellus 47A:1751–1796.

Bengtsson L, Botzet M, Esch M (1996) Will greenhouse gas-induced warming over the next 50 years lead to higher frequency and greater intensity of hurricanes? Tellus 48A:57–73.

Botkin DB, Janak JF, Wallis JR (1972) Some ecological consequences of a computer model of forest growth. Journal of Ecology 60:849–872.

Bugmann HKM, Solomon AM (1995) The use of a European forest model in North America: a study of ecosystem response to climate gradients. Journal Biogeography 22:477–484.

Bugmann HKM, Xiaodong Y, Sykes MT, Martin P, Linder M, Desanker P, et al. (1996) A comparison of forest gap models: model structure and behaviour. Climatic Change 34:289–313.

Clark JS (1989) Ecological disturbance as a renewal process: theory and application to fire history. Oikos 56:17–30.

Cody ML (1975) Towards a theory of continental species diversities. In: Cody ML, Diamond JM (eds) *Ecology and Evolution of Communities*, pp. 214–257. Belknap, Cambridge, MA.

Cramer W, Bondeau A, Woodward FI, Prentice IC, Betts RA, Broukin V, et al. (2001) Global response of terrestrial ecosystem structure and function to CO_2 and climate change: results from six dynamic global vegetation models. Global Change Biology 7:357–374.

Currie DJ (1991) Energy and large-scale patterns of animal and plant species richness. American Naturalist 137:27–49.

Currie DJ, Paquin V (1987) large-scale biogeographical patterns of species richness of trees. Nature 329:326–327.

Desanker PV, Prentice IC (1994) MIOMBO—a vegetation dynamics model for the Miombo woodlands of Zambian Africa. Forest Ecology and Management 69:87–96.

Foley JA (1994) Net primary productivity in the terrestrial biosphere: the application of a global model. Journal of Geophysical Research 99:20773–20783.

Foley JA, Prentice IC, Ramankutty N, Levis S, Pollard D, Sitch S, Have Hine, A (1996) An integrated biosphere model of land surface processes, terrestrial carbon balance, and vegetation dynamics. Global Biogeochemical Cycles 10:603–628.

Foley JA, Levis S, Prentice IC, Pollard D, Thompson SL (1998) Coupling dynamic models of climate and vegetation. Global Change Biology 4:561–579.

Harrison SP (1990) *An Introduction to General Circulation Modelling Experiments with Raised CO_2*. WP-90-27. International Institute of Applied Systems Analysis, Laxenburg, Austria.

Hawksworth DL, Kalin-Arroyo MT (1995) Magnitude and distribution of Biodiversity. In: Heywood VH, et al. (eds) *Global Biodiversity Assessment*, pp. 107–173. Cambridge University Press, Cambridge, UK.

Haxeltine A, Prentice IC (1996) BIOME3: an equilibrium terrestrial biosphere model based on ecophysiological constraints, resource availability, and competition among plant functional types. Global Biogeochemical Cycles 10:693–709.

Haxeltine A, Prentice IC, Cresswell ID (1996) A coupled carbon and water flux model to predict vegetation structure. Journal of Vegetation Science 7:651–666.

Heywood VH (ed) (1979) Flowering plants of the world. Oxford University Press, Oxford, UK.

Hora B (ed) (1981) *The Oxford Encyclopaedia of Trees of the World*. Oxford University Press, Oxford, UK.

Huntley B, Berry PM, Cramer W, McDonald P (1995) Modelling present and potential future ranges of some European higher plants using climate response surfaces. Journal of Biogeography 22:967–1001.

Kienast F (1987) *FORECE—A Forest Succession Model for Southern central Europe*. Oak Ridge National Laboratory, Oak Ridge, ORNL/TM-10575, 69 pp.

Lantham RE, Ricklefs RE (1993) Continental comparisons of temperate-zone tree species diversity. In: Ricklefs RE, Schluter D (eds) *Species Diversity: Historical and Geographical Perspectives*, pp. 178–184. University of Chicago Press, Chicago.

Larcher W (1983) *Physiological Plant Ecology*. Second ed. Springer-Verlag, Berlin.

Leemans R (1991) Sensitivity analysis of a general forest succession model. Ecological Modelling 53:247–262.

Leemans R, Cramer W (1991) *The IISAS climate database for land area on a grid of 0.5 resolution*. WP-41 International Institute of Applied Systems Analysis, Laxenburg, Austria.

Melillo JM, McGuire AD, Kicklighter DW, Moore B III, Vorosmarty CJ, Schloss AL (1993) Global climate change and terrestrial net primary production. Nature 363:234–240.

Melillo JM, Prentice IC, Schulze E-D, Farquhar GD, Sala OE (1996) Terrestrial ecosystems: biotic feedbacks to climate. In: Houghton J, et al. (eds) *Climate Change: The IPCC 1995 Assessment*. Cambridge University Press, Cambridge, UK.

Mooney HA, Lubchenco J, Dirzo R, Sala OE (1995) Biodiversity and ecosystem functioning: basic principles. In: Heywood VH (ed) *Global Biodiversity Assessment*, pp. 279–325. Cambridge University Press, Cambridge, UK.

Neilson RP (1995) A model for predicting continental scale vegetation distribution and water balance. Ecological Applications 5:362–386.

Neilson RP, King GA, Koerper G (1992) Towards a rule-based biome model. Landscape Ecology 7:27–43.

Nix HA, Switzer MA (1991) Rainforest animals: atlas of vertebrates endemic to Australia's wet tropics. Kopwari 1:112.

Ojima DS, Galvin KA, Turner BL II (1994) The global impact of land-use change. BioScience 44:300–304.

Parton WJ, Scurlock JMO, Ojima DS, Gilmanov TG, Scholes RJ, Schimel DS, et al. (1993) Observations and modelling of biomass and soil organic matter dynamics for the grassland biome worldwide. Global Biogeochemical Cycles 7: 785–809.

Pianka ER (1967) On lizard species diversity: North American flatland deserts. Ecology 48:333–351.

Pitelka L, the Plant Migration Workshop Group (Ash J, Berry S, Bradshaw RHW, Brubaker LB, Clark J, Davis MB, et al.) (1997) Plant migration and climate change. American Scientist 85:464–473.

Prentice IC, Leemans R (1990) Pattern and process and the dynamics of forest structure: a simulation approach. Journal of Ecology 78:340–355.

Prentice IC, Sykes MT (1995) Vegetation geography and global carbon storage changes. In: Woodwell GM, Mackenzie FT (eds) *Biotic Feedbacks in the Global Climatic System: Will the Warming Speed the Warming?* pp. 304–312. Oxford University Press, New York.

Prentice IC, Sykes MT, Cramer W (1991) The possible dynamic response of northern forests to global warming. Global Ecology and Biogeography Letters 1:129–135.

Prentice IC, Sykes MT, Cramer W (1993) A simulation model for the transient effects of climatic change on forest landscapes. Ecological Modelling 65:51–70.

Prentice IC, Cramer W, Harrison SP, Leemans R, Monserud RA, Solomon AM (1992) A global biome model based on plant physiology and dominance, soil properties and climate. Journal of Biogeography 19:117–134.

Prentice IC, Sykes MT, Lautenschlager M, Harrison SP, Dennisenko O, Bartlein PJ (1993) Modelling global vegetation patterns and terrestrial carbon storage at the last glacial maximum. Global Ecology and Biogeography Letters 3:67–76.

Price DT, Apps MJ, Kurz WA, Prentice IC, Sykes MT (1993) Simulating the carbon budget of the Canadian boreal forest using an integrated suite of process-based models. In: Chhun-Huor Ung (ed) *Forest Growth Models and their Uses*, pp. 251–264. Canadian Forest Service.

Richerson PJ, Lumm K (1980) Patterns of plant species diversity in California: relation to weather and topography. American Naturalist 116:504–536.

Rosenzweig ML, Abramsky Z (1993) How are diversity and productivity related? In: Ricklefs RE, Schluter D (eds) *Species Diversity: Historical and geographical perspectives*, pp. 52–65. University of Chicago Press, Chicago.

Running SW, Coughlan JC (1988) A general model of forest ecosystem processes for regional applications. I. Hydrologic balance, canopy gas exchange and primary production processes. Ecological Modelling 42:125–154.

Sisk TD, Launer AE, Switky KR, Ehrlich PR (1994) Identifying extinction threats. BioScience 44:592–604.

Sitch S (2000) The role of vegetation dynamics in the control of atmospheric CO_2 content. PhD Thesis, Lund University, Lund, Sweden.

Sykes MT (1997) The biogeographic consequences of forecast changes in the global environment: Individual species' potential range changes. In: Huntley B, Cramer W, Morgan AV, Prentice HC, Allen JRM (eds) *Past and Future Rapid Environmental Changes: The Spatial and Evolutionary Responses of Terrestrial Biota*, pp. 427–440. NATO ASI series, Springer-Verlag, Berlin.

Sykes MT, Prentice IC (1995) Boreal forest futures: modelling the controls on tree species range limits and transient responses to climate change. Water, Air, Soil and Pollution 82:415–428.

Sykes MT, Prentice IC (1996) Climate change, tree species distributions and forest dynamics: a case study in the mixed conifers/northern hardwoods zone of northern Europe. Climatic Change 34:161–177.

Sykes MT, Prentice IC, Cramer W (1996) A bioclimatic model for the potential distribution of northern European tree species under present and future climates. Journal of Biogeography 23:203–233.

Tilman D (1982) *Resource Competition and Community Structure*. Princeton University Press, Princeton.

Tilman D, Pacala S (1993) The maintenance of species richness in plant communities. In: Ricklefs RE, Schluter D (eds) *Species Diversity: Historical and Geographical Perspectives*, pp. 13–25. University of Chicago Press, Chicago.

Turner JRG, Lennon JJ, Lawrenson JA (1988) British bird distributions and the energy theory. Nature 335:539–541.

Urban DL (1990) *A Versatile Model to Simulate Forest Pattern: A Users Guide to Zelig 1.0*. University of Virginia, Dept. of Environmental Sciences, Charlottesville, VA. 108 pp.

VEMAP members 1995. Vegetation/ecosystem mapping and analysis project (VEMAP): a comparison of biogeography and biogeochemistry models in the context of global change. Global Biogeochemical Cycles 9:407–437.

Watt AS (1947) Pattern and process in the plant community. Journal of Ecology 35:1–22.

Woodward FI (1987) *Climate and Plant Distribution*. Cambridge University Press, Cambridge, UK.

Woodward FI, Rochefort L (1991) Sensitivity analysis of vegetation diversity to environmental change. Global Ecology and Biogeography Letters 1:7–23.

Woodward FL, Smith TM (1994) Global photosynthesis and stomatal conductance: Modelling the controls by soil and climate. Botanical Research 20:1–41.

Woodward FL, Smith TM, Emanuel WR (1995) A global land primary productivity and phytogeography model. Global Biogeochemical Cycles 9:471–490.

Wright DH, Currie DJ, Maurer BA (1993) Energy supply and patterns of species richness on local and regional scales. In: Ricklefs RE, Schluter D (eds) *Species Diversity: Historical and Geographical Perspectives*, pp. 66–74. University of Chicago Press, Chicago.

3. The Use of Global-Change Scenarios to Determine Changes in Species and Habitats

Rik Leemans

The alarming decline of the world's biological diversity, or *biodiversity*, is perceived by many as an irreversible process that could eventually obliterate the very foundation of human existence. Humans strongly depend on biological resources for food, construction materials, medicine, and energy. Humans currently directly or indirectly use most of the available biospheric resources (Vitousek et al. 1997). These resources are in principle renewable. With proper management, they can be used sustainably; unfortunately, human use often exceeds the renewal capacity, after which the biological resource base becomes degraded. Resource management ultimately thus defines the fate of biodiversity. Such degradation could also increase the vulnerability of biodiversity to other stresses (e.g., environmental change).

Biodiversity is the collection of genes, species, communities, and ecosystems that constitutes the living component of the Earth System. It responds to changes in the physical and biological environment in many complex ways, although the resilience in these responses is poorly known. The question, "Can biodiversity decline continue without curtailing ecosystem functioning?" has generated in-depth discussions (Schulze and Mooney 1993; Walker 1995; Chapin et al. 1997; Tilman et al. 1997) and led to several innovative ecological experiments with new insights (Naeem et al. 1994; Tilman et al. 1996; McGradysteed et al. 1997; van der Heijden et al. 1998). These discussions and insights suggest that, with increasing biodiversity, several ecosystem properties (e.g., carbon uptake and productivity) increase and that ecosystem response becomes more predictable and less variable. Most of these studies, however, have used simple model and experimental ecosystems; therefore, the results are difficult to extrapolate to determine the actual responses to environmental change of the wide variety and diversity of ecosystems on earth.

Leemans (1996) has evaluated changes in the focus of biodiversity research immediately after the UNCED conference in 1992 and few years later (in 1995). He classified many scientific publications with "Diversity" or "Biodiversity" in the title and found that the majority of papers focused at the genetic and population levels of biodiversity (i.e., within species diversity). Most of the remaining papers emphasized species and communities and described or compared geographic differences. Very few papers focused on biodiversity at the ecosystem level and its interactions with the environment (i.e., ecosystems functioning). Only a few papers actually described changes in biodiversity. Habitat loss and alien species were the most common threat discussed. No comprehensive linkages were made to the socioeconomic driving forces. His analysis further showed that there was only a small shift toward the higher organizational levels (ecosystems) after the adoption of the Convention on Biodiversity. Many more papers involving ecosystems and biodiversity have appeared; however, the skewed distribution of biodiversity issues in the literature still reflects the "genetic to species" bias of the Convention. Beyond that, almost all of the papers are descriptive and qualitative; quantitative approaches are rare.

These biases unfortunately do not provide the solution to the central biodiversity questions listed in the introductory chapter. Only some aspects of human influence on biodiversity are addressed. Different approaches are needed to comprehend the width of the central questions fully. Comparing the Framework Convention on Climate Change (FCCC) and the biodiversity convention yield different notions. The Objective of FCCC (Article 2) is to stabilize the atmospheric concentrations of greenhouse gases (GHGs) at nondangerous levels. These levels should allow ecosystems to adapt to climate change in a natural way, guarantee food-security, and let economic development proceed sustainably. On a very general but highly conceptual level, this FCCC objective implicitly defines targets and time paths for GHG emissions (Swart et al. 1998). This much more explicit objective resulted over the last decade in the development of approaches that analyze, evaluate, and synthesize major socioeconomic, physical, chemical, and biological aspects of climate change. Such approaches are currently labeled *integrated assessment* (IA). The first (Houghton et al. 1990; Izrael et al. 1990), which was published prior to the UNCED conference, and the second (Bruce et al. 1996; Houghton et al. 1996; Watson et al. 1996) assessment of the Intergovernmental panel on Climate Change (IPCC) are good examples of such an approach. These assessments have led to a much more scientifically integrated way of looking at climate change. The IA approaches, first comprehensively developed for acidification (Alcamo et al. 1996), have since become mature and potentially bridge the interests of scientists and policy makers (Weyant et al. 1996).

A central approach within IA is scenario development. Scenarios include a description of the current situation and of a possible or desirable future state, as well as of the series of events that lead from the current to the future state. Scenarios require a consistent and coherent set of assumptions on the phenomena and processes analyzed, their determining factors and expected future development. Scenario analysis is thus not a simple sensitivity analysis or trend extrapolation because scenarios should explicitly include system-level feedbacks, interactions, and inertia. Scenario analysis generally involves one or more no-policy or reference scenarios (Carter et al. 1994; Alcamo et al. 1995; Alcamo et al. 1996) and scenarios including different measures. By comparing the results of all these scenarios, the effectiveness of each measure can be evaluated (Alcamo and Kreileman 1996). Scenario analysis actually deals with many aspects of the different UNCED conventions and could be of help to generate understanding, could define the timing and level of policy action, and could, beforehand, indicate the effectiveness of policy measures.

Scenario studies have been instrumental in the scientific and policy discussions concerning climate change. The earliest scenarios defined different reference GHG emissions assuming different levels of population, economic, and technological development (Leggett et al. 1992). Later, different aspects of the objective of FCCC were evaluated [e.g., stabilization (Enting et al. 1994), mitigation (Ishitani et al. 1996), and other policy scenarios (Alcamo and Kreileman 1996; Swart et al. 1998)]. These studies influenced the viewpoints of different parties in the FCCC negotiations. For example, the important notion that stabilization of GHG concentrations minimally requires a 50% reduction of current global GHG emissions resulted from these scenario studies. This conclusion has contributed to the first agreement in history to start reducing GHG emissions at the FCCC Conference of Parties in Kyoto (1997).

Comprehensive scenario analysis is thus a powerful tool in bridging science and policy (van Daalen et al. 1998), and it has already been frequently used in assessing environmental problems (Alcamo et al. 1996). The Global Biodiversity Assessment (GBA; Heywood and Watson 1995) has unfortunately not applied scenario analysis for future projection of biodiversity levels, although its objective was to provide an independent, critical, peer-reviewed, scientific analysis of current issues, theories, and views of major global aspects of biodiversity and its threats. Most of the concern emerging from the GBA is that current extinction rates are dramatically higher than background extinction rates. This would eventually reduce the resilience and functioning of ecosystems, jeopardizing continued availability of all goods, services, and values provided by the biosphere. Much emphasis in the GBA is on preservation, conservation, and sustainable use of biodiversity. Only 10 pages deal with possible future developments (pp. 792–802), but only a limited inventory is presented. The notion that societal growth increases pressures on biodiversity, however, is well presented and discussed in this section. Further, possible accelerating effects of climate change on extinction are introduced, but not elaborated upon. Despite the section's conclusion that ecosystem management should focus on adaptability, variability, and resilience to allow for future environmental change, it does not comprehensively determine future dynamics and levels of biodiversity, nor does it link quantitatively to other environmental problems (e.g., acidification, climate change, deforestation and land degradation) and socioeconomic issues (e.g., land-use change, food security, and sustainable development). This research presented in this book is the first to develop more detailed and comprehensive quantitative approaches. Such quantitative scenario analysis could well bridge the gap between climate change, societal dynamics, resource use, and biodiversity.

I will go one step beyond the qualitative descriptions in the GBA in this chapter and identify major problems, obstacles, and challenges in developing future biodiversity scenarios. To do so, I will use one of the most advanced global IA models to date, the IMAGE 2 model (Alcamo 1994; Alcamo et al. 1998), which has been developed over the last decade at RIVM. This model was originally aimed at projecting GHG emissions stemming from energy use, land use, and industrial activities, then at calculating the build-up of atmospheric concentrations, and finally at determining climate change and assessing its effects. To do so, the model dynamically simulates changes in biospheric carbon uptake and release, shifts in crop and vegetation zones under climate change, and expanding or contracting agricultural areas. These aspects are also applicable and important for biodiversity assessments.

In order to represent important interactions within the society–biosphere–climate system, adequate regional and spatial and temporal resolutions had to be defined in IMAGE 2. This resulted in the determination of socioeconomic trends, which are often externally defined as *scenario assumptions*, and energy use for 13 socioeconomic regions and the simulation of land-use and climate-change impacts on a $0.5 \times 0.5°$ longitude and latitude grid. The realistic simulation of land-use and environmental change processes has expanded the application of the model toward food security, land degradation, and biodiversity. Some of these broader results were published in UNEP's Global Environmental Outlook (GEO) (United Nations Environment Programme 1997).

First, I will briefly summarize the processes and activities and their characteristics that influence terrestrial biodiversity. Second, I will summarize some of the relevant aspects of the IMAGE 2 model and define scenarios, mainly differing in population and economic growth assumptions. The results of these scenarios will be presented and possible changes in biodiversity will be discussed. Finally, I will discuss the limitation and uncertainties imbedded in this approach and conclude with some recommendations for improvements and future developments.

Human Influences on Biodiversity

Throughout history humans have influenced and altered biodiversity in many ways (McNeely et al. 1995). In many regions biodiversity has increased through the domestication of plants and animals. Human activities have traditionally supported the maintenance of species and genetic biodiversity. For example, shifting cultivation systems throughout the world have had a profound effect on biodiversity. Certain areas (e.g., sacred groves) were kept permanently out of rotation. Here, the original vegetation and wildlife flourished. In the cultivated areas a high diversity of domestic plants and animals was generally maintained, whereas the fallow fields provided productive habitat and feeding grounds for many nondomestic species. The farmers often grew crop varieties and cultivars well adapted to local conditions in order to reduce risk to pests and extreme weather conditions.

With the increasing population pressures and the development of modern agriculture, the sustainable use of biodiversity has lost its role in these systems. This trend has rapidly led to the destruction of local and regional biodiversity in agricultural systems and marginalized natural vegetation and wildlife as a natural resource (Table 3.1). Modern agriculture relies on fewer crop varieties. The intensification through increased dependence on artificial fertilizer, irrigation, and pesticides for pest and weed control has lead to high and stable yields, but it has also led to a significant reduction in the genetic diversity of common crops and livestock. This trend is now also accelerating in developing countries. In India, for example, it is estimated that more than 95% of the traditional varieties have disappeared.

Many factors currently threaten biodiversity, whereas others help to maintain these levels (Fig. 3.1). The most obvious threatening factors are those that alter habitats of species, introduce new exotic species, overexploit species and habitats, and/or change environmental conditions. Many of the causal

Table 3.1. Habitat and percentage human disturbance by continent

Continent	Area (millions km^2)	Undisturbed	Partially disturbed	Human dominant
Europe	5.8	15.6	19.6	64.9
Asia	53.3	42.2	29.1	28.7
Africa	34.0	48.9	35.8	15.4
North America	26.1	56.3	18.8	24.9
South America	20.1	62.5	22.5	15.1
Australasia	9.5	62.3	25.0	12.0
Antarctica	13.2	100.0	0.0	0.0
World total	162.0	51.9	24.2	23.9
World total minus rock and barren land	134.9	47.0	26.7	26.3

Source: Hannah et al. 1994.

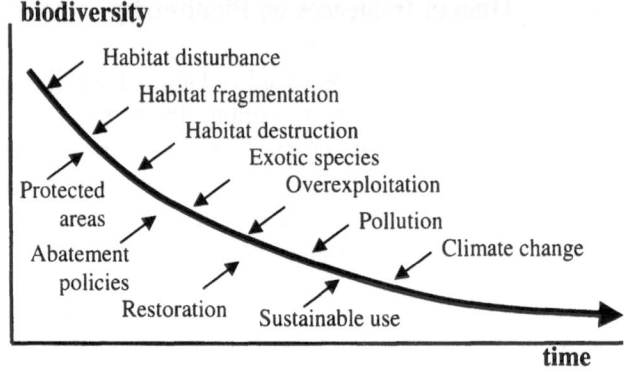

Figure 3.1. Major factors that influence biodiversity gains and losses.

factors impacting biodiversity are demographic, cultural, and socioeconomic. The establishment of protected areas can compensate some of these aspects. Further restoration, the development of more sustainable uses, and other abatement measures could enhance biodiversity levels. The influences on biodiversity depend strongly on the long-term evolutionary and shorter-term historic context, local population structure, species composition, and habitat availability. Large regional differences therefore exist in the threats to biodiversity and the possibilities to maintain or restore them.

The systemic biological and ecological responses with respect to ecosystem and biospheric functioning are becoming more predictable with advances in understanding of ecosystem functioning (Walker et al. 1997). The translation of these responses to the organizational levels (genes, species, and communities) and thus the complete scope of biodiversity, however, still remain a major challenge. Local and regional differences in environmental heterogeneity, species and population characteristics, and cultural and socioeconomic conditions will probably always lead to rather general translations that are tempered with many uncertainties, but they can at least sketch the emerging trends in a transparent way.

For many researchers, such uncertainty and generality would imply that the development of scenarios for changes in biodiversity would be a trivial and unrealizable exercise and, therefore, not worth the effort. Despite the inherent limitations of scenarios, the comprehensive analysis of plausible linkages between the societal developments (i.e., the causes of biodiversity decline), the subsequent resource use, and its consequences for ecosystem functioning and biodiversity is more than just describing an alarming trend of biodiversity decline. It puts convincingly the many local and regional examples of biodiversity in a societal perspective, which is much more appealing for policy makers and their advisors.

In the next section, I will describe the IMAGE 2 model, which is developed to make this linkage especially for terrestrial systems. To emphasize biodiversity, several coarse biodiversity indicators are used. These indicators are only coarse proxies for biodiversity. The implemented scenarios, however, will show that the trends in biodiversity decline can be relieved with the development of proper land-use management and the mitigation of climate change.

The Structure and Assumptions of the IMAGE 2 Model

Maintaining consistency in scenarios is a major challenge. This is partly solved by using IA Models for generating these scenarios. These models are tools for accomplishing a measure of harmony between the many disparate components of the scenarios. I will use IMAGE 2 for this discussion. Its goal is to provide a disciplinary and geographic overview of global environmental change. The model is fully documented elsewhere (Alcamo 1994; Alcamo et al. 1998), so a brief overview suffices here.

Assumptions about the regional demographic, technological, and economic developments are the major driving forces for the scenarios. These assumptions allow IMAGE 2 to compute future changes in energy and land use. These changes lead to emissions from fuel combustion, industrial production, and land-use change, as well as changes in carbon fluxes. The atmospheric concentration of various greenhouse gases consequently changes, as does the flux of heat and moisture between the terrestrial, oceanic, and atmospheric compartments. This regionally affects climate. These climatic changes then feedback to the biosphere in different ways, for example, by changing the productivity of crops and consequently the required amount of future agricultural land.

The IMAGE 2 model consists of several individual global submodels organized into three fully linked subsystems: Energy–Industry, Terrestrial Environment, and Atmosphere–Ocean (Fig. 3.2). The Energy–Industry models compute the emissions of greenhouse and other gases from five sectors in 13 large world regions based on estimates of industrial production and energy consumption. The Terrestrial Environment models simulate changes in global land use and cover on a grid-scale that takes into account shifts in the demand and potential productivity of land. These models also compute the subsequent fluxes of gases between the biosphere and the atmosphere. The Atmosphere–Ocean models calculate the changes in atmospheric composition of greenhouse and other gases, changes in the heat and moisture balance of the earth, and subsequent shifts in temperature and precipitation patterns. Each submodel has been tested either with data from 1970 to 1990 or with long-term averages, depending on suitability and availability of data. Simulations of IMAGE 2 result in scenarios for the period up to 2100.

Scenario Description

The main driving forces of global change in these scenarios are population and economic growth and activities. I have selected an intermediate and a

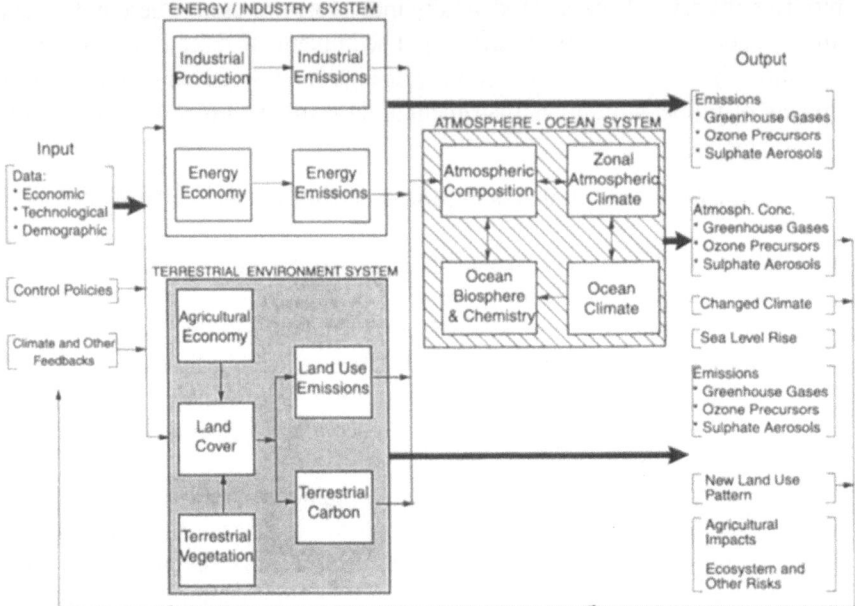

Figure 3.2. A flow diagram of IMAGE 2.

low scenario for this chapter (Scenario A and B, respectively). Scenario A was intended to reflect IPCC's medium estimate of global CO_2 emissions (Leggett et al. 1992). This "realistic" scenario is the basis of biodiversity scenarios presented in this book. Scenario B was intended to give lower overall consumption levels, with what are hoped to be lower impacts on biodiversity. Both scenarios, however, depict business-as-usual or reference developments. No global change or biodiversity policies are assumed.

Both scenarios use the population and GDP estimates of IPCC (Table 3.2). The higher population estimates in scenario A are probably more likely than the lower estimates in B (Lutz et al. 1997). The GDP assumptions are based on the estimates of the World Bank (World Bank 1991). These assumed regional figures imply a substantial increase in GDP per capita in both scenarios (Table 3.2). For example, GDP per capita in Latin America and East Asia will exceed current levels in OECD Europe in constant dollars. Nevertheless, a large income gap will remain between industrialized and developing regions.

To calculate a scenario of energy consumption and the consequent emissions for each of 13 world regions, IMAGE 2 takes four main factors into account:

1. Changes in the level of activity in different economic sectors as a function of changes in income and population.

Table 3.2. Assumptions for population (millions) and Gross Domestic Product (US $ per capita)

| | Population | | | | | | Gross Domestic Product | | | | | |
| | Observed | | Scenario A | | Scenario B | | Observed | | Scenario A | | Scenario B | |
Region	1970	1990	2050	2100	2050	2100	1970	1990	2050	2100	2050	2100
Canada	21	27	32	32	23	15	13,001	21,273	65,523	11,545	46,102	64,815
USA	205	250	298	295	235	166	15,931	21,866	65,531	11,417	48,209	66,522
Latin America	284	446	820	873	771	773	2,024	2,569	8,425	25,048	5,198	10,762
Africa	360	639	2,198	2,862	1,621	1,611	613	646	1,956	6,553	1,205	2,803
OECD Europe	351	377	394	388	323	218	12,268	19,065	58,722	10,347	41,317	58,088
Eastern Europe	108	123	149	148	129	97	1,213	1,913	9,584	16,768	6,047	7,278
CIS	243	289	350	347	302	228	1,452	2,476	7,666	13,413	4,777	5,749
Middle East	115	202	726	932	439	345	2,883	2,823	7,018	19,773	4,166	7,893
India + S. Asia	739	1,171	2,375	2,644	1,897	1,479	220	327	1,907	7,436	1,185	3,240
China + C.P. Asia	899	1,242	1,886	1,953	1,390	950	127	369	3,481	15,226	2,117	6,552
East Asia	240	368	746	831	596	465	569	1,508	8,795	34,293	5,465	14,941
Oceania	16	21	23	23	17	12	11,670	15,579	58,690	10,309	42,862	59,305
Japan	104	124	132	130	101	69	12,088	23,734	8,9411	15,705	65,299	90,349
World	3,686	5,280	10,129	11,455	7,844	6,427	3,073	3,971	9,473	21,319	6,566	10,453

2. Structural change of the economy that leads to changes in energy intensity of the different sectors.
3. Technological change that improves the performance of devices and appliances used to deliver energy services and emission levels.
4. Changes in fuel prices that stimulate energy conservation and shifts in fuel mix and their emission factors.

The precise energy and industry assumptions are presented in Leemans et al. (1998). Land cover change is an essential aspect to determine changes in biodiversity. The underlying scenario assumptions leading to changes in land use will therefore be discussed in more detail.

IMAGE 2 computes changes in land cover by taking into account the need for agricultural land, including pasture, cropland, and managed forests. The model computes land-use change by computing the growing demand in the different regions for livestock, crops, and forest products and the amount of crop, pasture, and forest land required to provide these products.

To define a land-use scenario, IMAGE 2 takes such factors into account as agricultural trade between regions, potential productivity, livestock, cropping intensity, and yield improvements. The need for agricultural land will depend, first and foremost, on regional agricultural demands; however, the amount of agricultural land for some regions will also depend on the amount of food traded with other regions. The trade patterns must be specified for each scenario. In scenario A and B the trade patterns are similar: Currently importing countries maintain their current dependence on imports in the future. The other factors, except potential productivity, are also scenario assumptions. Potential productivity influences the amount of food that can be produced per hectare of land and is computed internally by the model using local climate and soil characteristics and global CO_2-concentrations.

Agricultural demand consists of the need for all agricultural commodities, specifically meat and crops consumed by humans, and feed required by livestock (i.e., demand for forest products are computed separately in the model, whereas the demand for modern biomass[1] is generated by the Energy Economy model). To compute regional demands, the model multiplies per capita consumption of food times population estimates. IMAGE 2 computes this consumption under the main premise that people eat more food as their income increases, up to a particular "preferred" consumption level. Prices increase when the availability of agricultural land decreases, which shifts the consumption levels away from the preferred (often meat) toward achievable consumption levels. IMAGE 2 thus computes food consumption based on (1) population, (2) income, (3) land productivity and availability, and (4) preferred level of food consumption. The first two factors are taken from the

---

[1] Modern biomass is frequently proposed as a carbon-neutral energy source, but its production requires land. Many energy scenarios assume that after 2020 this source becomes abundant and will provide about 20% of the global energy demand.

population and GDP assumptions (Table 3.2). The third factor is computed internally in the IMAGE 2 model. The last factor, preferred level of consumption, is difficult to specify because it varies greatly from region to region, and it depends on difficult-to-quantify cultural and geographical factors; hence, we have taken a pragmatic approach and run the IMAGE 2 model "backward" from 1970 to 2010 in order to obtain the trend of this factor. This is done by specifying what the model is supposed to compute for this period—per capita consumption of different foods from 1970 to 2010. Data for food consumption comes from AGROSTAT (FAO 1999) for 1970 to 1990, and from trend estimates of IFPRI (Rosegrant et al. 1995) from 1990 to 2010. We then extrapolate these trends from 2010 to 2100. The same estimates of preferred consumption are used for both scenarios. An important variable that affects the overall land needed in a region for cropland is the cropping intensity (i.e., the number of crops grown per hectare of land over a calendar year). This must be specified for each scenario and region over the scenario period. There has historically been an upward trend, and both scenarios assume that this trend will continue up to a region-specific maximum. Cropping intensities of temperate cereals sharply increase for most developing regions up to the second half of the next century, but they level off in the early part of the century in industrialized regions. Improvements in management, fertilizer use and irrigation, high-yielding crop varieties, and machinery have contributed to a steady increase in crop yields throughout the world. The future rate of technological improvement in crop yield must be specified for each scenario and region. The industrialized regions are assumed to have already passed the "green revolution," although yields will continue to improve at somewhat lower rates due to biotechnology (Table 3.3). Scenarios for these regions are based on a slowing down of the 1970 to 1990 trends. For the developing regions, trends up to 2010 are taken from the FAO data for 1970–1990 (FAO 1999) and the IFPRI projections (Rosegrant et al. 1995) up to 2010. This rapid improvement rate levels off after 2030. Different rates of improvement are assigned to the two scenarios, depending on their rate of economic growth (Tables 3.2 and 3.3).

The assumptions concerning livestock also have an important influence on estimating future land requirements (for feed-crops and rangeland). First, animal productivity (i.e., the amount of meat produced per animal) strongly determines the amount of land used. It is assumed that industrialized countries are already close to their maximum value and that other regions will reach the current OECD Europe level when their GDP per capita reaches the current OECD income level; hence, the trend of this factor varies from scenario to scenario along with economic assumptions of the scenarios. Second, two different rangeland systems can be distinguished from the AGROSTAT statistics: intensive and extensive rangelands. The first is located in the European countries and Asia, whereas the latter is found in the Americas, Africa, and Australia or mainly the arid areas. Productivity on the intensively managed rangelands is about five times higher than it is on the extensive

Table 3.3. Assumptions for yield improvements relative to 1990 in temperate cereals (other crops are defined in a comparable way, see Leemans et al.)

	FAO agrostatiation		IFPRI	Scenario A		Scenario B	
	1970	1990	2010	2050	2100	2050	2100
Canada	1.01	1.00	1.10	1.28	1.45	1.24	1.38
USA	0.77	1.00	1.27	1.60	1.82	1.52	1.68
Latin America	0.59	1.00	1.37	2.17	2.80	1.90	2.32
Africa	0.73	1.00	1.37	2.17	2.80	1.90	2.32
OECD Europe	0.64	1.00	1.10	1.21	1.21	1.21	1.21
Eastern Europe	0.63	1.00	1.27	1.47	1.61	1.44	1.56
CIS	0.75	1.00	1.27	1.60	1.82	1.52	1.68
Middle East	0.71	1.00	1.27	1.60	1.82	1.52	1.68
India + S. Asia	0.60	1.00	1.37	1.92	1.92	1.92	1.92
China + C.P. Asia	0.39	1.00	1.27	1.53	1.53	1.53	1.53
East Asia	0.85	1.00	1.10	1.28	1.45	1.23	1.35
Oceania	0.62	1.00	1.27	1.60	1.82	1.52	1.68
Japan	0.62	1.00	1.10	1.28	1.45	1.24	1.38

Source: Leemans et al. 1998.

rangelands due to differences in stocking density. Although regional differences are initially apparent, this productivity gap between regions is assumed to close during the time period of the simulations, but the differences between intensive and extensive stocking densities remain. The result of this set-up is that much less land is required in the intensive rangeland regions to produce a kilogram of meat than in the regions with extensive rangelands.

The agricultural demand for food, fodder, fuelwood, timber, and modern biomass (as an efficient emerging renewable energy source; Johansson et al. 1993) is satisfied by continued production on current agricultural land, by intensification, and by expansion of agricultural land. If demand is less than the current production capacity, land is taken out of production and returned to its natural vegetation. The simulation of land use and land cover therefore results in a consistent pattern between regional demand and production potential. We use a heuristic scheme to incorporate some of the spatial interactions, inertia, and societal preferences in defining future agricultural patterns (Alcamo et al. 1998). For example, expansion predominantly occurs in areas close to current agricultural land and rivers to mimic the availability of infrastructure and labor. Of these "close-by" areas, those with highest potential productivity are used first. Expansion of agricultural land can lead to deforestation. Some of the cleared trees are used to satisfy timber demand. The amount of wood stemming from deforestation is unfortunately much larger than the demand, so a large part is burned, which leads to emissions of greenhouse gases. The land-cover simulations lead to frequently updated land cover patterns with arable land, rangeland, regrowth forests, and natural vegetation on a 0.5° longitude and latitude grid.

All the emissions stemming from energy use, industrial activities, and land use enter the atmosphere. IMAGE 2 then calculates by accounting for atmospheric chemistry, carbon uptake by the oceans, and the biosphere, the final atmospheric GHG concentrations. These are used in a simple climate model to determine climate change (both temperature and precipitation). Regional and seasonal climate patterns are defined by a simple downscaling procedure using the results of an advanced climate model, an observed climatology, and the determined climate change of IMAGE 2.

Scenario Results

The simulated climate patterns are used in subsequent years to calculate potential crop productivity and vegetation patterns, which are also corrected for the increases in CO_2 concentration (cf. Table 3.4). Climate change thus influences the land-use and vegetation patterns directly in the model, and these influences are presented as impacts. The vegetation models included in IMAGE are the same, which led to the alarming vegetation shifts discussed by Watson et al. (1996), Peters and Lovejoy (1992) and Huntley et al. (1997) (Table 3.5). The only difference is that we have developed a simple adaptation and migration scheme (Alcamo et al. 1998), which allows some ecosystems to adapt.

With the simulated climate change in these business-as-usual scenarios, natural adaptation unfortunately occurs only in a small percentage of the impacted vegetation. Forests generally do not adapt effectively. Another analysis with IMAGE 2 (Swart et al. 1998) indicated that effective adaptation is only possible at a very slow rate of climate change (less than 0.1°C per decade) and limited absolute climate change (less than 1°C in total). Such

Table 3.4. Direct and indirect effects of elevated CO_2 and climate

Direct effects	Possible indirect effects
Changes of elevated CO_2	
Changes in plant growth	Changes in species abundance
Changes in water use efficiency	Changes in ecosystem patterns and biogeochemical cycling
Changes in nutrient use efficiency	Changes in competitive abilities of species
Changes in climate	
Changes in plant growth	Changes in species abundance
Changes in decomposition	Changes in ecosystem structure and biogeochemical cycling
Changes in ecosystem patterns	Changes in biogeochemical cycling
	Changes in disturbance regimes
	Changes in competitive abilities of species and ecosystem structure and function

Table 3.5. Relative impact levels on natural vegetation (vs. total area) and nature reserves (vs. total reserve area) of differences of increases in global mean annual temperatures

	0.5°C	1.0°C	1.5°C	2.0°C	2.5°C	3.0°C
Natural vegetation	11	19	26	32	37	43
Nature reserves	9	17	24	32	37	42

rates and levels would limit impacts on vegetation. To reach this, global emission levels should be reduced immediately by 1–2% annually.

The simulated changes in vegetation patterns (Fig. 3.3; see color insert) indicate some concrete impacts on biodiversity. The regional impacts on vegetation are largest in the boreal regions, where higher-than-average temperature changes also occur. Tundra vegetation declines rapidly and boreal forests shift polewards. At the end of the present century in both scenarios much of the European boreal and alpine forest is replace by temperate broad-leaved forests, but because migration rates are too slow to cope with such rapid change, depauperate or degraded forest types result that are dominated by opportunistic species with wide distributions and rapid spread. The abundance of specialized and late successional species with small niches will decline. Most extinctions will occur in this group of species. In boreal regions the shifts are strongly determined by changes in temperature; in temperate regions, more by moisture. The simulations show that in the more maritime regions precipitation simultaneously increases with temperature. Here only small changes are calculated. In continental areas, however, precipitation remains the same or decreases. The resulting drought kills many forests, and grasslands spread. These changes are also detrimental for biodiversity.

Simulated natural vegetation change in the tropics is much smaller (Fig. 3.3). The average increase in temperature is also much smaller than it is in the higher latitudes. Much of the simulated changes in the tropics are generated by increases in aridity, not in temperature. Some of the current rainforest areas will experience more frequent drought periods in the future. This could increase fire occurrence (Fig. 3.4) and change the structure and species composition of this for biodiversity-important biome.

Much of the land is currently managed by humans. Although the shifts in vegetation zones make clear that climate change will result in a major change in biodiversity patterns and that some of the changes have already been observed (Grabherr et al. 1994; Myneni et al. 1997; Parmesan et al. 1999), many policy makers remain skeptical that these rapid changes are possible. Many believe that ecosystems are resilient and will evolve; if not, adaptive management will mitigate the negative effects. To extract managed land from natural lands, I have analyzed the impact on nature reserves (Table 3.5). A database with many large nature reserves from the World Conservation Monitoring Centre (Groombridge 1992) was used. The elegance of vegetation shifts in nature reserves is that they are legally protected to conserve specific species, their habitats and their ecosystems. The expansion of nature reserves are at the heart of the abatement measures specified in the Biodiversity Convention. They cannot move and therefore cannot adapt naturally. Under

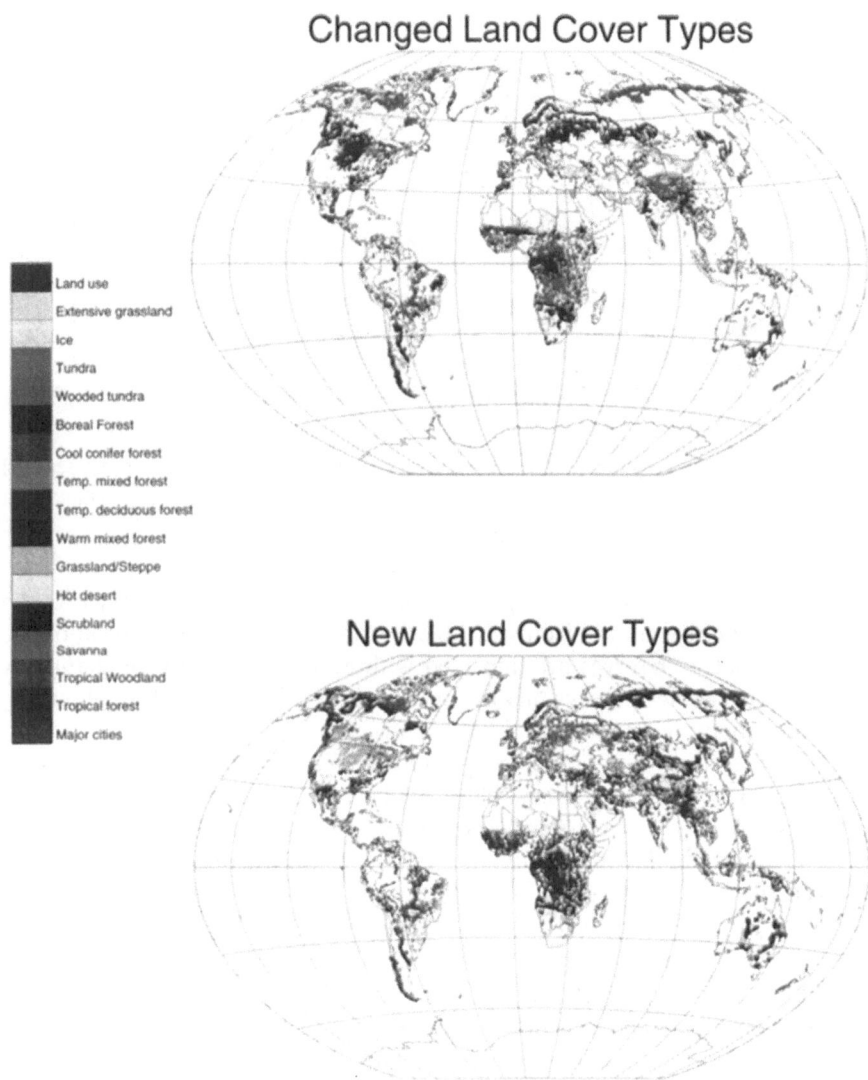

Figure 3.3. Change in actual land cover in 2100 for scenario A. The top panel presents the changes of the original vegetation. Red denotes agricultural land that is abandoned. The lower panel presents the future land cover. Red denotes the expansion of agricultural land, whereas orange depicts regrowth after clear-cut or abandonment (see color insert).

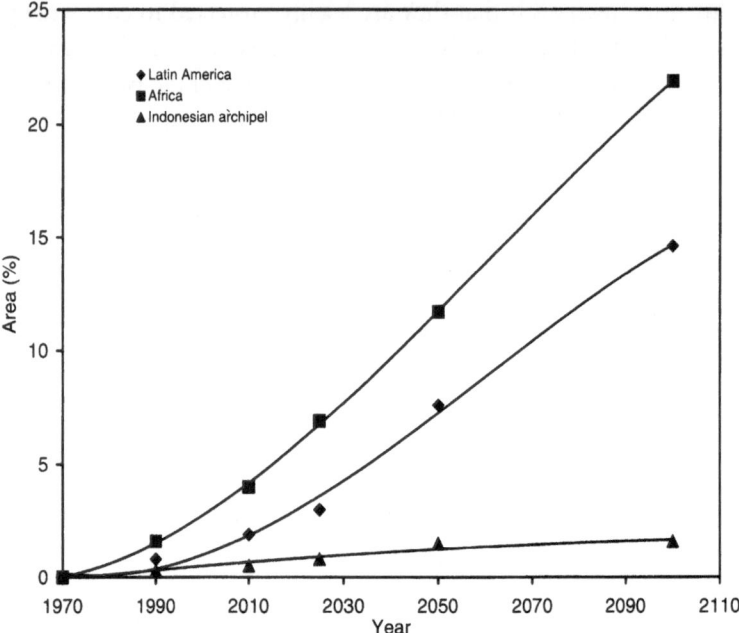

Figure 3.4. Changes in potential fire occurrence in the wet tropics (i.e., no dry season) of Latin America, Africa, and Asia. Fire occurrence is defined for the percentage area where a structural drought period of at least 2 weeks appears.

climate change, they just change and will probably no longer comply with the original conservation objectives. Table 3.5 shows that a large portion of the selected nature reserves are impacted with even a small climate change. A geographic analysis shows that the impacted sites are located throughout the world and are not restricted to specific regions.

The preceding analysis shows that changes on biodiversity due to climate change can be expected everywhere. Similar results are obtained for changes in land use (Fig. 3.3). The consumption of food crops, animal products (and also rangeland), and modern biomass globally increase threefold. Part of this increase is due to increases in population; part of it is due to changed consumption patterns. The result is that over the period 1990–2100, the global extent of cropland (not grassland) increases from 9 to 14 million km^2 (Fig. 3.5). The production of modern biomass additionally requires about 6 million km^2 in 2100 (30% of all arable land). The global extent of pastureland increases more rapidly from 34 to 44 million km^2 in 2030, after the total extent decreases somewhat. This is most apparent in Africa, where currently around 9 million km^2 are used for grazing. This increases to 15 million km^2 in 2025 they in scenario A and somewhat slower in scenario B. In scenario B the upper level is reached in 2050. The arable land in Africa increases from 1 million km^2 in 1990 to 3 million km^2 in 2100. This increase in arable land mostly stems from converting pasture land. The total increase in land use saturates in 2025 and 2050 for scenarios A and B, respectively, because of the lack of available and suitable land. Deforestation rates are also different for both scenarios (Fig. 3.5). The initial deforestation rates in scenario B are lower than they are in scenario A. After 2030 there is even a slight increase in forest extent.

These changes in land use lead to sustained global deforestation rates up to 2025, after which the total forest area tends to stabilise; however, regionally there are large differences. In the simulations, there is a shift of major deforestation areas away from Latin America toward Africa. Asian deforestation just continues, although in Latin America in the second half of next century grazing and other agricultural land is abandoned and returns to the natural vegetation. In Europe and North America the decrease of agricultural land starts immediately after the start of the simulation. Here agricultural land is converted to natural vegetation and the forest areas increase. This can have potentially positive effects on biodiversity, but the simulations also show that timber production increases. Although reforested, a large proportion of these forests remain managed and are "harvested" at least once during the simulated period.

The impacts of climate change on land use are also pronounced (Fig. 3.3). Changes in potential crop yields are very regional specific. At the warmer edge of a crop's distribution, yields often decline, whereas extent and

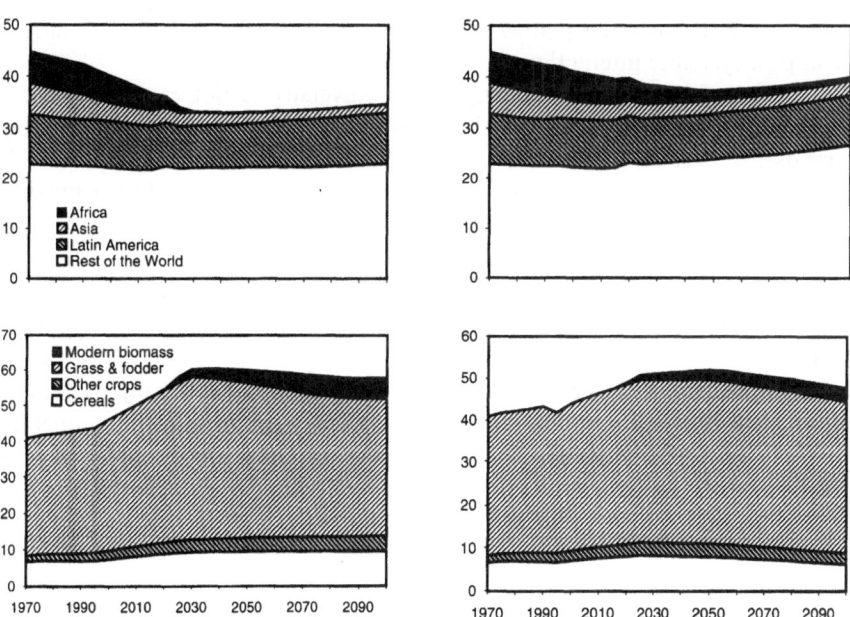

Figure 3.5. Changes in the extent (million km²) of forests (top) and land use (bottom). The left graph depicts scenario A and the right one scenario B.

productivity increase at the colder edge. New opportunities emerge for cereals in Scandinavia, Canada, and Russia, whereas yields decline in southern China and on the African plateau. Moisture availability also has a significant effect, especially in the central Great Plains. The simulated drier conditions here result in productivity losses. The model then simulates a shift of these major agricultural regions toward the southeast of the United States with consequences for biodiversity in the new land-use regions.

These scenario results show that it is important to include climate change and land-use change simultaneously. Both strongly determine the future land-cover patterns and thus biodiversity. One should be cautious, however, not to take these results too literally. IMAGE 2 is far from perfect and does not include all the complex drives of land-use change. It sketches a broad picture. Some of the results are a direct consequence of the assumptions. For example, if animal production in Africa was assumed to be more intensive, less land would be required, and more forest would probably be saved; however, the sensitivity of the model to changes in assumptions can easily be tested, which I did. Some interesting feedbacks and interactions could be observed in scenario A. The availability of suitable land in Africa became limited. This led to changes in diets, away from the preferred diets with more meat, toward a basic diet with more cereals, pulses, and roots. Assuming a rapidly increasing animal productivity initially relieved some of the land constraints, but it simultaneously saturated the preferred diet. In the end of the simulation the higher productivity made little differences in these simulations, mainly because of the lack of suitable land to fulfill the dietary preferences. This example, however, shows that these models can be used to analyse complex systemic interactions.

The impact of land use change on biodiversity is not easy to analyze. Changes in habitats of individual species should ideally be evaluated. This is impossible because of the coarseness of the model. We have therefore developed a proxy or indicator: the total amount of used land (Table 3.6), defined as the sum of the extent of arable land, pastures, and managed forests. This area is strongly dominated by humans, which leaves little room for biodiversity. The original biodiversity probably cannot sustain itself under such intense land use. If we look at the simulated changes, a clear pattern emerges. In tropical areas the extent of used land increases, although it largely remains stable in temperate regions. This means that over the coming decades the largest stresses from land use will probably occur in the tropics and not in the temperate regions. The differences between scenarios is also apparent. Scenario A with high resource use leaves less room for biodiversity than does scenario B.

Concluding Remarks

Developing regions have both the highest population and relative income growth in the scenarios presented here (Table 3.2). This results in the largest increases in consumption and thus resource use. The resulting expansion of

Table 3.6. Nondomesticated land as a percentage of total regional land area for scenarios A and B

Region	1990	2025 A	2025 B	2050 A	2050 B	2100 A	2100 B
Canada	91	91	91	90	91	86	91
USA	62	61	62	59	64	60	70
Latin America	68	62	66	63	65	71	71
Africa	72	50	63	48	55	50	50
OECD Europe	69	68	68	67	69	63	71
Eastern Europe	32	30	31	31	33	25	35
CIS	76	77	78	76	78	76	82
Middle East	91	74	82	71	79	69	81
India + S. Asia	59	46	52	40	52	42	62
China + other C.P. Asian countries	59	47	48	44	46	37	44
East Asia	74	56	58	46	50	34	48
Oceania	62	63	61	65	71	68	85
Japan	60	50	49	35	35	27	24
World	71	62	67	61	65	61	67

rangelands and arable land will lead to additional losses of biodiversity. This changed land-use could easily lead to overexploitation, additional pollution, habitat destruction and fragmentation, and thus strong declines in biodiversity. Direct human-induced changes in biodiversity and additional pressure on nature reserves will thus be largest in those regions. Indeed, many examples from GBA illustrate this; however, problems also arise in developed regions. First, there is the direct land-use component: Agriculture does not expand, but forestry does. The higher proportion of managed forests could have their toll on the original biodiversity. Further, the impacts of climate change caused by the anthropogenic emissions of GHGs are severe and influence large parts of the developed world. In the near future, rapid climate change will probably be the largest threat to biodiversity in these regions.

An additional lesson that can be learned from these scenarios is the connectivity between components of the society–climate–biosphere system. Most future energy scenarios strongly depend on the use of modern biomass as a major energy supply. This fuel has no net CO_2 emissions, but it requires land. This scenario analysis has already shown that the additional land requirements will be substantial (Fig. 3.4). Production of biomass for energy both competes with food production and, through its land-use consequences, could also pose an additional threat to biodiversity. The analysis presented here illustrates that an obvious solution for one environmental problem can create additional obstacles for other problems. Comprehensive IA is essential to analyze and advise on optimal solutions for a series of environmental problems.

I have shown that integrated assessment and scenario development can help to address and probably answer the questions posed in the introduction. In this chapter, however, I have not tried to make the full circle, and I limited the presentation and discussion to the "How?", "Why?" and "What?" questions. We have seen that if unabated the dynamic interactions between socioeconomic developments and natural resources will lead to a severe decline in biodiversity. The ultimate questions on "How will a decline in biodiversity affect the biosphere and society" must still be addressed. The emerging understanding of biodiversity and ecosystem functioning, however, supplies some clues to its answer: Ecosystems with lowered biodiversity levels are likely to be less productive, less resilient, and, when stressed additionally, respond in a highly unpredictable way. This could well erode the certainty with which we rely on the continued supply and availability of natural resources that we need for our daily lives.

The redundancy discussion in the biodiversity debate (Schulze and Mooney 1993; Walker 1995; Chapin et al. 1997; Tilman et al. 1997) has resulted in the definition of the dynamic role of biodiversity, *The Insurance Hypothesis*: Species that are redundant under normal, relatively constant environmental conditions may be critical for the maintenance of ecosystem processes in the face of environmental variation and change. The two business-as-usual scenarios have illustrated that environmental change will control future conditions. In line with the "insurance hypothesis," this means that high levels of biodiversity should be maintained and protected worldwide and that the causes of its decline should be strongly abated. This means both the elimination of alien species, and connecting and restoring species' habitats, as well as (and probably more important) a reduction in the rapid increase of land use and emissions of greenhouse gases and other pollutants. The scenario analysis illustrate that, although population growth contributes to the problem, rapidly increasing consumption patterns, rapid expansion of rangelands to support changing diets, and slow rates of technological innovation are the major drivers of environmental change. This makes solving the environmental-change problems not simply a technical environmental issue, but particularly a developmental issue.

The scenario studies clearly highlight the complex systemic interactions and feedback between society and the other components of the earth system and provide a broad outline of possible future developments. As such, these scenarios already contribute to the understanding of trends in biodiversity: however, many aspects of biodiversity are not included in these scenarios. First, the genetic and species levels of biodiversity are not considered. Second, the expansion and intensification of agricultural land only approximates habitat disturbance, fragmentation, and destruction. Finally, the influence of alien species is neglected. Some of these issues are dealt with qualitatively in other chapters in this book. Much research, therefore, has to be directed to the development of proper indicators that can bridge the coarseness of the global scenarios and the detailed aspects of the biodiversity issues that should be subject of the study.

The analysis finally shows that the advantages of such comprehensive scenario studies should not be denied. Although these business-as-usual scenarios sketch a grim future for biodiversity, the integrated characteristics of these scenarios interest many scientist and policy makers. This could help to initiate the development and analysis of effective abatement policies. The often-heard argument that such grim or doomsday scenarios do not become reality is not because they are unrealistic, but rather that it is the result of their convincing cases. Their main lesson is that a continuation of the business-as-usual track is harmful. Their insights will help to look at alternatives again and select more appropriate and successful tracks.

References

Alcamo J (ed) (1994) IMAGE 2.0: integrated modeling of global climate change. Kluwer Academic Publishers, Dordrecht.

Alcamo J, Bouwman A, Edmonds J, Grübler A, Morita T, Sugandhy A (1995) An evaluation of the IPCC IS92 emission scenarios. In: Houghton JT, Meira Filho LG, Bruce J, Lee H, Callander BA, Haites E, et al. (eds) Climate change 1994: radiative forcing of climate change and an evaluation of the IPCC IS92 emission scenarios, pp. 247–304. Cambridge: Cambridge University Press.

Alcamo J, Kreileman E (1996) Emission scenarios and global climate protection. Global Environmental Change 6:305–334.

Alcamo J, Kreileman E, Leemans R (1996) Global models meet global policy—how can global and regional modellers connect with environmental policy makers? what has hindered them? what has helped? Global Environmental Change 6:255–259.

Alcamo J, Leemans R, Kreileman E (1998) Global change scenarios of the 21st century. Results from the IMAGE 2.1 model. Pergamon and Elseviers Science, London.

Alcamo J, Kreileman GJJ, Bollen JC, van den Born GJ, Gerlagh R, Krol MS, et al. (1996) Baseline scenarios of global environmental change. Global Environmental Change 6:261–303.

Bruce JP, Lee H, Haites EF (eds) (1996) Climate Change 1995. Economic and social dimensions of climate change. Cambridge, Cambridge University Press.

Carter TR, Parry ML, Harasawa H, Nishioka S (1994) IPCC Technical guidelines for assessing impacts of climate change. Intergovernmental Panel on Climate Change, WMO and UNEP, Geneva. IPCC Special Report CGER-1015-'94. December 1994.

Chapin FS, Walker BH, Hobbs RJ, Hooper DU, Lawton JH, Sala OE, et al. (1997) Biotic control over the functioning of ecosystems. Science 277:500–504.

Enting IG, Wigley TML, Heimann M (1994) Future emissions and concentrations of carbon dioxide. Mordialloc, Australia: CSIRO, Australian Division of Atmospheric Research. Technical Paper 0 643 05256 9. November 1994.

FAO (1999) FAOSTAT database collections. Rome: Food and Agriculture Organization of the United Nations. Database http://aps/fao.org/default.html.

Grabherr G, Gottfried M, Pauli H (1994) Climate effects on mountain plants. Nature 369:448–448.

Groombridge B (ed) (1992) Global biodiversity, status of the Earth's living resources. Chapman and Hall, London.

Hannah L, Lohse D, Hutchinson C, Carr JL, Lankerani A (1994) A preliminary inventory of human disturbance of world ecosystems. Ambio 23:246–250.

Heywood VH, Watson RT (eds) (1995) Global biodiversity assessment. Cambridge University Press, Cambridge, UK.

Houghton JT, Jenkins GJ, Ephraums JJ (eds) (1990) Climate Change: The IPCC Scientific Assessment. Cambridge University Press, Cambridge, UK.

Houghton JT, Meira Filho LG, Callander BA, Harris N, Kattenberg A, Maskell K (eds) (1996) Climate Change 1995. The Science of Climate Change. Cambridge University Press, Cambridge, UK.

Huntley B, Cramer WP, Morgan AV, Prentice HC, Allen JRM (eds) (1997) Past and future rapid environmental changes: the spatial and evolutionary responses of terrestrial biota. Springer, Berlin.

Ishitani H, Johansson TB, Al-Khouli S, Audus H, Bertel E, Bravo E, Edmonds JA, et al. (1996) Energy supply mitigation options. In: Watson RT, Zinyowera MC, Moss RH (eds) *Climate Change 1995. Impacts, Adaptations and Mitigation of Climate Change: Scientific-Technical Analysis*, pp. 587–647. Cambridge University Press, Cambridge, UK.

Izrael YA, Hashimoto M, Tegart WJM (eds) (1990) *Climate Change: The IPCC Impact Assessment*. Australian Government Publishing Service, Canberra. 358 pp.

Johansson TB, Kelly H, Reddy AKN, Williams RH (eds) (1993) *Renewable Energy: Sources for Fuels and Electricity*. Island Press, Washington, D.C.

Leemans R (1996) Biodiversity and global change. In: Gaston KJ (ed) *Biodiversity: A Biology of Numbers and Difference*, pp. 367–387. Blackwell Science, London.

Leemans R, Kreileman E, Zuidema G, Alcamo J, Berk M, van den Born GJ, et al. (1998) *The IMAGE User Support System: Global Change Scenarios from IMAGE 2.1*. National Institute of Public Health and the Environment, Bilthoven. RIVM Publication 481508 006. October 1998.

Leggett J, Pepper WJ, Swart RJ (1992) Emissions scenarios for the IPCC: an update. In: Houghton JT, Callander BA, Varney SK (eds) *Climate Change 1992. The Supplementary Report to the IPCC Scientific Assessment*, pp. 71–95. Cambridge University Press, Cambridge, UK.

Lutz W, Sanderson W, Scherbov S (1997) Doubling of world population unlikely. Nature 387:803–805.

McGradysteed J, Harris PM, Morin PJ (1997) Biodiversity regulates ecosystem predictability. Nature 390:162–165.

McNeely JA, Gadgil M, Levèque C, Padoch C, Redford K, Arden-Clarke C, et al. (1995) Human influence on biodiversity. In: Heywood VH, Watson RT (eds) *Global Biodiversity Assessment*, pp. 711–821. Cambridge University Press, Cambridge, UK.

Myneni RB, Keeling CD, Tucker CJ, Asrar G, Nemani RR (1997) Increased plant growth in the northern high latitudes from 1981 to 1991. Nature 386:698–702.

Naeem S, Thompson LJ, Lawler SP, Lawton JH, Woodfin RM (1994) Declining biodiversity can alter the performance of ecosystems. Nature 368:734–737.

Parmesan C, Ryrholm N, Stefanescu C, Hill JK, Thomas CD, Descimon H, et al. (1999) Poleward shifts in geographical ranges of butterfly species associated with regional warming. Nature 399:579–583.

Peters RL, Lovejoy TE (eds) (1992) Global Warming and Biological Diversity. Yale University Press, New Haven. 386 pp.

Rosegrant MW, Agcaoili-Saombilla M, Perez ND (1995) *Global Food Projections to 2020: Implications for Investment. Food, Agriculture and the Environment*. International Food Policy Research Institute (IFPRI), Washington, D.C. Discussion Paper July 1995.

Schulze E-D, Mooney HA (eds) (1993) *Biodiversity and Ecosystem Function*. Springer-Verlag, Berlin.

Swart R, Berk MM, Janssen M, Kreileman E, Leemans R (1998) The safe landing analysis: risks and trade-offs in climate change. In: Alcamo J, Leemans R, Kreileman E (eds) *Global Change Scenarios of the 21st Century. Results from the IMAGE 2.1 Model*, pp. 193–218. Elsevier Science, London.

Tilman D, Wedin D, Knops J (1996) Productivity and sustainability influenced by biodiversity in grassland ecosystems. Nature 379:718–720.

Tilman D, Lehman CL, Thomson KT (1997) Plant diversity and ecosystem productivity: theoretical considerations. Proceedings of the National Academy of Sciences USA 94:1857–1861.

United Nations Environment Programme (1997) *Global Environment Outlook*. Oxford University Press, New York.

van Daalen CE, Thissen WAH, Berk MM (1998) The Delft process: experiences with a dialogue between policy makers and global modellers. In: Alcamo J, Leemans R, Kreileman E (eds) *Global Change Scenarios of the 21st Century. Results from the IMAGE 2.1 Model*, pp. 267–285. Elseviers Science, London.

van der Heijden MGA, Klironomos JN, Ursic M, Moutoglis P, Streitwolf-Engel R, Boller T, et al. (1998) Mycorrhizal fungal diversity determines plant biodiversity, ecosystem variability and productivity. Nature 396:69–72.

Vitousek PM, Mooney HA, Lubchenco J, Melillo JM (1997) Human domination of earth's ecosystems. Science 277:494–499.

Walker B (1995) Conserving biological diversity through ecosystem resilience. Conservation Biology 9:747–752.

Walker B, Steffen W, Bondeau A, Bugmann H, Campbell B, Canadell PTC, et al. (1997) The terrestrial biosphere and global change. Implications for natural and managed ecosystems. A synthesis of GCTE and related research. The International Geosphere-Biosphere Programme, Stockholm. IGBP Science November 1997.

Watson RT, Zinyowera MC, Moss RH (eds) (1996) Climate Change 1995. Impacts, adaptations and mitigation of climate change. Cambridge University Press, Cambridge, UK.

Weyant J, Davidson O, Dowlabathi H, Edmonds J, Grubb M, Parson EA, et al. (1996) Integrated assessment of climate change: an overview and comparison of approaches and results. In: Bruce JP, Lee H, Haites EF (eds) *Climate Change 1995. Economic and Social Dimensions of Climate Change*, pp. 367–396. Cambridge University Press, Cambridge, UK.

World Bank (1991) *World Development Report 1991*. Oxford University Press, New York.

4. Soil Biodiversity

Diana H. Wall, Gina Adams, and Andrew N. Parsons

There is considerable uncertainty as to the effect of global change on soil bio-diversity. This is primarily due to a tremendous lack of knowledge of soil organisms. The sheer abundance of species in the soil (millions·m^{-2}), our naïvete of soil biodiversity at the species or molecular level for groups such as bacteria, fungi, and microinvertebrates, the complexity of relationships of soil biodiversity to vegetation type and ecosystem functioning, and the limited studies on the effect of long-term elevated atmospheric CO_2 and soil warming on soil species diversity contribute to our uncertainty of the impact of global change drivers on soil biodiversity. Nevertheless, there is sufficient evidence to indicate that the composition and abundance of many groups of soil biota are affected, at local and regional scales, by changes in vegetation, soil physical and chemical habitat, climate, and invasive species (Bongers 1990; Ruess 1995). These changes in soil biota are important because they are linked to critical ecosystem processes that sustain life. Soil degradation has accelerated globally as human populations have expanded, threatening the stability of Earth's ecosystems. Determining how soil species diversity will change under global change drivers will help scientists, policy makers, and managers devise and implement strategies to preserve and maintain our terrestrial ecosystems for the long term.

We developed global scenarios of soil biodiversity change and the possible effects on ecosystem functioning for the year 2050 by using the current knowledge on, (1) the wealth of soil biodiversity and its functions in ecosystems (Table 4.1, 4.2), (2) the factors that determine soil biodiversity (Fig. 4.1),

Table 4.1. Major phyla found in soils, in order of increasing body width; their functions in ecosystems, number of species described to date in soil habitats and estimated total number of species that exist in all habitats (terrestrial and aquatic). Species in litter and decaying wood are included in estimates of soil-dwelling species

Body width	Taxonomic group	Function in ecosystem	Described soil species	Reference for described species	Estimated number of species in all habitats	Reference for estimated number of species
1–2 μm	Bacteria	decay; trace gas producers & consumers; N fixers; pathogens; biocontrol	13,000	(Torsvik et al. 1994) (Akimov and Hattori 1996)	1,000,000	(Hammond et al. 1995)
3–100 μm	Fungi	decay; pathogens; plant symbionts; biocontrol	18,000–35,000	(Brussaard et al. 1997)	1,500,000	(Hammond et al. 1995)
15–100 μm	Protozoa	regulate bacterial growth; increase N availability; predators	1,500	(Brussaard et al. 1997)	200,000	(Hammond et al. 1995)
5–120 μm	Nematoda (roundworms)	increase N availability; feed on bacteria, fungi, soil algae, small fauna, plant & animal parasites, & pathogens enhance microbe growth; disperse microbes	5,000	(Brussaard et al. 1997)	400,000–10,000,000	(Hammond et al. 1995)
80 μm–2 mm	Acari (mites)	feed on bacteria, fungi plants, small fauna; parasites; enhance microbe growth; disperse microbes	20,000–30,000	(Brussaard et al. 1997)	900,000	(www1)
150 μm–2 mm	Collembola (springtails)	fungivores; predators; detritivores; feed on algae	6,500	(Brussaard et al. 1997)	24,000	(www2)
300 μm–1 mm	Diplura	fungivores; predators; detritivores	660, 800[a]	(www3; www4)	1,600	(www3)

Size	Group	Function	Described species	Reference	Estimated total	Reference
500 µm–4 mm	Symphyla	detritivores; plant-feeders	160[b]	(Scheller 1982)	No estimate	(Behan-Pelletier pers. comm.)
	Enchytraeidae (pot worms)	detritivores; bioturbators; enhance microbial growth; disperse microbes	600	(Brussaard et al. 1997)	1,200	(Brussaard et al. 1997)
500 µm–4 mm	Isoptera (termites)	bioturbators; aid soil structure; soil and wood feeders; enhance microbial growth; mounds are habitat for biota	1,600	(Bignell pers. comm.; Bignell and Eggleton 1998)	3,000	(Bignell pers. comm.; Bignell and Eggleton 1998)
	Formicoidea (ants)	bioturbators; enhance microbial growth; hill are habitat for biota; aid soil structure	8,800	(Brussaard et al. 1997)	15,000	(Brown 1982)
2–20 mm	Isopoda (sowbugs, pillbugs)	bioturbators; predators; detritivores; herbivores; microbivores	5,000	(www5)	No estimate	
1–50 mm	Chilopoda (centipedes)	predators; mix soil & litter	2,500[c]	(Hoffman 1982)	No estimate	
1–50 mm	Diplopoda (millipedes)	detritivores; calcium cycling; mix soil & litter	10,000[d]	(Hoffman 1990)	60,000	(Hoffman 1990)
	Oligochaete (earthworms)	consume, mix soil & litter; reconstitute litter aid soil structure	3,600	(Brussaard et al. 1997)	No estimate	
	Diptera larvae (flies, blackflies)	plant & animal parasites; predators; decay dung	60,000[e]	(McAlpine 1990)	240,000	(www6)

[a] The majority of Dipluran species are in soil (Maddison 1997). Thus, the number of described soil-dwelling Diplura species described was assumed to be equal to the total number described.

[b] Symphyla live in soil and litter (Scheller 1982); thus, the number of described soil-dwelling species was assumed to be equal to the total number described.

[c] Chilopoda live in soil, leaf litter, rotting woods and caves (Maddison 1997); thus, the number of described soil and litter dwelling species was assumed to equal the total number of described species.

[d] Diplopoda are characteristic of the upper soil and litter horizons (Hoffman 1990); thus, the number of described soil and litter-dwelling species was assumed to be equal to the total number described.

[e] We calculated the number of described species in soils based on McAlpine's estimate (McAlpine 1990) for the United States that 50% of Diptera species are in soils.

Table 4.2. Ecosystem services provided by soil biota[a]

Regulation of major biogeochemical cycles
Retention and delivery of nutrients to plants
Generation and renewal of soil structure and soil fertility
Bioremediation of wastes and pollutants
Provision of drinking water
Modification of the hydrological cycle
   Mitigation of floods & droughts
   Erosion control
Translocation of nutrients, particles and gases
Regulation of atmospheric trace gases (e.g., CO_2, NO_x) (production and
   consumption)
Modification of anthropogenically driven global change (e.g., carbon sequestration
   modifiers of plant responses)
Regulation of animal and plant populations
Control of potential agricultural pests
Contribution to all plant production for food, fuel and fiber

Determinants of landscape heterogeneity
Vital component of habitats important for recreation and natural history

[a] Modified from Bengtsson et al. 1997; Daily 1997; Wall and Virginia 1999.

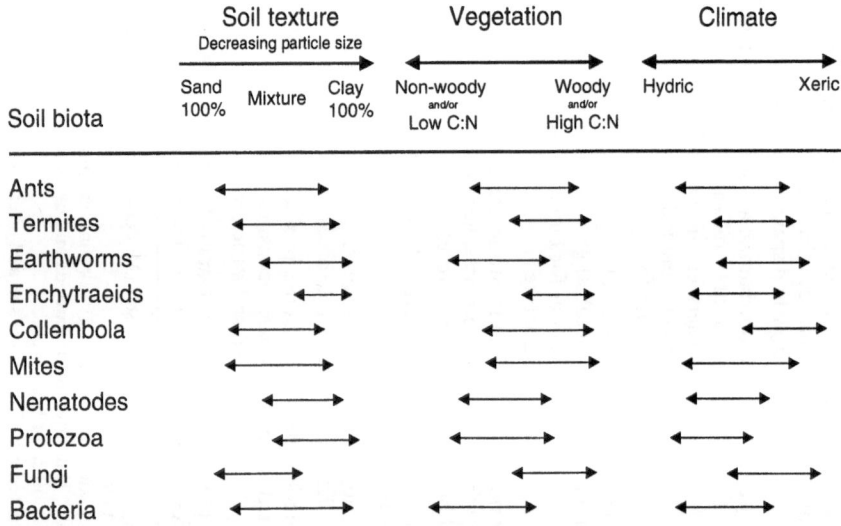

Figure 4.1. Species diversity for most groups of soil biota is strongly affected by global change drivers altering the three determinants of soil biodiversity, soil texture, vegetation, and climate. For each group arrows represent the range where highest biodiversity occurs at a global scale under soil texture [pure sand-100%; a mixture of particle sizes (e.g., sand, silt, and clay), and almost 100% pure clay], vegetation (low to high C:N and nonwoody-to-woody) and climate (hydric to xeric).

(3) the range of conditions over which the highest biodiversity occurs for selected soil groups (Fig. 4.1), and (4) the effects of five drivers of global change—changes in land use, climate, atmospheric CO_2 concentration, nitrogen deposition and acid rain, and introduced species—on soil biodiversity or the factors that determine it. We then ranked the drivers by the impact they would have on the highest biodiversity. This provides the global change scenarios most likely to affect soil biodiversity at the species level (Fig. 4.2) that would have an impact on ecosystem processes.

Soil Biodiversity

Representatives of all above-ground phyla, from chordates [e.g., gophers, prairie dogs (Andersen 1987; Holland and Detling 1990) and caecilians

Soil biota	Δ Land use	Elevated CO_2 Δ Climate (Vegetation change)	Biotic introductions
Ants	★★★★	★★★	★★★★
Termites	★★★★★	★★★★★	★★★★
Earthworms	★★★★★	★★★	★★★★★
Enchytraeids	★★★★★	★★★★★	★★★★★
Collembola	★★★	★★★★	★★★
Mites	★★★★	★★★★	★★★
Nematodes	★★★★	★★★★	★★★★
Protozoa	★★	★★★	★★★★
Fungi	★★★	★★★★	★★★★
Bacteria	★★	★★★	★★★

Figure 4.2. Land use change, elevated CO_2, and biotic exchange will have the most profound impacts on soil biodiversity, but the impacts will differ with the group of organisms shown. Land use change has a radical impact on soil biodiversity because it affects the three determinants of soil biodiversity shown in Figure 4.1. Land use change will have the most visible impact on the entire soil biota. The greatest effect will be seen more quickly on the macrofaunal groups (e.g., termites, earthworms, enchytraeids) because of their effects on ecosystem function (e.g., change in soil structure, increased erosion, change in rate of wood decomposition, changes in hydrology, changes in carbon, and nitrogen fluxes). The effects of elevated CO_2 will be mediated indirectly through changes in one determinant of soil biodiversity, vegetation (Fig. 4.1). Organisms restricted by climate and vegetation will be most affected and include termites, and enchytraeids, with lesser effects on soil mesofauna and fungi. Biotic introductions of plants and other soil organisms will have critical effects on two determinants of soil biodiversity (i.e., soil texture and vegetation) and will cause the greatest changes in the diversity of earthworms and enchytraeids.

(Wake 1993)], to molluscs, arthropods, nematodes, mites and microbes, and from vascular plant roots to algae, are found in subsurface habitats. Some of these groups are listed in Table 4.1. The number of described species in soil represents a small fraction of the estimated total number of discovered soil species. Many scientists predict that there are a greater number of species in soil than above ground (Anderson 1995; Hammond et al. 1995). The high number of species, the complexity of their interactions below ground, and linkages to above ground (Bardgett et al. 1988; Hooper et al. in press), and the difficulty in determining species' contributions to ecosystem processes, has made predictions of their response to environmental change at local and regional scales a tremendous challenge for soil ecologists.

Stotzky and Bollag (1992) noted, "Soil is the most complex of microbial habitats." There is, as yet, no location with a complete species list of all soil organisms, or even a location with a species list of all soil invertebrates. Identifying the total number of species within a given soil for any ecosystem has been hampered by many of the same problems found in assessing biodiversity for above-ground systems (Blair et al. 1996; Wall and Virginia 1999). For example, many identification keys are based exclusively on the adult morphological and biological (life cycle) characteristics; thus, juveniles may be misidentified as a "new" species. As with above ground, many species within a taxonomic group are rare (Bills and Polishook 1994; Price and Siddiqi 1994), the species associations with habitat and other organisms are poorly known, making "hunting" for a species difficult, and migratory and dispersal mechanisms vary with species and taxonomic group.

Determining the influence of global change drivers on soils and soil biodiversity must include consideration of the spatial and temporal scales the organisms use. The spatial habitat varies with the organism's biology and size. Local scales defined for smaller species and the molecular level (e.g., bacteria, fungi, protozoa, nematodes) as a gram of soil may be, for larger species (e.g., gophers, molluscs, ants, millipedes, ectomycorrhizae), meters of soil. The scientist's choice of sampling and extraction methods can affect the biodiversity of the animals retrieved from soil; therefore, methods should be considered when comparing biodiversity across larger scales. Temporal scale considerations (Moore and de Ruiter 1991), such as whether the species migrates centimeters to meters vertically within the soil profile on a diurnal or seasonal basis, or whether it is a transient soil resident (e.g., an invertebrate with part of its life cycle in soil), can influence species diversity estimates. Changes in season, climate, plant species composition, and phenology of plants similarly contribute to changes in the temporal patterns of soil biodiversity.

Major limitations to determining species diversity in soils have been the minimal, and still declining, taxonomic expertise as well as the amount of labor involved in sampling, extraction, and identification of the soil biota. In a Cameroon soil study (Lawton et al. 1998) more than 90% of the nematodes were unidentified because of the labor, time, and cost constraints. Despite these difficulties, melding of technology (e.g., GIS maps of soil types, soil carbon, moisture and vegetation, stable isotope techniques, molecular biology, and microscopic video images for www display) along with interdisciplinary advances are accelerating our understanding of soil biodiversity across temporal and spatial scales (Behan-Pelletier and Newton 1999). Molecular techniques, often combined with morphological techniques, expand our ability to identify species and molecular types for microbial, fungal, and invertebrate groups, and to determine their distribution across geographic scales. These tools also result in discoveries of new key groups of micoorganisms (Dedysh et al. 1998) and relationships among invertebrates (Blaxter 1998; Blaxter et al. 1998). Direct field measurements relating ecosystem processes to soil species diversity [e.g., use of stable isotopes to measure carbon and nitrogen fluxes in individual soil invertebrates (Hobbie et al. 1998), electron probe microanalysis, and laser microprobe microanalysis] are emerging as necessary tools for soil ecology. These and other approaches, combined with global change models (Smith et al. 1998), increase our ability to predict the response of soil biodiversity and ecosystem processes to global change drivers.

Soil Species and Ecosystem Processes

Ecosystem services provided by soil organisms are numerous and critical to human existence (Table 4.2). Information on the role of soil species in global ecosystem processes derives from two disciplines: ecosystem science, where research focuses on the functional or trophic group level, and agricultural sciences, where research focuses at the species level. The synthesis of the two approaches is generating a greater understanding of the potential responses of soil species to global change. The services attributable to soil biota (Table 4.2) include: maintenance of soil fertility, decomposition of organic matter, cycling of nitrogen and carbon, influencing plant fitness through symbiotic, mutualistic, and parasitic associations (Freckman and Caswell 1985; Schimel and Gulledge 1998; Wall and Moore 1999), channeling and moving soil organic matter and other biota horizontally and vertically in the soil, thus affecting hydrology, aggregation of soil particles and soil stability (Lynch and Bragg 1985; Hendrix et al. 1990; Gupta 1994; Lavelle et al. 1994b; Linden et al. 1994; Lavelle et al. 1997), and contributing to clean air and water by degrading pollutants (Coleman and Crossley 1996; Daily et al. 1997; Freckman et al. 1997; Hooper and Vitousek 1998). There has been increasing interest in how soil species diversity structures above-ground plant communities (van der Heijden et al. 1998). In addition, the soil biota serve as food sources, predators, parasites, and biological control agents for plants and animals above and below ground (Gaugler and Campbell 1991; Beckage 1998; Strong 1998). In performing these functions, soil biota are an integral link to other organisms, vertically to above-ground systems, and horizontally to sediments of freshwater and marine ecosystems (Groffman and Bohlen 1988; Freckman et al. 1997; Wall Freckman et al. 1997; Wagener et al. 1998).

Ecosystem research has simplified the vast diversity of the soil biota in order to clarify their role in ecosystems by categorizing them into functional groups, each containing many species of similar morphologies, physiologies, and food sources. Functional groups in soil foodwebs can include primary producers, plant parasites and pathogens, fungal and bacterial decomposers, fungal symbionts, methanogens, nitrogen-fixing bacteria, fungal and bacterial feeders, predators, and omnivores. A consequence of this simplification is that all species of a functional group (e.g., fungal-feeding mites) are considered as processing their food base (carbon, nitrogen) at the same rate, excreting at the same rate, and metabolizing at the same rate.

The functional group approach has permitted exploration of the overall role of soil organisms in ecosystem processes (e.g., decomposition and nutrient flux). Although it ignores species differences, it has yielded valuable findings. Results of the International Biological Program (IBP) showed soil fauna were minor players in the energetics of global terrestrial ecosystems, contributing less than 5% to total soil respiration (Petersen and Luxton 1982). Since then, numerous experimental studies combined with below-ground foodweb models (Hunt et al. 1987; Paustian and Schnurer 1987b; de Ruiter et al. 1993; Smith et al. 1998) involving earthworms and other soil fauna, altogether have shown important effects on other ecosystem processes such as nutrient cycling, herbivory, and trace gas fluxes (Brussaard et al. 1997). Hunt et al. (1987) showed in the shortgrass steppe that bacterial feeding nematodes and amoebae, previously thought to be less significant than other biota in assessments of energetics of the soil community, contributed 83% of the total N mineralization attributed to soil invertebrates. Lauenroth and Milchunas (1991) estimated that one third of the herbivory in the same ecosystem was due to plant parasitic nematodes.

Thus, ecosystem science has provided a backbone for clarifying the role of soil functional groups in ecosystem processes and shown the pivotal linkages among functional groups within the soil foodweb to above-ground systems (Freckman et al. 1997). As with above-ground biota, disturbance of one species may have an indirect or "domino" effect on other soil species within the same, or across different, functional groups. This confounds our ability to predict the effect of global change drivers on soil biodiversity, and, subsequently, changes in ecosystem processes (Coleman et al. 1992; Anderson 1995; Ettema 1997; Giller et al. 1997; Lavelle et al. 1997; Swift 1997; Ettema 1998; Ingram and Wall-Freckman 1998; Swift et al. 1998; Wardle et al. 1998).

Just as the use of functional groups was an important step for evaluating the functions of groups of species in ecosystems, agricultural science also made significant contributions to ecosystem research. This has revealed that species within the same functional group and even within the same genus have varying morphologies, life strategies, and, consequently, varying responses to microhabitat changes. These studies disclose the extent of differences in species' responses within a functional group and the direct (Table 4.3) and feedback effects they have on plant communities. For example, trophically

Table 4.3. Direct effects of some soil organisms on plants

Root Organism	Effect on Plant
Bacterial symbionts (*Rhizobia*, actinorhizas)	Symbiotic nitrogen fixation; sink for photosynthate
Bacterial pathogens	
Pseudomonas solanacearum	Causes bacteria wilt of solanaceous crops (Walker 1969)
Agrobacterium tumefaciens	Causes crown gall of apple, peach, apricot, grape, etc. (Walker 1969)
Streptomyces scabies	Causes common scab of potato (Walker 1969)
Fungal symbionts (ecto- and endomycorrhizae)	Symbiotic resource capture—increases plant tolerance to water and osmotic stress; increased P uptake; sink for photosynthate: (Bowen 1978; Allen 1991) may protect from pathogenic fungi (Newsham et al. 1995)
Fungal pathogens (e.g., *Rhizoctonia* and *Phytophthora*)	Kill root seedlings, infect vascular system, decrease plant growth, cause root rot, increase exudation (Walker 1969; Van Gundy et al. 1977)
Fungal endophytes	Causes enhanced tillering and root growth grasses, increases drought tolerance, protect against pathogens and herbivory, (beneficial to plants) (Schardl and Phillips 1997).
Nematode pathogens	
Pratylenchus penetrans on alfalfa	Changes root architecture, plant nutrient and water uptake and translocation, decreases root biomass. Plants may exhibit no apparent damage, or crop loss.
Bursaphelenchus xylophilus on pine trees	Increases proliferation of lateral roots, decreases tree growth, causes pine wilt, and kills trees.
Aphelenchoides fragariae	Foliar parasite, discolors leaves, kills cells.
Arthropods	
Beetles (Scarabaeidae, Elateridae)	
Diabrotica spp.	Feeds on roots (corn rootworm) (Gerson 1996)
Scale insects	
Daktulosphaira vitifoliae	Grape phylloxera, root sucking insect, feeds on vine roots, can cause galls (Bournier 1977)

similar nematode species respond differently to soil disturbance because of variable ingestion and metabolic rates (minutes to hours) (Schiemer 1983; Ferris et al. 1995; Ferris et al. 1996; Ferris et al. 1997), and generation times (weeks to months) (Ettema and Bongers 1993; Venette and Ferris 1997; Ettema 1998). Soil species within the same genus have differing effects on plants (Table 4.3), and thus different effects on crop production and economics. The response of a plant species to soil pathogens, even within the same genus, can be very specific and is used as a tool for diagnosing plant pathogens and symbionts and determining management alternatives (Cook and Baker 1983; Yeates 1999). Root architecture of the same plant species will change depending on the species of nematode parasitizing the root. The plant response to a root pathogen may also be noted above ground with changes in plant morphology (e.g., size of fruit, plant growth), plant metabolism (e.g., leaf respiration, nutrient status), and plant death.

Agricultural science has provided evidence that: (1) soil species within the same trophic group respond differently to changes in the above- or below-ground environment, (2) trophically similar species can induce different responses in plants, and (3) changing plant species (e.g., in crop rotation) and plant litter inputs (e.g., in no-tillage systems) can affect both individual plant root pathogens, or biocontrol organisms, and the structure of the entire soil community. Thus, we suggest that for global change scenarios predicting changes in vegetation (e.g., species composition) metabolism (NPP, carbon allocation), litter quality, quantity, and morphology (root architecture), soil biodiversity will be affected at the species and molecular level. Further, because of the differential effects at the soil species and DNA level, scenarios of global change influencing climate, plants and the soil physicochemical environment cannot be accurately predicted by examining soil biodiversity only at the functional or trophic group level. Moreover, the magnitude of these global change effects on soil species diversity will be interlocked with the time scale (e.g., the abruptness of the vegetation change), which adds to the difficulty in predicting the response of soil biodiversity to global change drivers. Studies determining the effect of future scenarios of soil biodiversity under global change will need to incorporate knowledge of species and their contribution to ecosystem processes, across temporal and spatial scales.

Geographical Patterns

Geographical patterns of soil biodiversity are poorly known (Brussaard et al. 1997). The International Biological Program (Petersen and Luxton 1982) was perhaps the closest to a global soil biodiversity synthesis (i.e., an estimate of how many and which taxonomic and functional groups exist in biomes across the globe). The objective of the IBP, however, was to obtain results on the contribution of soil fauna functional groups to ecosystem energetics rather than a taxonomic species diversity comparison across biomes. Syntheses of species biogeography have been developed for several years for some above-ground taxa (e.g., for flowering plants, see Williams 1964, Ricklefs et al. 1995), but they are generally lacking for soil biota. Assessments of soil bio-diversity exist for some countries (Szegi 1984; Dighton and Jones 1994), but the few global assessments of soil biodiversity are mostly for only a few components of the vast diversity in soils, generally the larger soil organisms (e.g., ants, termites, earthworms) (Pearce and Waite 1994; Blair et al. 1995; Brussaad et al. 1997; Folgarait 1998).

Assumptions that species richness declines at the poles ignores below-ground complexity (Mooney et al. 1995; Ricklefs et al. 1995; Brussaard et al. 1997; Boag and Yeates 1998). Global biogeographic patterns based on limited data must be evaluated carefully for each group of soil biota. Termite and ant diversity declines with distance from the equator, but this pattern is not assured for earthworms, enchytraeids, mites, and nematodes (Lavelle 1993; Lavelle et al. 1994b; Brussaard et al. 1997; Folgarait 1998). Earthworms vary along a thermolatitudinal gradient, and are absent from dry ecosystems (Lavelle et al. 1997). The number of mite species was similar in tropical South American and temperate North American sites (Stanton 1979; Heneghan et al. 1998). For nematodes, where there often are more than 100 rare species in a soil sample, Procter (1984, 1990) and Boag and Yeates (1998) suggested a greater diversity in temperate than tropical latitudes; however, other data show greater species richness in a Cameroon site (Lawton et al. 1996; Bloemers et al. 1997; Lawton et al. 1998). Even ant diversity that follows the latitudinal pattern declines with altitude and aridity (Folgarait 1998).

Establishing biogeographical patterns for soil organisms is difficult. Species overlap between soil samples taken within the same ecosystem can be low, particularly for microscopic organisms (e.g., mites, Collembola, nematodes, tardigrades, protozoa, bacteria, fungi). For example, less than 6% of the total nematode species were common to 50 cores collected within the same ecosystem (Price and Siddiqi 1994). Ettema (1998) notes that because nematode diversity is significantly patchy at the local scale of soil cores, as well as at regional and ecosystem levels, patchiness occurs across all scales (see also Robertson and Freckman 1995). Communication via the internet increases opportunities for syntheses of geographical distribution patterns (e.g., Collembola; http://www.geocities.com/~fransjanssens/).

The lack of syntheses on distribution patterns for the entire soil biota (Table 4.1) limits our ability to predict how soil species diversity and soil communities will vary with biome, climate, and vegetation on regional to global scales. Many species are cosmopolitan, and many regions have endemic species (daGama et al. 1997; Wall Freckman and Virginia 1998), but we have been unable to identify traits that allow a species to be either cosmopolitan or endemic. A greater understanding of biogeographic patterns and the mechanisms determining habitat ranges will allow better predictions of species loss, species introductions, the identities of species needed for restoration and maintenance of soils, and the effect of global change on subsurface biota that are involved in transfers of materials between ecosystems. As we currently probe the impact of global change scenarios on species diversity and biome shifts and develop management plans for sustainable soils, we rely on limited knowledge of ca. 5–10% of the soil species (Brussaard et al. 1997; Wall and Virginia 1999).

Determinants of Soil Biodiversity

The abundance and structure of the soil community at the global scale are determined by: (1) vegetation, (2) soil physical and chemical properties, (3) microclimate, and (4) the interactions among soil organisms (Swift et al. 1979; Anderson 1995; Giller et al. 1997). These factors are integrated over geologic time, making it difficult to determine whether a change in one factor will result in a corresponding change in soil biodiversity.

On a global and regional scale, vegetation is the primary determinant of soil biodiversity. Numerous studies have shown decreasing soil organism abundance and diversity away from plants (Ingham et al. 1985; Freckman and Mankau 1986; Yeates 1987; Ingham et al. 1989; Yeates and Orchard 1993; McSorley and Frederick 1996). Plants influence the composition of the soil community through both root and above-ground organic inputs and changes in soil microclimate (e.g., moisture, temperature, oxygen content, etc.). Plant biomass, the quantity and quality of the litter (C, N, lignin) (Fig. 4.1), the plant species community composition, and the functional composition of plants (C_3, C_4) all contribute to structuring soil communities (Lawton et al. 1998; Hooper et al. 2000). For example, input of organic matter from vegetation with lower C:N ratios and less lignin results in decomposition by a bacterial-based foodweb, whereas a presence of a greater amount of plant structural material (e.g., lignin and cellulose) results in decay by a fungal-based foodweb (Paustian and Schnurer 1987a,b; Christensen 1989). Plant phenology, root architecture, depth of rooting, and plant metabolism, including reallocation of carbon within the plant, add to the multiplicity of factors that affect soil biodiversity. On a global and regional basis, the vegetation shifts predicted by global change models will indirectly impact soil species diversity, and the feedback effects on global ecosystem processes are unknown.

Soil physical and chemical properties provide the habitat for soil organisms and include soil texture (the combination of sand, silt, and clay) (Fig. 4.1), moisture, salinity, pH, organic carbon, and N, P, K. Soil physical structure determines both the space and boundaries for organisms and the habitat temperature and moisture dynamics for the species that coexist in that habitat (Wallace 1963; Sohlenius 1985; Robertson and Freckman 1995). Organic matter, soil chemistry, and legacies of soil changes due to historical land use modify the range of soil textures and thus influence soil biodiversity. On a global scale, for example, mite diversity is generally higher in well-structured soils with high organic matter, and lower in soils with a sandy texture of limited organic matter (Walter 1999). Based on literature and personal communication with taxonomic experts, we hypothesize that on a global basis, the diversity of invertebrates (e.g., ants, termites, earthworms, and enchytraeids) will be higher in soils that have a greater soil textural diversity as compared with soils of nearly pure sand or pure clay (Fig. 4.1).

Adding to the complexity of factors that influence geographic patterns of soil biodiversity at the regional level is the influence of geologic history on soils. Earthworm distribution in North America has been defined by glaciation, with species moving into areas of the Wisconsinian Glaciation (Gates 1982; James 1990, 1995). Enchytraeids have replaced earthworms in many tundra, boreal forests, and northern parts of temperate zones. Soil biodiversity in the Antarctic Dry Valleys appears to be structured by legacy carbon left from ancient lakes, soil salinity, and soil moisture (Wall Freckman and Virginia 1998; Burkins et al. 2000; Virginia and Wall 1999).

Microscale soil texture characteristics at the patch and microscale can influence trophic dynamics. Robertson and Freckman (1995) used geostatistics to examine the spatial variation of nematode trophic groups across cultivated ecosystems and found that soil texture and pH explained the patchy distribution of all groups except plant parasites at meters to hectare scales. At the microhabitat scale, Elliott et al. (1980) noted little competition between bacterial-feeding protozoa and nematodes for their food source, bacteria, because protozoa access bacteria in soil textures with smaller-sized soil pores than do nematodes. Thus, a mixture of textures would allow for greater diversity of biota.

Physical disturbance to soils decreases soil biodiversity (Paoletti et al. 1992; Freckman and Ettema 1993; Lavelle et al. 1994a). Physical disturbances that change soil structure (e.g., plowing, loss of topsoil through water and wind erosion) have a rippling effect on other factors defining the organism's habitat, such as soil moisture, oxygen availability, and soil chemistry. For example, soil texture compaction differentially affects invertebrates (Whalley et al. 1995) and vertebrates (e.g., caecilians, gophers, prairie dogs) (Andersen 1987; Ducey et al. 1993). The loss of species and resulting impact on ecosystem processes is more obvious for larger invertebrates (e.g., earthworms, termites, and ants), whose diversity is narrowly defined by soil parameters such as texture (Sochtig and Larink 1992) (Fig. 4.1).

Regional climate patterns (e.g., water, temperature, wind) contribute to the distribution patterns of soil organisms (Swift et al. 1979). Enchytraeids and earthworms are absent in dry and desert ecosystems, and in cold, acid waterlogged regions of the world (Blair et al. 1995; Lavelle et al. 1997). Climate changes that result in drier surface soils affect the seasonal distribution patterns of enchytraeids in soils, and increase their densities at depth (Springett et al. 1970; Briones et al. 1997). The distribution of some plant parasitic nematode species in the UK can be explained by their relationship to isotherms (Boag et al. 1991) and plant type (e.g., deciduous, coniferous, grasses) (Boag 1974; Boag and Williams 1976). Young et al. (1998) suggest that freezing–thawing events predicted by global-change models affect soil particle size distribution and thus have an influence on local-scale soil species diversity.

Rainfall patterns could shift soil community composition dramatically on a regional basis, directly or indirectly, through changes in vegetation. Flooding can be essential for the life cycle of many species, even if it occurs rarely. In the playas of the southwestern United States deserts, brine shrimp that inhabit the soil in an egg state for years, suddenly appear active with flooding (Crawford 1981); alternatively, flooding events can create anaerobic soil conditions that affect species composition.

Effects of Global Change Drivers

Although soil biodiversity may be determined by factors other than, or in addition to, the three major factors we have selected (Fig. 4.1), we considered the soil biota holistically and in the context of those factors that determine distribution and diversity. We asked, under what conditions (i.e., soil structure, vegetation, and climate) will there be the greatest phyla and species diversity and which of these groups might be more vulnerable to global change drivers? Representative groups that are known to affect ecosystem processes were used in Figure 4.1. We approached this task cautiously because for many groups, such as nematodes, protozoa, bacteria, and fungi, there are few data on the relationship of diversity to soil texture, vegetation type, or climatic regime.

Those global change drivers that affect the soil physical and chemical properties, vegetation, and climate and shift them between conditions that support high versus low biodiversity over the largest land area (e.g., regional, biome, landscape, and continental) will have the greatest influence on future soil biodiversity. These will probably be land use change, elevated CO_2 and climate change (through vegetation change), and biotic exchange (Fig. 4.2). In comparison, nitrogen deposition and soil warming will affect species-level diversity to a lesser extent. We are aware that interactions among these factors (e.g., among water and vegetation) can be simultaneous and additive, yet subject to a time lag due to migration and dispersal limitations, and the like. This adds to the challenge of predicting the effect of global change drivers on soil biodiversity at large spatial scales.

Effects of global change drivers on soil biodiversity are described in greater detail later, but we expect land use change to have the greatest impact because it radically alters the three most important determinants of diversity (e.g., soil structure, vegetation, and microclimate). Evidence strongly suggests that the species diversity of all soil taxa decreases precipitously with land use change. Land use change is a physical disturbance that results in vegetation change and the disruption of soil habitats, thus changing the organic inputs to soil and altering the decomposition foodweb. This, in turn, affects plant roots and the herbivorous foodweb. Effects on climate at micro-, local, and regional scales are linked with these changes. We would expect the diversity of termites, earthworms, and enchytraeids to be most affected because their diversity is confined to a small range of soil textures. Bacteria would be the least affected because their diversity, although just beginning to be discovered, appears to occur across a broad spectrum of soil textures, plant communities, and climate.

Elevated CO_2 and climate change will have major effects on soil biodiversity indirectly through vegetation change (e.g., plant metabolism and plant community composition). Based on experiments at the functional group level, we expect that, if below-ground plant resources (no change in plant species) increase, the abundance of bacterial and fungal foodwebs, root pathogen populations, and herbivory will increase. It is unclear if these increased carbon inputs will affect biodiversity; however, root pathogen diversity (e.g., obligate parasites:fungi, bacteria, plant–parasitic nematodes) and symbiont diversity (mycorrhizae, rhizobia), is expected to decline with any sudden change in plant composition that results in nonhost plants. Those organisms (e.g., ants, termites, enchytraeids) whose highest diversity is constrained by climate and/or confined to an extreme of vegetation types (either woody or non woody) (Fig. 4.1) would be expected to have greater changes in diversity, with perhaps a decline in diversity, if vegetation types were switched to the other under elevated CO_2 and climate change (Fig. 4.2).

As a global change driver, biotic introductions of animals, plants, and microbes have had, and will continue to have, a devastating effect on soil biodiversity (Fig. 4.2). Introductions of plant species change the soil habitat, root-symbiotic and pathogenic relationships, hydrology, and organic matter inputs to the soil. For plant introductions, we would predict the greatest changes in diversity for termites, enchytraeids, bacterial and fungal symbionts (e.g., rhizobia, mycorrhizae), and root pathogens (Fig. 4.2). Because of their impacts on soil physical and chemical habitats invasive soil macroinvertebrates (e.g., ants, termites, earthworms) are predicted to have major repercussions on the diversity of soil mesofauna, bacteria, and fungi.

Studies on atmospheric deposition (primarily sulfate and nitrate) showed effects on soil chemistry and diversity (Huhta 1984; Bewley and Parkinson 1986; Persson et al. 1989; Baath et al. 1990; Hyvonen and Persson 1990). We would expect aquatic soil animals confined to nonporous, less drained and more clay soils (earthworms, enchytraeids, nematodes, protozoa), as well as bacteria and fungi, to have decreased biodiversity globally because of changes in water chemistry due to increased atmospheric nitrogen deposition.

Soil warming experiments have shown differential effects on groups of soil taxa at local scales. The mechanisms by which warming effects taxa include direct climatic effects (e.g., increased temperature on soil habitats) and altered wet-drying frequency cycle that affects migration, dispersal and life histories. Because warming can affect both hydric and xeric (Fig. 4.1) habitats, we made no prediction as to which biotic group will be most affected at larger scales. Polar regions, however, may lose a key group—enchytraeids—with warming, but they may also gain in overall soil biodiversity.

Land Use Change

Accelerating destruction of soils through conversion of land for human use will be the primary driver that impacts soil biodiversity because of its strong effect on soil physical and chemical properties, vegetation, and microclimate (Sochtig and Larink 1992; Murphy et al. 1995; Brussaard et al. 1997; Lavelle et al. 1997). In general, we expect that land conversion to regrowth forests, agriculture and cities will destroy lineages of soil species and the multiplicity of their interactions at the regional scale. The disturbance to soil taxa and soil community structure has been recorded in many ecosystems (Freckman and Ettema 1993; Brussaard et al. 1997; Lavelle et al. 1997; Folgarait 1998). Altered diversity and composition of earthworms, nematodes, and mites have been used to indicate the changes in soil quality (Bongers 1990; Blair et al. 1996).

The Images model (this volume) predicts the types of land conversion from 1990 to 2050 for each continent (Table 4.4). The North American continent is projected to lose agricultural land, tundra, forests, woodlands, and deserts and gain substantially in regrowth forests, grassland/steppe, grassland, ice, scrub and savanna (Table 4.4). North America's loss of agricultural land is typical of other developed regions (e.g., Europe, Australia, and New Zealand). In contrast, the developing countries are predicted to increase their agricultural land, driving the global increase in agroecosystems. The types of agricultural management practices selected in the future (e.g., slash and burn, rotations, no-tillage with minimal chemical inputs, conventional, intensive monoculture agriculture with high chemical inputs) will determine the extent of impact of this land use change on soil biodiversity (Anderson 1994; Brussaard et al. 1997; Giller et al. 1997; Swift 1997).

Whether agricultural management is based on intensive, high-chemical input or no-till, minimal-chemical input, all types of management result in disturbance to the soil and the biota at local to regional scales. Cultivation, pesticides, increased herbicide use on no-till crops, increases in pests and invasive species, depletion of soil organic matter through erosion, and overuse of land can all affect trophic structure and composition, species abundance, richness, and biomass of soil organisms (Brussaard et al. 1997; Niles and Wall Freckman 1997; Mando 1998). In some groups, however, disturbance results in effects other than a decline in biodiversity. For example, following a disturbance, ant diversity may decline with or without a corresponding increase in individuals, remain unchanged, or increase as opportunists from nearby land invade (Folgarait 1998). The effects of these changes on ecosystem processes and services can be expected to be profound at the regional and continental scales.

Table 4.4. The IMAGES (this volume) projection of land use change from 1990 to 2050. The net land conversion was determined as the area (km²) by which the land type will increase or decrease (–) from 1990 to 2050. The upper figure shows the net land conversion in area ($\times 10^3$ km²); the lower figure shows the net land conversion as % of 1990 area

Each cell shows: upper figure (net land conversion in area, $\times 10^3$ km²) / lower figure (net land conversion as % of 1990 area).

	Agricultural land	Extensive grassland	Regrowth forest	Ice	Tundra	Wooded tundra	Boreal forest	Cool conifer forest	Temperate mixed forest	Temperate deciduous forest	Warm mixed forest	Grassland/steppe	Hot desert	Scrubland	Savanna	Tropical woodland	Tropical forest
North America	-554.6 / -14.2	110.7 / 9.4	834.0 / 250.6	70.2 / 36.6	-623.0 / -25.3	-157.8 / -16.2	-52.7 / -1.1	-216.5 / -27.7	-262.7 / -28.9	-316.3 / -59.2	-277.3 / -54.7	1,381.4 / 84.2	-11.2 / -5.2	86.9 / 104.6	8.3 / n.a.	-19.3 / -54.0	0.0 / n.a.
Mexico, Central America, and Caribbean	71.7 / 6.1	0.0 / n.a.	0.0 / n.a.	0.0 / n.a.	0.0 / n.a.	0.0 / n.a.	0.0 / n.a.	0.0 / n.a.	0.0 / n.a.	0.0 / n.a.	0.0 / n.a.	-58.1 / -22.7	-117.4 / -53.8	146.6 / 173.2	-15.6 / -5.6	-15.1 / -7.3	-12.1 / -28.6
South America	564.1 / 11.1	-40.9 / -4.5	364.4 / n.a.	0.0 / n.a.	-47.1 / -43.3	11.2 / 46.8	-48.2 / -29.1	0.0 / n.a.	40.1 / 89.1	4.4 / 37.1	-80.7 / -32.0	-60.1 / -5.7	165.5 / 301.8	-77.0 / -14.6	-621.1 / -25.3	-55.1 / -1.4	-119.3 / -4.3
Europe	-601.5 / -11.8	26.8 / 8.0	430.4 / 89.1	-10.5 / -20.9	-40.4 / -27.1	-54.7 / -42.7	-249.5 / -20.2	24.0 / 3.1	470.7 / 89.7	7.2 / 1.8	1.4 / 1.7	-30.0 / -21.6	2.5 / n.a.	23.6 / 18.3	0.0 / n.a.	0.0 / n.a.	0.0 / n.a.
Africa	7,127.1 / 88.6	2,472.9 / 88.5	-89.8 / -97.1	n.a. / n.a.	n.a. / n.a.	n.a. / n.a.	n.a. / n.a.	n.a. / n.a.	0.0 / n.a.	n.a. / n.a.	-56.9 / -100.0	-695.4 / -56.4	-419.7 / -4.8	-1,496.9 / -98.7	-4,364.4 / -100.0	-1,509.3 / -100.0	-967.5 / -100.0
Asia	3,975.2 / 44.3	2,011.0 / 36.4	584.2 / 131.0	-62.0 / -62.3	-1,548.3 / -49.7	-269.0 / -23.9	179.3 / 2.1	135.1 / 31.1	-23.7 / -11.1	-235.8 / -79.5	-356.8 / -92.1	-1,351.3 / -29.2	-1,993.7 / -43.4	-244.1 / -50.0	-576.7 / -51.5	-171.1 / -65.2	-52.3 / -53.1
Indonesia and Pacific Islands	565.4 / 122.4	9.1 / 12.8	132.3 / 214.9	0.0 / n.a.	0.0 / n.a.	0.0 / n.a.	0.0 / n.a.	0.0 / n.a.	0.0 / n.a.	0.0 / n.a.	-24.6 / -80.0	0.0 / n.a.	0.0 / n.a.	0.0 / n.a.	-12.0 / -66.8	-45.7 / -51.8	-624.5 / -39.0
Australia and New Zealand	-243.8 / -8.5	318.5 / 16.9	42.2 / 365.0	0.0 / n.a.	-4.4 / -100.0	-2.2 / -50.0	-17.4 / -100.0	0.0 / n.a.	-2.3 / -100.0	-2.5 / -100.0	-36.3 / -100.0	-83.9 / -6.3	-113.0 / -11.3	152.1 / 38.1	-7.1 / -7.4	0.0 / 0.0	0.0 / 0.0
Greenland	0.0 / n.a.	0.0 / n.a.	0.0 / n.a.	95.6 / 4.8	-95.6 / -61.2	0.0 / n.a.	0.0 / n.a.	0.0 / n.a.	0.0 / n.a.	0.0 / n.a.	0.0 / n.a.	0.0 / n.a.	0.0 / n.a.	0.0 / n.a.	n.a. / n.a.	n.a. / n.a.	n.a. / n.a.
Global	10,903.4 / 30.6	4,908.0 / 37.9	2,297.7 / 158.0	93.3 / 4.0	-2,358.9 / -39.3	-472.4 / -21.0	-188.5 / -1.3	-57.4 / -2.9	222.0 / 13.1	-543.0 / -43.4	-831.1 / -61.3	-897.3 / -8.7	-2,487.1 / -16.8	-1,408.9 / -43.6	-5,588.6 / -67.1	-1,815.5 / -30.5	-1,775.6 / -32.3

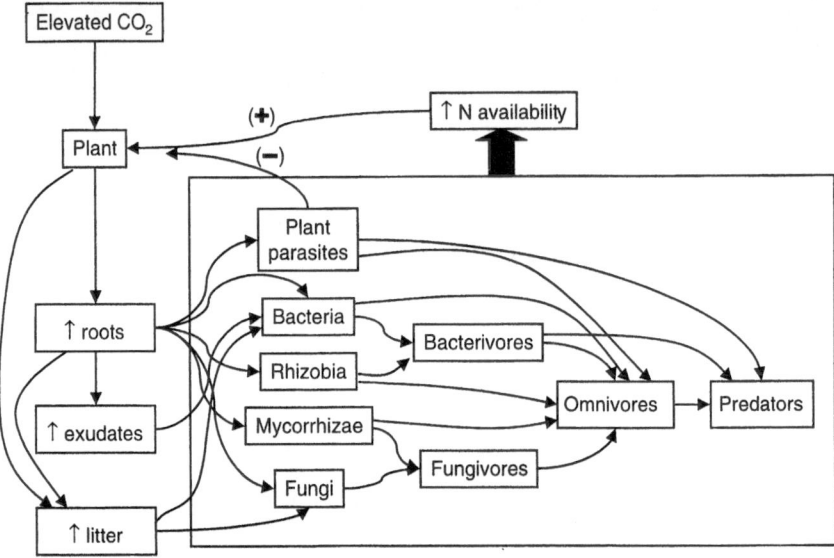

Figure 4.3. Elevated atmospheric CO_2 is likely to influence soil foodwebs indirectly via changes in plant physiology. Figure 4.2 shows the components of soil foodwebs that may be influenced by increases in litter production and root activity. The response of biodiversity of such foodwebs is virtually unstudied. Consequences for biomass may be more reliably predicted (Norby and Cotrufo 1998) and stimulation of micro-bial-based foodwebs (with positive feedbacks to plant productivity) or of plant parasites (with negative feedbacks to plant productivity) may occur.

Elevated CO_2

Effects of elevated CO_2 on soil biodiversity may be largely indirect via altered vegetation (Yeates et al. 1997; Folgarait 1998; Wolters et al. 2000). Because of the existing high CO_2 concentrations within the soil, direct effects of ele-vated CO_2 on soil biodiversity have been considered less important than indi-rect effects (van Veen et al. 1991; O'Neill 1994; Coleman and Crossley 1996; Lavelle et al. 1997; Wolters et al. 2000). This assumption, however, requires further investigation (Sustr and Simek 1996) because gradients as low as 100 ppm CO_2 in the soil atmosphere attract plant parasitic nematodes to roots over distances of several meters (Dusenbery 1987; Pline and Dusenbery 1987). CO_2 also attracts entomopathogenic nematodes to their insect hosts in soil (Barbercheck and Kaya 1991). Indirect effects of elevated CO_2 on soil foodwebs, via an altered supply of plant-based resources, have been studied in more detail than direct effects and could occur with (1) changes in plant species composition, and (2) changes in plant physiology (Fig. 4.3).

The composition of plant communities could be impacted by elevated CO_2 through differential physiological responses of plant species, depending on plant functional group (e.g., C_3/C_4) and capacity for additional carbohydrate utilization. Plant community composition could change within ecosystems (Korner 1996) or across biomes (Cure and Acock 1986; Oechel et al. 1994; Drake et al. 1996) and could indirectly influence soil biodiversity and ecosystem processes. There are few data, however, on the effects of altered plant species composition due to elevated CO_2 on soil biodiversity.

The effect of elevated CO_2 on plant physiology has been studied in more detail. Elevated CO_2 consistently enhances photosynthesis, at least initially, for a wide variety of plant species (Bazzaz 1990; Korner 1996). The plant physiological changes that may have greatest impact on soil biodiversity are increased net primary productivity (biomass) (Yeates et al. 1997), including an increase in above-ground biomass with increased litter inputs to soil and increased below-ground inputs (noted later); changes in plant tissue quality (Lincoln et al. 1993; Penuelas and Estiarte 1998) increased water use efficiency (Casella et al. 1996; Drake et al. 1997; Jackson et al. 1998) and altered phenology (Bazzaz et al. 1989; Newton et al. 1995). Below-ground plant responses most frequently noted in elevated CO_2 experiments are reallocation of enhanced photosynthate manifested as an enhanced root biomass (especially under water and nutrient limited conditions), increased root length, and rhizodeposition (Stulen and den Hertog 1993; van Noordwijk et al. 1998; Wardle et al. 1998) and root respiration (Boone et al. 1998). There is insufficient evidence to determine whether root turnover increases (Pregitzer et al. 1995; van Noordwijk et al. 1998).

Most data on the response of soil organisms to plant physiological responses has been at the functional group level (O'Neill 1994; Yeates et al. 1997; van Noordwijk et al. 1998; Wardle et al. 1998). These studies indicate that increased below-ground plant resources affect both herbivory and decomposition foodwebs (Zak et al. 1996). Decomposition foodweb responses include an increase (O'Neill et al. 1987; Klironomos et al. 1998) to no change (Yeates and Orchard 1993; Lussenhop et al. 1998) in microbial biomass, increased fungal mass, hyphal length, percent colonization and/or spores, and an increased abundance of fungal feeding microarthropods (Klironomos et al. 1996; Jones et al. 1998; Lussenhop et al. 1998; Rillig et al. 1999), bacterial feeding nematodes (Yeates and Orchard 1993), protozoa (Lussenhop et al. 1998), and enchytraeids (Yeates et al. 1997). The differences in microbial biomass response to elevated CO_2 may in part reflect differences between fungal and bacterial foodwebs. For example, fungal biomass appears to be more consistently enhanced than bacterial biomass, and Diaz (1996) has suggested that this may favor mycorrhizal plant species over nonmycorrhizal species.

Although it has generally been assumed that increases in below-ground resources driving shifts in soil biotic communities under elevated CO_2 are due to reallocation of enhanced photosynthate, other mechanisms should be considered. In non-CO_2 experiments other abiotic and biotic factors have been shown to increase root exudation and cause shifts in microbial composition. Increased leakiness of roots and a substantial increase in the amount of root exudates can be due to waterlogging and the decreased oxygen availability in soils (Kuan and Erwin 1980), the nematode *Meloidogyne* parasitizing and inciting gall formation in roots (Van Gundy et al. 1977), and "normal" root microflora (Prikryl and Vancura 1980; Jones and Darrah 1993). Elevated CO_2 could potentially alter these factors (e.g., by altering water use efficiency of plants, host susceptibility to parasitism or composition of root microflora, respectively) leading to nonphotosynthate-driven effects on root exudates with consequences for soil biota.

There are few studies of the indirect effect of elevated CO_2 on soil biodiversity at the species level, and the limited data make it difficult to extrapolate findings to regional and global levels. A number of studies have noted shifts in soil species composition under elevated CO_2. Jones et al. (1998) found significant and consistent shifts in soil fungal species composition between Ecotron experiments maintained at ambient and elevated CO_2. Collembola, which are major consumers of fungi and graze selectively on different fungal species, also showed dramatic shifts in species composition.

In a pot experiment Klironomos et al. (1998) found differential responses of mycorrhizal fungi to elevated CO_2, with consequences for the host plant *Artemisia tridentata*. The mycorrhizal species *Glomus intraradices* and *Glomus etunicatu* increased intraradicle structures and spore production, and their host plants showed increased growth under elevated CO_2. Other mycorrhizal fungi, *Acaulospora sp.* and *Scutellospora calospora*, increased extraradicle hyphal length and the host plants showed increased leaf phosphate content, but not growth, under elevated CO_2. The authors suggest that host plants of mycorrhizal species having enhanced phosphate supply under elevated CO_2 may have a longer-term competitive advantage, which is a hypothesis that could be readily tested in field experiments in different biomes.

The increased plant carbon:nitrogen ratio often observed under elevated CO_2 has been hypothesized to reduce plant litter quality and thus alter the soil decomposer community and rates of decomposition. This hypothesis has generally been unsupported (Norby and Cotrufo 1998). There are exceptions, however, such as the reduced quality of ash litter under elevated CO_2 being less palatable to the woodlouse *Oniscus assellus* (Cotrufo et al. 1998). Changes under elevated CO_2 in the amount of litter or in plant phenology (e.g., relative proportions and timing of leaf, stem, root, fruit, and woody litter) may be more likely than altered litter quality and could have profound effects on soil species and decomposition rates (Andrén et al. 1995; Norby and Cotrufo 1998).

CO_2-induced changes in above-ground, living plant tissue chemistry has significant effects on above-ground herbivores (see Lincoln et al. 1993; Penuelas and Estiarte 1998); below ground, however, there is little evidence of CO_2—induced changes in root quality (Van Ginkel et al. 1996; Jones et al. 1998) affecting herbivores (van Noordwijk et al. 1998). Differences in root quality in nonelevated CO_2 experiments, however, have had significant effects on plant pathogenic soil species, and there is strong circumstantial evidence for co-evolution of root chemistry and root herbivores, which suggests that CO_2-induced changes in root chemistry might have significant impacts on herbivorous soil biota (Wallace 1963; van Noordwijk et al. 1998; Wardle et al. 1998).

Biotic Introductions

Introductions of above-ground or below-ground plant or animal species can change soil structure, soil chemistry, soil biotic communities (Mamiya 1983; Rutherford et al. 1990; Boag et al. 1997; Burtelow et al. 1998; Folgarait 1998; Fraser and Boag 1998; Ehrenfeld 1999), and nutrient cycling (Vitousek et al. 1987). Dispersal of soil-invasive species into new habitats occurs with transport of soil attached to plants, vertebrates (e.g., cattle, birds, zoo animals) and invertebrates, field and construction machinery, and wind erosion of soils. In addition, movement of soil through irrigation and flooding is a source of soil organism dispersal on a local to regional scale.

The introduction of exotic plant species can have unexpected effects on soils and soil biodiversity. Invasion of nitrogen-fixing symbiotic bacteria with the tree, *Myrica faya*, changed the nutrient cycling and organic matter status of the Hawaiian ecosystem (Vitousek et al. 1987). The introduction of two exotic plant species, *Berberis thunbergii* and *Microstegium vimineum* in the northeastern United States, resulted in soils with higher pH, a thinner litter layer and organic horizon, and higher populations of earthworms (Kourtev et al. 1998). Kourtev et al. (1998) speculated that the exotic plants created a more favorable soil environment for earthworms than uninvaded soils. The resulting larger populations of earthworms may have incorporated more litter that resulted in the decrease in organic matter.

Introductions of exotic soil macroinvertebrates can cause major changes to soil habitat characteristics (Folgarait 1998) and will change the species composition of mesofauna (e.g., mites, Collembola, nematodes, and protozoa). Ant and earthworm species, with different biologies than the indigenous community, can alter distribution of organic matter, soil aeration, water infiltration, and soil chemistry (Lee 1985; Lavelle et al. 1997; Burtelow et al. 1998; Folgarait 1998), affecting soil biodiversity and ecosystem processes. Lee (1961) recognized that the major threats to earthworms were habitat degradation and invasion of exotic competitors. Lee has more recently implored scientists and others to preserve whole ecosystems and to gather scientific experimental data on impacts of invasive species on soils and ecosystem processes (Lee 1999; see also James 1995). Invasions of Australasian planarian flatworms, which have no known beneficial effects on soils but are obligate predators of earthworms, are causing great concern in the United Kingdom and northern Europe (Boag et al. 1997; Fraser and Boag 1998). All flatworm species in North America are introduced (Ogren and Kawakatsu 1998), but there is little information on their ecology and effect on ecosystem processes.

The invasive species in soils have had major effects on economies of nations. Some microscopic species of soil bacteria, fungi and plant parasitic nematodes have been subject to state, national and international quarantine regulations for decades. For example, the nematode species *Bursaphelenchus xylophilus* (Table 4.3) introduced from North America to Japan and then Portugal (Mota 1999), is capable of killing pine tree plantations and forests in less than 6 weeks and is subject to quarantine regulations (Mamiya 1983; Rutherford et al. 1990). These illustrations show that the introductions of plants and soil biota can have important, costly impacts on soil biodiversity, soil structure and fertility, and ecosystem processes at regional and global scales. Initiatives for global programs on species invasions must consider the impacts on soil systems as a priority.

Nitrogen Deposition

Nitrogen deposition can have major effects on the biodiversity of soil communities because of its indirect effects on vegetation (e.g., changes in plant productivity, plant quality, and plant composition) and direct effects through acidification of the soil habitat. Major taxonomic groups have differential responses to acid deposition, and some evidence also exists for shifts in diversity at the functional group and species level. We expect enchytraeids, earthworms, mesofaunal groups, and microbial groups to be most influenced because of the acidification of the soil. Nematode and microarthropod functional groups are viewed as good indicators of the effects of acid deposition on soils (Van Straalen et al. 1988).

At a regional and global scale, the effect of nitrogen deposition may be strongly modified by the production potential of the ecosystem (Vitousek et al. 1997). When ecosystems have a high productivity potential, nitrogen deposition may increase the productivity of existing plant species rather than directly altering plant species composition (Vitousek et al. 1997); thus, effects on soil biota will be through altered amounts and quality of resource inputs from already existing plant species. In ecosystems adapted to low nitrogen availability, increased N deposition may dramatically alter plant species composition (Wedin and Tilman 1996; Vitousek et al. 1997) with consequences for soil biotic composition (Berg and Verhoef 1998).

Studies of acidification effects on soil biota have often considered acidification effects due to sulphate, as well as nitrogen deposition. Many studies are from forest ecosystems where declines in abundance and shifts in community composition of earthworms, enchytraeids, nematodes, bacteria, and fungi have been observed (Huhta 1984; Bewley and Parkinson 1986; Van Straalen et al. 1988; Persson et al. 1989; Baath et al. 1990; Hyvonen and Persson 1990; Ruess 1995). Long-term studies in the Tatra Mountains of Slovakia showed that vegetation and soil biodiversity changed with acid deposition (Rusek 1993). Collembollan species that were rare in 1977 became abundant and widespread in distribution by 1990. In contrast, some previously abundant collembolan species have totally disappeared from the region. Vegetation patterns shifted as a result of acid deposition, and collembolan species composition became more homogenous and less diverse at the landscape scale.

Insights on the effect of nitrogen deposition on soil species diversity may also be gained from studies of additions of inorganic fertilizer to soils, although fertilization, in contrast to atmospheric nitrogen deposition, involves sudden and often larger nitrogen inputs. At the functional level, nitrogen fertilization appears to stimulate root herbivory– and bacterial-based foodwebs (Gupta 1994). At the species level, inorganic N fertilization increased (Lehle and Funke 1989) or had no effect on protozoan diversity (Berger et al. 1986). It also decreased the diversity of microarthropods (Siepel and Van de Bund 1988).

Soil Warming

Predictions of future temperature and precipitation are less certain than predictions of atmospheric CO_2 concentrations and land use change. In general, predictions suggest an increase in mean global temperature of 1–3.5°C, increased precipitation (with regional variation), and increased severity and frequency of catastrophic climatic events (Acosta et al. 1998). The predicted climate changes may affect soil biotic diversity directly by altering soil temperature, moisture, and wet-drying and freeze-thawing cycles, or indirectly by altering vegetation community and productivity, and the amount and timing of decomposition (Hodkinson et al. 1998; Sohlenius and Bostrom 1999). Young et al. (1998) predict that these changes in vegetation may alter soil biodiversity through changes in soil carbon content and spatial heterogeneity of soil structure.

These climatic effects would be predicted to affect most strongly organisms that presently live at extremes of climate tolerance ranges. For example, enchytraeid species, which are cold tolerant and key taxa in acid soils of polar regions, might decline at high latitudes with increased frequency of climatic events that affect soil temperature, moisture, and freeze-thaw cycles (Somme and Birkemoe 1997; Huhta et al. 1998).

The soil biotic response to warming may strongly interact with moisture availability, at both regional and local scales. At regional scales, under warmer drier conditions, grasslands may increase, and their rhizome-associated, aggregating microorganisms may increase soil aggregates and biodiversity. In contrast, warmer wetter conditions may increase rainforests, and the soil aggregating, biodiversity-promoting microorganisms of grasslands may be lost (Young et al. 1998). At a local scale, shifts in species composition of soil biota in response to warming may also be strongly influenced by moisture. Enchytraeid species that were able to migrate through the soil profile to moist areas (*Cognettia sphagnetorum*) or which could tolerate dry conditions (*Achaeta eiseni*) generally increased at a +2.5°C warmer site, whereas one other species (*Cernosvitoviella atrata*) was unable to avoid dry conditions and suffered severe mortality at the increased temperature (Briones et al. 1997). The more tolerant and mobile species could also withstand the "extreme" summer temperatures at the warmer site. Biodiversity at higher taxonomic levels than species was also influenced by temperature and moisture interactions. Briones et al. (1997) demonstrated that Dipteran larvae died under warmer and drier conditions, but tardigrades survived, probably in an anhydrobiotic survival state. Studies of Collembola and oribatid mites also demonstrated that, when moisture is limiting, differential responses to temperature at species and higher taxonomic levels can become apparent. In dry conditions in both the laboratory (Hodkinson et al. 1996) and arctic field sites (dry semi-polar desert as opposed to moist tundra) (Coulson et al. 1996) increased temperature resulted in a decline in Collembola, but not in oribatid mite densities. Differential species responses also occurred within these groups.

Effects of warming may be large in those ecosystems that are currently limited by temperature such as the arctic tundra and semi-polar deserts (Swift et al. 1998). For most soil micro- and mesofaunal groups, arctic regions currently contain less than 15% of the species biodiversity of temperate ecosystems. It might therefore be predicted that an increase in temperature would alleviate much of this limitation (Swift et al. 1998); however, the studies of Collembola and oribatid mites descibed earlier (Coulson et al. 1993; Coulson et al. 1996; Hodkinson et al. 1996; Hodkinson et al. 1998) reveal that, as long as there is sufficient moisture, cryotolerant organisms in arctic regions may be well adapted to fluctuating temperatures, and population densities will be little affected by soil warming. Even small microscopic invertebrates have mechanisms for survival of temperature extremes (Crawford 1981; Coleman and Crossley 1996).

Few studies have investigated the effects of soil warming on biota in temperate regions, especially at the species level. One warming study in a Northern Michigan sugar maple forest noted shifts in microbial function and composition at higher taxonomic levels that may have important consequences for the organic matter substrate pool and ecosystem functioning (Zogg et al. 1997). The potential effects of such dramatic responses on soil species will require further investigation.

The preceding studies reveal considerable differences in the response of soil organisms to soil warming, but we presently have too little data to determine how these will impact soil biodiversity and ecosystem functioning at large spatial scales. It is also clear that many interacting factors (e.g., the microclimate produced by vegetation, and the structure of the soil that determines its capacity to provide refuges from drought and/or to maintain moisture) will strongly influence an organism's response to soil warming. These interacting factors will in turn be strongly influenced by the other drivers of global change; we must strive to consider their effects simultaneously.

Conclusions

Land use change is expected to be the primary global change driver affecting soil biodiversity. Elevated CO_2 and biotic introductions will have major impacts on soil species and ecosystem services. All drivers, including nitrogen deposition and warming, are likely to show strong interactions with each other. Few studies have directly investigated the effects of global change drivers on soil biota, compared with the evidence accumulating for aboveground biota. Nevertheless, there is sufficient evidence from microcosm and field experiments investigating either effects of single global change drivers, or effects of disturbances on soils similar to those predicted under global change to predict that global change drivers will disrupt current species assembledges within and across functional groups.

To ascertain whether these changes in soil biodiversity will have long-lasting effects on ecosystem functioning will require testing of hypotheses on the relationship between soil species and ecosystem function, collaboration of scientists in many disciplines, and incorporation of technologies at microsite to regional scales. GIS maps of soil textures, soil carbon, vegetation, and moisture can be used to predict areas where there is high soil biodiversity and productivity, where soils and biota are vulnerable to change, and where plant pathogens may spread. We believe syntheses of available results, and future experiments across soil disciplines of agriculture and ecosystem science, should be designed to provide us, as quickly as possible, the information necessary for judicious decisions on the future sustainability of the world's soils and the world's life in soils.

Acknowledgments. The authors appreciate the help of Dan Bumbarger, Mark St. John, and Nicole DeCrappeo, and Drs. Debra Coffin, Bill Parton, J. Ole Becker, and Dorota Porazinska for their helpful discussion, reviews, and comments. Among the taxonomic experts who contributed, we especially appreciate the enthusiasm of: Drs. R. Scheffrahn, V. Behan-Pelletier, D. Walter, and B. Kondriatieff. DHW acknowledges Don Strong and the UC Davis Bodega Marine Laboratory and staff for providing a great environment to write this paper. We acknowledge the support of NSF DEB 9708596, DEB 9626813 and NSF OPP 9211773 and OPP 9624743. This work was conducted as part of the Future Scenarios of Global Change Working Group supported by the National Center for Ecological Analysis and Synthesis, a Center funded by NSF (DEB 9421535), the University of California—Santa Barbara, the California Resources Agency, and the California Environmental Protection Agency.

References

Acosta R, Allen M, Cherian A, Granich S, Mintzer I, Suarez A, et al. (1998) Climate Change Information Kit: United Nations Environment Programme's Information Unit for Conventions, UNEP/IC, Châtelaine, Switzerland.

Akimov V, Hattori T (1996) Towards cataloguing soil bacteria: preliminary note. Microbes and Environments 11:57–60.

Allen MF (1991) The ecology of mycorrhizae. Cambridge University Press, Cambridge, UK.

Andersen DC (1987) Below-ground herbivory in natural communities: a review emphasizing fossorial animals. The Quarterly Review of Biology 62(3):261–286.

Anderson J (1995) The soil system. In: Heywood V (ed) *Global Biodiversity Assessment*, pp. 405–411. Cambridge University Press, Cambridge, UK.

Anderson JM (1994) Functional attributes of biodiversity in land use systems. In: Greenland DJ, Szabolcs I (eds) *Soil Resilience and Sustainable Land Use*, pp. 267–290. CAB International, Wallingford.

Andren O, Bengtsson J, Clarholm M (1995) Biodiversity and species redundancy among litter decomposers. In: Collins HP, Robertson GP, Klug MJ (eds) The significance and regulation of soil biodiversity, pp. 141–151. Kluwer Academic Publishers, Boston.

Baath E, Berg B, Lohm U, Lundgren B, Lundkvist H, Rosswall T (1990) Effects of experimental acidification and liming on soil organisms and decomposition in a Scots pine forest. Pedobiologia 20:85–100.

Barbercheck ME, Kaya HK (1991) Competitive interactions between entomopathogenic nematodes and *Beauvaria bassiana* (Deuteromycotina: Hyphomycetes) in soilborne larvae of *Spodoptera exigua* (Lepidoptera: Noctuidae). Environmental Entomology 20(2):707–712.

Bardgett RD, Wardle DAW, Yeates GW (1988) Linking above-ground and below-ground interactions: how plant responses to foliar herbivory influence soil organisms. Soil Biology and Biochemistry 30:1867–1878.

Bazzaz FA (1990) The response of natural ecosystems to the rising global carbon dioxide levels. Annual Review of Ecology and Systematics 21:167–196.

Bazzaz FA, Garbutt K, Reekie EG, Williams WE (1989) Using growth analysis to interpret competition between a C_3 and a C_4 annual under ambient and elevated carbon dioxide. Oecologia 79(2):223–235.

Beckage NE (1998) Parasitoids and polydnaviruses—an unusual mode of symbiosis in which a DNA virus causes host insect immunosuppression and allows the parasitoid to develop. BioScience 48:305–311.

Behan-Pelletier V, Newton G (1999) Computers in biology: linking soil biodiversity and ecosystem function—the taxonomic dilemma. BioScience 49:140–149.

Bengtsson J, Jones H, Setala H (1997) The value of biodiversity. Trends in Ecology and Evolution 12(9):334–336.

Berg MP, Verhoef HA (1998) Ecological characteristics of a nitrogen-saturated coniferous forest in The Netherlands. Biology and Fertility of Soils 26(4):258–267.

Berger H, Foissner W, Adam H (1986) Field experiments on the effects of fertilizers and lime on the soil microfauna of an alpine pasture. Pedobiologia 29(4):261–272.

Bewley RJF, Parkinson D (1986) Sensitivity of certain soil microbial processes to acid decomposition. Pedobiologia 29:73–84.

Bignell DE, Eggleton P (1998) Termites. In: Calow P (ed) Encycolpedia of ecology and environmental management. Blackwell Scientific, Oxford. pp. 742–744.

Bills GF, Polishook JD (1994) Abundance and diversity of microfungi in leaf litter of a lowland rain forest in Costa Rica. Mycologia 86:187–198.

Blair JM, Parmelee RW, Lavelle P (1995) Influences of earthworms on biogeochemistry. In: Hendrix PF (ed) *Earthworm Ecology and Biogeography in North America*, pp. 125–156. CRC Press, Boca Raton, FL.

Blair JM, Bohlen PJ, Freckman DW (1996) Soil invertebrates as indicators of soil quality. In: Doran JW, Jones AJ (eds) *Methods for Assessing Soil Quality*. Soil Science Society of America, pp. 273–291. Madison, WI.

Blaxter M (1998) *Caenorhabditis elegans* is a nematode. Science 282:2041–2046.

Blaxter MLP, De Lay LP, Garey JR, Liu LX, Scheldeman P, Vierstraete A, et al. (1998) A molecular evolutionary framework for the phylum Nematoda. Nature 392:71–75.

Bloemers GF, Hodda M, Lambshead PJD, Lawton JH, Wanless FR (1997) The effects of forest disturbance on diversity of tropical soil nematodes. Oecologia (Berlin) 111(4):575–582.

Boag B (1974) Nematodes associated with forest and woodland trees in Scotland (1974) Annals of Applied Biology 77:41–50.

Boag B, Williams KJO (1976) The criconematidae of the British Isles. Annals of Applied Botany 84:361–369.

Boag B, Yeates GW (1998) Soil nematode biodiversity in terrestrial ecosystems. Biodiversity and Conservation 7(5):617–630.

Boag B, Crawford JW, Neilson R (1991) The effect of potentail climatic changes on the geographical distribution of the plant-parasitic nematode *Xiphinema* and *Longidorus* in Europe. Nematologica 37:312–323.

Boag B, Jones HD, Neilson R (1997) The spread of the New Zealand flatworm (*Artioposthia triangulata*) within Great Britain. European Journal of Soil Biology 33(1):53–56.

Bongers T (1990) The maturity index: an ecological measure of environmental disturbance based on nematode species composition. Oecologia 83:14–19.

Boone RD, Nadelhoffer KJ, Canary JD, Kaye JP (1998) Roots exert a strong influence on the temperature sensitivity of soil respiration. Science 396:570–572.

Bournier A (1977) Grape insects. Annual Review of Enotmology 22:355–376.

Bowen GD (1978) Dysfunction and shortfalls in symbiotic responses. In: Horstall JG, Coowling EB (eds) *Plant Disease*, pp. 231–256. Academic Press.

Briones MJI, Ineson P, Piearce TG (1997) Effects of climate change on soil fauna: responses of enchytraeids, Diptera larvae and tardigrades in a transplant experiment. Applied Soil Ecology 6:117–134.

Brown WL Jr (1982) Hymenoptera. In: Parker SP (ed) *Synopsis and Classification of Living Organisms*, pp. 652–680. McGraw-Hill Book Company, New York.

Brussaard L, Behan-Pelletier VM, Bignell DE, Brown VK, Didden WAM, Folgarait PI, et al. (1997) Biodiversity and ecosystem functioning in soil. Ambio 26(8):563–570.

Burkins MB, Virginia RA, Chamberlain CP, Wall DH (2000) Origin and distribution of soil organic matter in Taylor Valley, Antarctica. Ecology 81:2377–2391.

Burtelow A, Bohlen PJ, Groffman PM (1998) Influence of exotic earthworm invasion on soil organic matter, microbial biomass and denitrification potential in forest soils of the northeastern US. Applied Soil Ecology (In press).

Casella E, Soussana JF, Loiseau P (1996) Long-term effects of CO_2 enrichment and temperature increase on a temperate grass sward. 1. Productivity and water use. Plant and Soil 182(1):83–99.

Christensen M (1989) A view of fungal ecology. Mycologia 81:1–19.

Coleman DC, Crossley DA (1996) Fundamentals of Soil Ecology. San Diego, Academic Press.

Coleman DC, Odum EP, Crossley DA Jr (1992) Soil biology, soil ecology, and global change. Biology and Fertility of Soils 14:104–111.

Cook RJ, Baker KF (1983) The nature and practice of biological control of plant pathogens. American Phytopathological Society, St. Paul, MN.

Cotrufo MF, Briones MJI, Ineson P (1998) Elevated CO_2 affects field decomposition rate and palatability of tree leaf litter: importance of changes in substrate quality. Soil Biology and Biochemistry 30(12):1565–1571.

Coulson S, Hodkinson ID, Strathdee A, Bale JS, Block W, Worland MR, et al. (1993) Simulated climate change: the interaction between vegetation type and microhabitat temperatures at Ny Alesund, Svalbard. Polar Biology 13:67–70.

Coulson SJ, Hodkinson ID, Webb NR, Block W, Bale JS, Strathdee AT, Worland MR, Wooley C (1996) Effects of elevated temperature on high-arctic soil microarthropod populations. Polar Biology 16(2):147–153.

Crawford CS (1981) *Biology of Desert Invertebrates.* Springer-Verlag, New York.

Cure JD, Acock B (1986) Crop responses to carbon dioxide doubling: a literature survey. Agricultural and Forest Meteorology 38(1–3):127–146.

daGama MM, Sousa JP, Ferreira C, Barrocas H (1997) Endemic and rare Collembola distribution in high endemism areas of South Portugal: a case study. European Journal of Soil Biology 33(3):129–140.

Daily GC (1997) *Natures Services: Societal Dependence on Natural Ecosystems.* Island Press, Washington DC.

Daily GC, Matson PA, Vitousek PM (1997) Ecosystem services supplied by soil. In: Daily GC (ed) *Natures Services: Societal Dependence on Natural Ecosystems*, pp. 113–132. Island Press, Washington DC.

Dedysh SN, Panikov NS, Liesack W, Grosskopf R, Zhou J, Tiedje JM (1998) Isolation of acidophilic methane-oxidizing bacteria from northern peat wetlands. Science 282(5387):281–284.

de Ruiter PC, Moore JC, Zwart KB, Bouwman LA, Hassink J, Bloem J, et al. (1993) Simulation of nitrogen mineralization in the belowground food webs of two winter wheat fields. Journal of Applied Ecology 30:95–106.

Diaz S (1996) Effects of elevated [CO_2] at the community level, mediated by root symbionts. Plant and Soil 187:309–320.

Dighton J, Jones HE (1994) A review of soil biodiversity. *Royal Commission on Environmental Pollution*, UK.

Drake BG, GonzalezMeler MA, Long SP (1997) More efficient plants: a consequence of rising atmospheric CO_2? Annual Review of Plant Physiology and Plant Molecular Biology 48:609–639.

Drake BG, Muehe MS, Peresta G, Gonzalez-Meler MA, Matamala R (1996) Acclimation of photosynthesis, respiration and ecosystem carbon flux of a wetland on Chesapeake Bay, Maryland to elevated atmospheric CO_2 concentration. Plant and Soil 187(2):111–118.

Ducey PK, Formanowicz DR Jr, Boyet L, Mailloux J, Nussbaum RA (1993) Experimental examination of burrowing behavior in caecilians (Amphibia: Gymnophiona): effects of soil compaction on burrowing ability of four species. Herpetologica 49(4):450–457.

Dusenbery DB (1987) Theoretical range over which bacteria and nematodes locate plant roots using carbon dioxide. Journal of Chemical Ecology 13(7):1617–1624.

Ehrenfeld JH (1999) Nitrogen mineralization and nitrification in suburban and undeveloped Atlantic white cedar wetlands. Journal of Environmental Quality 28:523–529.

Elliott ET, Anderson RV, Coleman DC, Cole CV (1980) Habitable pore space and microbial trophic interactions. Oikos 35(3):327–335.

Ettema CH (1997) Relating soil nematode diversity to ecosystem processes and disturbances. Journal of Nematology 29(4):577.

Ettema CH (1998) Soil nematode diversity: species coexistence and ecosystem function. Journal of Nematology 30(2):159–169.

Ettema CH, Bongers T (1993) Characterization of nematode colonization and succession in disturbed soil using the maturity index. Biology and Fertility of Soils 16(2):79–85.

Ferris H, Lau S, Venette R (1995) Population energetics of bacterial-feeding nematodes: respiration and metabolic rates based on CO_2 production. Soil Biology and Biochemistry 27(3):319–330.

Ferris H, Venette RC, Lau SS (1997) Population energetics of bacterial-feeding nematodes: carbon and nitrogen budgets. Soil Biology and Biochemistry 29(8):1183–1194.

Ferris H, Eyre M, Venette RC, Lau SS (1996) Population energetics of bacterial-feeding nematodes: stage-specific development and fecundity rates. Soil Biology and Biochemistry 28(3):217–280.

Folgarait PJ (1998) Ant biodiversity and its relationship to ecosystem functioning: a review. Biodiversity and Conservation 7:1221–1244.

Fraser PM, Boag B (1998) The distribution of lumbricid earthworm communities in relation to flatworms: a comparison between New Zealand and Europe. Pedobiologia 42(5–6):542–553.

Freckman DW, Caswell EP (1985) Ecology of nematodes in agroecosystems. Annual Review of Phytopathology 23:275–296.

Freckman DW, Mankau R (1986) Abundance, distribution, biomass and energetics of soil nematodes in a northern Mohave desert. Pedobiologia 29:129–142.

Freckman DW, Ettema CE (1993) Assessing nematode communities in agroecosystems of varying human intervention. Agriculture, Ecosystems and Environment 45:239–261.

Freckman DW, Blackburn TH, Brussaard L, Hutchings PA, Palmer MA, Snelgrove PVR (1997) Linking biodiversity and ecosystem functioning of soils and sediments. Ambio 26(8):556–562.

Gates GE (1992) Farewell to North American megadriles. Megadrilogica 4:12–77.

Gaugler R, Campbell JF (1991) Selection for enhanced host-finding resistance of scarab larvae (Coleoptera:Scarabaeidae) in an entomopathogenic nematode. Environmental Entomology 20(2):700–706.

Gerson U (1996) Arthropod root pests. In: Waisel Y, Eshel A, Kafkafi U (eds) *Plant Roots: the Hidden Half*, pp. 797–809. Marcel Dekker Publisher, New York.

Giller KE, Beare MH, Lavelle P, Izac AMN, Swift MJ (1997) Agricultural intensification, soil biodiversity and agroecosystem function. Applied Soil Ecology 6(1):3–16.

Groffman PM, Bohlen PJB (1988) Soil and sediment biodiversity: cross system comparisons and large scale effects. BioScience 49:139–148.

Gupta VVSR (1994) The impact of soil and crop management practices on the dynamics of soil microfauna and mesofauna. In: Pankhurst CE, Doube BM, Gupta VVSR, Grace PR (eds) *Soil Biota Management in Sustainable Farming Systems*, pp. 107–124. CSIRO, Melbourne, Australia.

Hammond PM, Hawksworth DL, Kalin-Arroyo MT (1995) Magnitude and distribution of biodiversity: 3.1. The current magnitude of biodiversity. In: Heywood VH (ed) *Global Biodiversity Assessment*, pp. 113–138. Cambridge University Press, Cambridge, UK.

Hendrix PF, Crossley DAJ, Blair JM, Coleman DC (1990) Soil biota as components of sustainable agroecosystems. In: *Sustainable Agricultural Systems*, pp. 637–654. Soil and Water Conservation Society, Ankeny, IA.

Heneghan L, Coleman DC, Zou X, Crossley DA Jr, Haines BL (1998) Soil microarthropod community structure and litter decomposition dynamics: a study of tropical and temperate sites. Applied Soil Ecology 9:33–38.

Hobbie EA, Rygiewicz PT, Moldenke AE, Griffis WL (1998) What does a mite bite? Insights on carbon and nitrogen movement through soil food webs. Bulletin of the Ecological Society of America 79:177.

Hodkinson ID, Coulson SJ, Webb NR, Block W (1996) Can high arctic soil microarthropods survive elevated summer temperatures? Functional Ecology 10(3):314–321.

Hodkinson ID, Webb NR, Bale JS, Block W, Coulson SJ, Strathdee AT (1998) Global change and arctic ecosystems: conclusions and predictions from experiments with terrestrial invertebrates on Spitzbergen. Arctic and Alpine Research 30(3):306–313.

Hoffman RL (1982) Chilopoda. In: Parker SP (ed) *Synopsis and Classification of Living Organisms*, pp. 681–688. McGraw-Hill Book Company, New York.

Hoffman RL (1990) Diplopoda. In: Dindal DL (ed) *Soil Biology Guide*, pp. 835–860. John Wiley and Sons, New York.

Holland EA, Detling JK (1990) Plant response to herbivory and belowground nitrogen cycling. Ecology 71(3):1040–1049.

Hooper D, Dangerfield M, Brussaard L, Wall DH, Wardle D, Bignell D, et al. (In press) Interactions between above- and belowground biodiversity in terrestrial ecosystems: patterns, mechanisms and feedbacks. BioScience.

Hooper DU, Vitousek PM (1998) Effects of plant composition and diversity on nutrient cycling. Ecological Monographs 68(1):121–149.

Huhta V (1984) Response of *Cognettia spagnetoroum* (Enchytraeidae) to manipulation of pH and nutrient status in coniferous forest soil. Pedobiologia 27:245–260.

Huhta V, Persson T, Setala H (1998) Functional implications of soil fauna diversity in boreal forests. Applied Soil Ecology 10:277–288.

Hunt HW, Coleman DC, Ingham ER, Elliott ET, Moore JC, Rose SL, et al. (1987) The detrital food web in a shortgrass prairie. Biology and Fertility of Soils 3:57–68.

Hyvonen R, Persson T (1990) Effects of acidification and liming on feeding groups of nematodes in coniferous forest soils. Biology and Fertility of Soils 9:205–210.

Ingham ER, Coleman DC, Moore JC (1989) An analysis of food-web structure and function in a shortgrass prairie, a mountain meadow, and a lodgepole pine forest. Biology and Fertility of Soils 8(1):29–37.

Ingham RE, Trofymow JA, Ingham ER, Coleman DC (1985) Interactions of bacteria, fungi, and their nematode grazers: effects on nutrient cycling and plant growth. Ecological Monographs 55(1):119–140.

Ingram J, Wall-Freckman D (1998) Soil biota and global change: preface. Global Change Biology 4:699–701.

Jackson RB, Sala OE, Paruelo JM, Mooney HA (1998) Ecosystem water fluxes for two grasslands in elevated CO_2: a modeling analysis. Oecologia 113(4):537–546.

James SW (1990) Oligochaeta: Megascolecidae and other earthworms from southern and midwestern North America. In: Dindal DL (ed) *Soil Biology Guide*, pp. 379–386. John Wiley and Sons, New York.

James SW (1995) Systematics, biogeography and ecology of Nearctic earthworms from eastern, central, southern and southwestern United States. In: Hendrix PF (ed) *Earthworm Ecology and Biogeography in North America*, pp. 29–52. Lewis Publishers, Boca Raton.

Jones DL, Darrah PR (1993) Re-Sorption of organic compounds by roots of Zeamays L. and its consequences in th rhizosphere II. Experimental and model evidence for simultaneous exudation and re-sorption of soluble carbon compounds. Plant and Soil 153:47–59.

Jones TH, Thompson LJ, Lawton JH, Bezemer TM, Bardgett RD, Blackburn TM, et al. (1998) Impacts of rising atmospheric carbon dioxide on model terrestrial ecosystems. Science 280(5362):441–443.

Klironomos JN, Rillig MC, Allen MF (1996) Below-ground microbial and microfaunal responses to *Artemisia tridentata* grown under elevated atmospheric CO_2. Functional Ecology 10(4):527–534.

Klironomos JN, Ursic M, Rillig M, Allen MF (1998) Interspecific differences in the response of arbuscular mycorrhizal fungi to *Artemisia tridentata* grown under elevated atmospheric CO_2. New Phytologist 138(4):599–605.

Korner C (1996) The response of complex multispecies systems to elevated CO_2. In: Walker B, Steffen W (eds) *Global Change and Terrestrial Ecosystems*, pp. 20–42. Cambridge University Press, Cambridge, UK.

Kourtev PS, Ehrenfeld JG, Huang WZ (1998) Effects of exotic plant species on soil properties in hardwood forests of New Jersey. Water Air and Soil Pollution 105(1–2):493–501.

Kuan TL, Erwin DC (1980) Predisposition effect of water saturation of soil on Phytophthora root rot of alfalfa. Phytopathology 70:981–986.

Lauenroth WK, Milchunas DG (1991) Short-grass steppe. In: Coupland RT (ed) *Ecosystems of the World 8A*, pp. 183–226. Elsevier, New York.

Lavelle P (1993) The structure of earthworm communities. In: Satchell JE (ed) *Earthworm Ecology: From Darwin to Vermiculture*, pp. 449–466. Chapman and Hall, London.

Lavelle P, Gilot C, Fragoso C, Pashanasi B (1994a) Soil fauna and sustainable land use in the humid tropics. In: Greenland DJ, Szabolcs I (eds) *Soil Resilience and Sustainable Land Use*, pp. 291–300. CAB International, Wallingford.

Lavelle P, Dangerfield M, Fragoso C, Eschenbrenner V, Lopez-Hernandez D, Pashanasi B, et al. (1994b) The relationship between soil macrofauna and tropical soil fertility. In: Swift MJ, Woomer P (eds) *Tropical Soil Biology and Fertility*, pp. 137–169. John Wiley-Sayce, New York.

Lavelle P, Bignell D, Lepage M, Wolters V, Roger P, Ineson P, et al. (1997) Soil function in a changing world: the role of invertebrate ecosystem engineers. European Journal of Soil Biology 33(4):159–193.

Lawton JH, Bignell DE, Bloemers GF, Eggleton P, Hodda ME (1996) Carbon flux and diversity of nematodes and termites in Cameroon forest soils. Biodiversity and Conservation 5:261–273.

Lawton JH, Bignell DE, Bolton B, Bloemers GF, Eggleton P, Hammond PM, et al. (1998) Biodiversity inventories, indicator taxa and effects of habitat modification in tropical forest. Nature (London) 391(6662):72–76.

Lee KE (1961) Interactions between native and introduced earthworms. Proceedings of the New Zealand Ecological Society 8:60–62.

Lee KE (1985) *Earthworms: Their Ecology and Relationships with Land Use*. Academic Press, Sydney.

Lee KE (1995) Earthworms and sustainable land use. In: Hendrix PF (ed) *Earthworm Ecology and Biogeography in North America*, pp. 215–234. Lewis Publishers, Ann Arbor.

Lehle E, Funke W (1989) The microfauna of forest soils. II. Ciliata (Protozoa: Ciliophora). Effects of anthropogenic influences. Verhandlungen Gesellschaft für Okologie 17:385–390.

Lincoln DA, Fajer ED, Johnson RH (1993) Plant-Insect herbivore interactions in elevated CO_2 environments. Trends in Ecology and Evolution 8(2):64–68.

Linden DR, Hendrix PF, Coleman DC, vanVliet PCJ (1994) Faunal indicators of soil quality. In: Doran JW, Coleman DC, Bezdicek DF, Stewart BA (eds) *Defining Soil Quality for a Sustainable Environment*, pp. 91–106. Soil Science Society of America and American Society of Agronomy, Madison, WI.

Lussenhop J, Treonis A, Curtis PS, Teeri JA, Vogel CS (1998) Response of the soil biota to elevated atmoshperic CO_2 in polar model systems. Oecologia 113(2):247–251.

Lynch JM, Bragg E (1985) Microorganisms and soil aggregate stability. In: Stewart BA (ed) *Advances in Soil Science*, pp. 135–171. Springer-Verlag, New York.

Maddison DR (1997) *The Tree of Life, Diplura*. ⟨http://phylogeny.arizona.edu/tree/eukayotes/animals/arthropoda/hexapoda/diplura/diplura.html⟩ (June 1998).

Mamiya Y (1983) Pathology of the pine wilt disease caused by *Bursaphelenchus xylophilus*. Annual Review of Phytopathology 21:201–220.

Mando A (1998) Soil-dwelling termites and mulches improve nutrient release and crop performance on Sahelian crusted soil. Arid Soil Research and Rehabilitation 12(2):153–163.

McAlpine JF (1990) Insecta: Diptera adults. In: Dindal DL (ed) *Soil Biology Guide*, pp. 1211-1252. John Wiley and Sons, New York.

McSorley R, Frederick JJ (1996) Nematode community structure in rows and between rows of a soybean field. Fundamental and Applied Nematology 19(3):251–261.

Mooney HA, Lubchenco J, Dirzo R, Sala OE (1995) Biodiversity and ecosystem functioning: basic principles. In: Heywood VH (ed) *Global Biodiversity Assessment*, pp. 275–326. Cambridge University Press, Cambridge, UK.

Moore JC, de Ruiter PC (1991) Temporal and spatial heterogeneity of trophic interactions within below-ground food webs. Agriculture, Ecosystems and Environment 34:371–397.

Mota M (1999) personal communication.

Murphy WM, Barreto ADM, Silman JP, Dindal DL (1995) Cattle and sheep grazing effects on soil organisms, fertility and compaction in a smooth-stalked meadowgrass-dominated white clover sward. Grass and Forage Science 59:191–194.

Newsham KK, Fitter AH, Watkinson AR (1995) Arbuscular mycorrhiza protect an annual grass from root pathogenic fungi in the field. Journal of Ecology 83(6):991–1000.

Newton PCD, Clark H, Bell CC, Glasgow EM, Tate KR, Ross DJ, et al. (1995) Plant growth and soil processes in temperate grassland communities and elevated CO_2. Journal of Biogeography 22:1239–1244.

Niles RK, Wall Freckman D (1997) From the ground up: nematode ecology in bioassessment and ecosystem health. In: Barker KR, Pederson GA, Windham GL (eds) *Plant-Nematode Interactions*. American Society of Agronomy, Crop Science Society of America and Soil Science Society of America, pp. 65–87. Madison WI.

Norby RJ, Cotrufo MF (1998) Global change—a question of litter quality. Nature 396(6706):17–18.

Oechel WC, Cowles S, Grulke N, Hastings SJ, Lawrence B, Prudhomme T, et al. (1994) Transient Nature of CO_2 Fertilization in Arctic Tundra. Nature 371(6497):500–503.

Ogren RE, Kawakatsu M (1998) American neartic and neotropical land planarians (Tricladida: Terricola) faunas. Pedobiologia 42:441–451.

O'Neill EG (1994) Responses of soil biota to elevated atmospheric carbon dioxide. Plant and Soil 165:55–65.

O'Neill EG, Luxmoore RJ, Norby RJ (1987) Increases in mycorrhizal colonization and seedling growth in *Pinus echinata* and *Quercus alba* in an enriched CO_2 atmosphere. Canadian Journal of Forest Research 17(8):878–883.

Paoletti MG, Pimentel D, Stinner BR, Stinner D (1992) Agroecosystem biodiversity—matching production and conservation biology. Agriculture Ecosystems and Environment 40(1–4):3–23.

Paustian K, Schnurer J (1987a) Fungal growth response to carbon and nitrogen limitation: application of a model to laboratory and field data. Soil Biology and Biochemistry 19:621–629.

Paustian K, Schnurer J (1987b) Fungal growth response to carbon and nitrogen limitation: a theoretical model. Soil Biology and Biochemistry 19:613–620.

Pearce MJ, Waite B (1994) A list of termite genera with comments on taxonomic changes and regional distribution. Sociobiology 23:247–262.

Penuelas J, Estiarte M (1998) Can elevated CO_2 affect secondary metabolism and ecosystem function? Trends in Ecology and Evolution 13(1):20–24.

Persson T, Lundkvist H, Wiren A, Hyvonen R, Wessen B (1989) Effects of acidification and liming on carbon and nitrogen mineralization and soil organisms in mor humus. Water, Air and Soil Pollution 45:77–96.

Petersen H, Luxton M (1982) A comparative analysis of soil fauna populations and their role in decomposition processes. Oikos 39:287–388.

Pline M, Dusenbery DB (1987) Responses of plant-parasitic nematode *Meloigogyne incognita* to carbon dioxide determined by video camera-computer tracking. Journal of Chemical Ecology 13(4):873–883.

Pregitzer KS, Zak DR, Curtis PS, Kubiske ME, Teeri JA, Vogel CS (1995) Atmospheric CO_2, soil nitrogen and turnover of fine roots. New Phytologist 129(4):579–585.

Price NS, Siddiqi MR (1994) Rainforest nematodes with particular reference to the Korup National Park, Cameroon. Afro-Asian Journal of Nematology 4:117–128.

Prikryl Z, Vancura V (1980) Root exudates of plants. VI. Wheat exudation as dependent on growth, concentration gradient of exudates and the presence of bacteria. Plant and Soil 57:69–83.

Procter DLC (1984) Towards a biogeography of free-living soil nematodes 1. Changing species richness diversity and densities with changing latitude. Journal of Biogeography 11(2):103–118.

Procter DLC (1990) Global overview of the functional roles of soil-living nematodes in terrestrial communities and ecosystems. Journal of Nematology 22(1):1–7.

Ricklefs RE, Hawksworth DL, Kalin-Arroyo MT (1995) Magnitude and distribution of biodiversity: 3.2. The distribution of biodiversity. In: Heywood VH (ed) *Global Biodiversity Assessment*, pp. 139–173. Cambridge University Press, Cambridge, UK.

Rillig MC, Field CB, Allen MF (1999) Soil biota responses to long-term atmospheric CO_2 enrichment in two California annual grasslands. Oecologia 119:572–577.

Robertson GP, Freckman DW (1995) The spatial distribution of nematode trophic groups across a cultivated ecosystem. Ecology 76:1425–1433.

Ruess L (1995) Studies on the nematode fauna of an acid forest soil: spatial distribution and extraction. Nematologica 41:229–239.

Rusek J (1993) Air-pollution-mediated changes in alpine ecosystems and ecotones. Ecological Applications 3(3):409–416.

Rutherford TA, Mamiya Y, Webster JM (1990) Nematode-induced pine wilt disease: factors influencing its occurrence and distribution. Forest Science 36:145–155.

Schardl CL, Phillips TD (1997) Protective grass endophytes: where are they from and where are they going? Plant Disease 81(5):430–438.

Scheller U (1982) Symphyla. In: Parker SP (ed) *Synopsis and Classification of Living Organisms*, pp. 688–689. McGraw-Hill Book Company, New York.

Schiemer F (1983) Comparative aspects of food dependence and energetics of freeliving nematodes. Oikos 41:32–42.

Schimel JP, Gulledge J (1998) Microbial community structure and global trace gases. Global Change Biology 4(7):745–758.

Siepel H, Van de Bund CF (1988) The influence of management practices on the microarthropod community of grassland. Pedobiologia 31(5–6):339–354.

Smith P, Andren O, Brussaard L, Dangerfield M, Ekschmitt K, Lavelle P, et al. (1998) Soil biota and global change at the ecosystem level: describing soil biota in mathematical models. Global Change Biology 4(7):773–784.

Sochtig W, Larink O (1992) Effect of soil compaction on activity and biomass of endogeic lumbricids in arable soils. Soil Biology and Biochemistry 24(12):1595–1599.

Sohlenius B (1985) Influence of climatic conditions on nematode coexistence: a laboratory experiment with a coniferous forest soil. Oikos 44:430–438.

Sohlenius B, Bostrom B (1999) Effects of climate change on soil factors and metzoan microfauna (nematodes, tardigrades, and rotifers) in a Swedish tundra soil—a soil transplantation experiment. Applied Soil Ecology 12:113–128.

Somme L, Birkemoe T (1997) Cold tolerance and dehydration in Enchytraeidae from Svalbard. Journal of Comparative Physiology B 167:264–269.

Springett JA, Brittain JE, Springett BP (1970) Vertical movement of Enchytraeidae (Oligochaeta) in moorland soils. Oikos 21:16–21.

Stanton NL (1979) Patterns of species diversity in temperate and tropical litter mites. Ecology 60:295–304.

Stotzky G, Bollag JM (eds) (1992) *Soils Plants and the Environment*. Marcel Dekker, New York.

Strong DR (1998) Predator control in the terrestrial ecosystems: the underground food chain of bush lupine. In: Olff H, Brown VK (eds) *Herbivores, Plants, and Predators*, pp. 577–602. Blackwell, Oxford, UK.

Stulen I, den Hertog J (1993) Root growth and functioning under atmospheric CO_2 enrichment. Vegetatio 104/105:99–115.

Sustr V, Simek M (1996) Behavioural responses to and lethal effects of elevated carbon dioxide concentration in soil invertebrates. European Journal of Soil Biology 32:149–155.

Swift MJ (1997) Special issue: soil biodiversity, agricultural intensification and agroecosystem function. Applied Soil Ecology 6(1):1–108.

Swift MJ, Heal OW, Anderson JM (1979) *Decomposition in Terrestrial Ecosystems*. Blackwell, Oxford, UK.

Swift MJ, Andren O, Brussaard L, Briones M, Couteaux MM, Ekschmitt K, et al. (1998) Global change, soil biodiversity, and nitrogen cycling in terrestrial ecosystems: three case studies. Global Change Biology 4(7):729–743.

Szegi J (ed) (1984) Soil biology and the conservation of the biosphere. Akadémiai Kiadó, Budapest.

Torsvik V, Goksoyr J, Daae FL, Sorheim R, Michalsen J, Salte K (1994) Use of DNA analysis to determine the diversity of microbial communities. In: Ritz K, Dighton J, Giller KE (eds) *Beyond the Biomass*, pp. 39–48. Wiley-Sayce Publication, New York.

van der Heijden MGA, Klironomos JN, Ursic M, Moutiglis P, Streitwolf-Engel R, Boller T, et al. (1998) Mycorrhizal fungal diversity determines plant biodiversity, ecosystem variability and productivity. Nature 396:69–72.

Van Ginkel J, Gorissen A, Van Veen J (1996) Long-term decomposition of grass roots as affected by elevated atmospheric carbon dioxide. Journal of Environmental Quality 25:1122–1128.

Van Gundy SD, Kirkpatrick JD, Golden J (1977) The nature and role of metabolic leakage from root-knot galls and infection by *Rhizoctonia solani*. Journal of Nematology 9:113–121.

van Noordwijk M, Martikainen P, Bottner P, Cuevas E, Rouland C, Dhillon S (1998) Global change and root function. Global Change Biology 4:759–772.

Van Straalen NM, Kraak MH, Denneman CAJ (1988) Soil microarthropods as indicators of soil acidification and forest decline in the Veluwe area, the Netherlands. Pedobiologia 32:47–55.

van Veen J, Lilijeroth E, Lekkerkerk L, van de Geijin S (1991) CO_2 fluxes in plant-soil systems at elevated atmospheric CO_2 levels. Ecological Applications 1:175–181.

Venette RC, Ferris H (1997) Thermal constraints to population growth of bacterial-feeding nematodes. Soil Biology and Biochemistry 29(1):63–74.

Virginia RA, Wall DH (1999) How soils structure communities in the Antarctic dry valleys. BioScience 49:973–983.

Vitousek PM, Walker LR, Whiteaker LD, Mueller-Dombois D, Matson PA (1987) Biological invasion by *Myrica faya* alters ecosystem development in Hawaii. Science 238:802–804.

Vitousek PM, Aber JD, Howarth RW, Likens GE, Matson PA, Schindler DW, et al. (1997) Human alteration of the global nitrogen cycle: Sources and consequences. Ecological Applications 7(3):737–750.

Wagener SM, Oswood MW, Schimel JP (1998) Rivers and soils: parallels in carbon and nutrient processing. BioScience 48(2):104–108.

Wake M (1993) The skull as a locomotor organ. In: Hanken J, Hall BK (eds) *The Skull: Functional and Evolutionary Mechanisms*, p. 240. The University of Chicago Press, Chicago.

Walker JC (1969) *Plant Pathology, Third edition.* McGraw-Hill Book Company, New York.

Wall DH, Virginia RA (2000) The world beneath our feet: soil biodiversity and ecosystem functioning. In: Raven PR, Williams T (eds) *Nature and Human Society: The Quest for a Sustainable World.* National Academy of Sciences, Washington, DC.

Wall DH, Moore JC (1999) Interactions underground: soil biodiversity, mutualsim and ecosystem processes. BioScience 49:109–118.

Wall Freckman D, Virginia RA (1998) Soil Biodiversity and Community Structure in the McMurdo Dry Valleys, Antarctica. In: Prisco JC (ed) Ecosystem Dynamics in a Polar Desert: The McMurdo Dry Valleys, Antarctica. pp. 323–325. American Geophysical Union, Washington, DC.

Wall Freckman D, Blackburn TH, Hutchings P, Palmer MA, Snelgrove PVR (1997) Linking biodiversity and ecosystem functioning of soils and sediments. Ambio 26:556–662.

Wallace HR (1963) *The Biology of Plant Parasitic Nematodes.* Edward Arnold Publishers, Ltd., London.

Walter D (1999) Personal communication.

Wardle DA, Verhoef HA, Clarholm M (1998) Trophic relationships in the soil microfood-web: predicting the responses to a changing global environment. Global Change Biology 4(7):713–728.

Wedin DA, Tilman D (1996) Influence of nitrogen loading and species composition on the carbon balance of grasslands. Science 274:1720–1723.

Whalley WR, Dumitru E, Dexter AR (1995) Biological effects of soil compaction. Soil and Tillage Research 35:53–68.

Williams CB (1964) *Patterns in the Balance of Nature.* Academic Press, London.

Wolters V, Silver WL, Bignell DE, Coleman DC, Lavelle P, van der Putten W, et al. (2000) Global change effects on above- and belowground biodiversity in terrestrial ecosystems: interactions and implications for ecosystem functioning. BioScience 50:1089–1099.

www1. http://phylogeny.arizona.edu/tree/eukaryotes/animals/arthropoda/arachnida/acari/acari.html.

www2. http://www.gypsymoth.ento.vt.edu/~ravlin/insect_orders/collembola.html.

www3. http://www.gypsymoth.ento.vt.edu/~ravlin/insect_orders/diplura.html.

www4. http://phylogeny.arizona.edu/tree/eukaryotes/animals/arthropoda/hexapoda/diplura/diplura.html.

www5. http://phylogeny.arizona.edu/tree/eukaryotes/animals/arthropoda/crustacea/isopoda/isopoda.html.

www6. http://www2.ncsu/unity/lockers/ftp/bwiegman/fly_html/diptera/html.

Yeates GW (1987) *How Plants Affect Nematodes. Advances in Ecological Research,* pp. 61–113. Academic Press, London.

Yeates GW (1999) Effects of plants on nematode community structure. Annual Review of Phytopathology 37:127–149.

Yeates GW, Orchard VA (1993) Response of pasture soil faunal populations and decompostition processes to elevated carbon dioxide and temperature—a climate chamber experiment. In: Prestidge RA (ed) Proceedings of the 6th Australasian grassland invertebrate ecology conference pp. 148–154. AgResearch, Hamilton, NZ.

Yeates GW, Tate KR, Newton PCD (1997) Response of the fauna of a grassland soil to doubling of atmospheric carbon dioxide concentration. Biology and Fertility of Soils 25:307–315.

Young IM, Blanchart E, Chenu C, Dangerfield M, Fragoso C, Grimaldi M, et al. (1998) The interaction of soil biota and soil structure under global change. Global Change Biology 4(7):703–712.

Zak DR, Ringelberg DB, Pregitzer KS, Randlett DL, White DC, Curtis PS (1996) Soil microbial communities beneath *Populus grandidentata* grown under elevated atmospheric CO_2. Ecological Applications 6(1):257–262.

Zogg GP, Zak DR, Ringelberg DB, MacDonald NW, Pregitzer KS, White DC (1997) Compositional and functional shifts in microbial communities due to soil warming. Soil Science Society of America Journal 61(2):475–481.

5. Scenarios of Biodiversity Changes in Arctic and Alpine Tundra

Marilyn D. Walker, William A. Gould, and
F. Stuart Chapin III

Arctic and alpine tundra, defined as those areas that lie to the north of the latitudinally or altitudinally controlled limits of tree growth (Gabriel and Talbot 1984), currently occupies approximately $8.3 \times 10^6 \text{km}^2$. The tundra biome is characterized by low biomass and species diversity relative to other biomes, and the spatial distribution of species of all groups is strongly structured by physical factors (Chapin and Körner 1995). About $2.3 \times 10^6 \text{km}^2$ (28%) of the total is occupied by ice, primarily the continental glaciers of Greenland and Antarctica. The remainder consists of a combination of shrub-dominated tundra (about 25%) and herbaceous-dominated tundra (about 47%). Herbaceous-dominated tundra is usually subdivided by ecologists into true tundra (or alpine tundra in the mountains) and polar desert (which is somewhat analogous to the alpine nival zone). The IMAGE model (Alcamo 1994) upon which this volume is based recognizes shrub tundra (i.e., woody tundra), tundra, and ice.

The arctic is functionally important for global carbon storage and as human and wildlife habitat. The arctic has three times as much soil carbon as does alpine tundra on a global basis, but it has only 13% of the floristic diversity found in alpine regions (Körner 1995). Thus, changes in alpine tundra will potentially have greater direct effects on diversity, but changes in the arctic could have greater feedback to the global system, at least in terms of carbon balance.

Present Patterns of Diversity

The species richness of the arctic tundra is the lowest of all terrestrial biomes (Table 5.1). The most complete information is available for vascular plant species. Animal diversity is extremely low in arctic tundra, except for birds, many of which nest there in summer, feeding on the abundant insect life. Although there is little information available on the global faunal diversity of alpine systems, the greater floristic diversity of alpine tundra is also likely to be reflected in its fauna.

There are two important distributional patterns in tundra that should influence changes in diversity. The first is that the number of species correlates with summer temperature—there are fewer species on higher mountains and at higher latitudes. The second is that the species of the highest latitudes and altitudes are the most widely distributed—the number of species with circumpolar distributions increases with colder temperatures. The majority of endemic species, on the other hand, are found in the more "favorable" regions—the low arctic and the lower altitudinal limit of alpine tundra (Körner 1995; Walker 1995).

Arctic floras and faunas are primarily a depauperate subset of the boreal system. For example, of approximately 30 terrestrial mammals known to occur in the Alaskan arctic, only four have uniquely arctic distributions (Bee and Hall 1956). In contrast, alpine areas, particularly in the southern hemisphere and equatorial regions, have a higher proportion of endemic species. These differences are due primarily to the fragmentation of alpine areas compared with the relatively continuous distribution of arctic tundra; however, endemics are not common in either system.

Regional species diversity numbers are strongly correlated with summer temperatures in the arctic (Young 1971; Rannie 1986; Chernov 1995), with diversity decreasing as mean summer temperature decreases to the north. The July mean temperature explains as much as 98% of the variance in species numbers for certain groups (i.e., vascular plants and some insect groups). Nonvascular plants have no such correlation with temperature. Certain

Table 5.1. Current species richness of major taxonomic groups in the tundra biome

	Arctic	Alpine
Vascular plants	1,000–1,500[a,b]	8,000–10,000[c]
Terrestrial mammals	61[d]	?
Birds	300[d]	?
Reptides	<10[d]	?—very few
Amphibians	<10[d]	?
Insects	3,000[d]	?
Microbes	Unknown[e]	

[a] Polunin 1959; [b] Murray 1995; [c] Körner 1995; [d] Chernov 1995; [e] Schimel 1995.

smaller taxonomic groups (e.g., sawflies) may show reverse patterns if they occur in habitats or are associated with food plants that are more abundant at high latitudes. At the scale of major taxonomic divisions, however, the relationship between species number and temperature is generally quite strong. Despite regional differences among alpine systems, Körner (1995) reported a very consistent "local flora" size of 200–300 species in alpine tundra (i.e., within a subsection of a mountain range with fairly uniform geology), suggesting that current conditions rather than evolutionary history most strongly govern alpine regional diversity.

Arctic diversity hotspots include polar oases, eskers, south-facing slopes, riparian areas, and migratory bird nesting areas (Walker 1995). Other areas (e.g., caribou calving grounds) are not necessarily hotspots of diversity, but are instead of critical concern. Diversity hotspots generally have "ameliorated" conditions, including warmer summer temperatures and a larger soil nutrient pool. They may also have intermediate-level disturbance regimes that result in high habitat diversity (Fox 1992; Hobbie et al. 1993; Naiman et al. 1993).

Past Patterns and Losses of Diversity

Arctic

Patterns of biodiversity in the arctic have been very dynamic over the past 100,000 years (Bartlein and Prentice 1989; Brubaker et al. 1995). Changes in the flora and vegetation have included large-scale invasions and retreats of trees and woody species, with the current tundra established somewhere between 6000 and 3000 BP (Brubaker et al. 1995). These invasions indicate that given adequate temperatures these landscapes could support woody species, and that trees can cause changes in rates of paludification and ion loss (Hu et al. 1993; MacDonald et al. 1993; Brubaker et al. 1995).

The extinction of the Pleistocene megafauna, which occurred throughout the temperate zone, also occurred in the arctic at the end of the Pleistocene (Guthrie and Stoker 1990), or as recently as 3000 years ago in isolated localities such as Wrangel Island (Vartanyan et al. 1993). The occurrence of abundant, large-bodied grazers and associated predators during the Pleistocene seems paradoxical and may have reflected more continental conditions at a time when extensive sea ice and continental glaciers blocked movement of moisture in the arctic. These drier conditions and disturbance by the megafauna apparently prevented the development of moss-dominated peat landscapes that now occupy the most extensive areas and have the lowest diversity in the arctic (Schweger et al. 1982; Zimov et al. 1995). Only a small proportion of the Pleistocene megafauna survived whatever changes caused the widespread extinctions of most other large mammals. The few large herbivores that remain in this system are either migratory and follow weather-driven changes in phenology and food availability (e.g., caribou, *Rangifer tarandus*) or specialize on moss-free disturbed portions of the landscape such as riparian corridors (e.g., muskox, *Ovibos moschatus*).

Alpine

The alpine flora has remained much more stable than its arctic counterpart, although mountain glacier systems have been equally dynamic as in the arctic (Ammann 1995). There have been rearrangements of vegetation on local scales, but regionally the species composition has remained fairly constant, and there have not been large-scale woody invasions of tundra (MacDonald 1989; Elias 1991). This may partially reflect a difference in scale between arctic and alpine systems. Changes in fragmented alpine areas are more difficult to detect than large-scale change in the arctic. Observed differences in paleoecological records, however, are likely to be real. The predominance of steeply sloped terrain in the alpine prevents large-scale peat development and associated loss of diversity (Körner 1995) such as what occurred in the arctic (Zimov et al. 1995; Walker et al. 1998a). High winds, snow patterns, and fire regimes may act alone or in concert to limit upward expansion of tree growth into alpine tundra (Shankman and Daly 1988; Carrarra et al. 1991; Holtmeier and Broll 1992).

Although vegetation is regionally more stable in the alpine than it is in the arctic, there is some evidence for the upward movement of certain plant species in response to climate warming and loss of some species from the highest altitudinal zone (Grabherr et al. 1994; Grabherr et al. 1995). Consequent changes in both temperature and precipitation could magnify this change, as many species' ranges would be simultaneously shifting and shrinking (Sholes 1994; Maxwell 1997).

There have been changes to alpine faunas associated with both climate and land-use change. For example, the alpine zone of the intermountain west of North America decreased in size during the late Pleistocene. *Ochotona princeps*, a small mammal whose habitat is restricted to the alpine zone, now has a patchy distribution, and extinctions appear to be somewhat at random in areas of near minimal size (Brown 1971; Hafner and Sullivan 1995). Extinctions of large grazers have occurred recently in some alpine regions, primarily in New Zealand and the Himalaya, often due to human occupation and increased use (Körner 1995).

Scenarios of Altered Diversity

Expected Changes in Tundra Distribution

The IMAGE model projects that tundra will undergo the largest projected climatically induced areal reduction of any biome on earth. There is minimal change to tundra from any other biome; almost all the change in tundra is loss (Table 5.2). The model projects that the area of ice will remain virtually unchanged within the next century, due to the persistence of continental ice

Table 5.2. Projected conversion of tundra to other biomes (and vice versa) between 1990 and 2100, as simulated by IMAGE

	From true tundra to other biomes	From shrub tundra to other biomes	From other biomes to tundra
Climate-induced changes			
Ice	176 (4%)	0 (0%)	90
Tundra	1994[a]—	4 (0%)	—
Shrub tundra	1476 (37%)	111[a]—	—
Boreal forest	674 (17%)	1640 (77%)	1
Grassland	1153 (29%)	138 (6%)	—
Land-use–induced changes			
Agriculture	398 (10%)	90 (4%)	—
Managed forests	125 (3%)	268 (13%)	—
Total change	4002 (100%)	2140 (100%)	91

Values are in 10^3 km^2, with percentage of total change in parenthesis. Figures marked as [a] are unchanged (i.e., areas that transition to themselves).

sheets (Fig. 5.1). Biome conversions will consequently add only about 1% of new land to the area of tundra (Alcamo 1994). In contrast, the area of current tundra is expected to shrink by 55% to a total of 3.7×10^6 km^2.

The projected changes in area of tundra differ dramatically among continents. Alpine tundra is expected to show the greatest reductions in area, with the present broad expanses of alpine tundra in the Himalayan and Tibetan

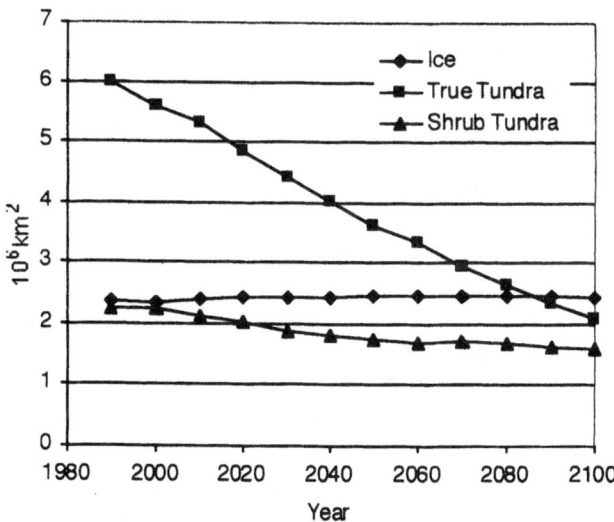

Figure 5.1. Changes in extent of ice, tundra, and shrub tundra as projected by the IMAGE model A1 scenario of climatic warming.

plateau to become restricted to rings of tundra beneath the nival zone (20% of their current area) (Alcamo 1994). In other alpine areas, the alpine zone is expected to move up in elevation and to be invaded by trees or converted to grassland at its lower boundary, depending on prevailing precipitation. In regions with relatively warm climate and/or low elevations, the alpine zone may disappear completely from mountaintops (Grabherr et al. 1994) (e.g., in New Zealand, New Guinea, and the western United States). These alpine zones harbor some of the highest concentrations of biodiversity because of geographic isolation and speciation into the alpine of low-elevation genera (Chabot and Billings 1972; Körner 1995).

Most expected changes in area of arctic tundra will be caused by climatically induced shifts to other biomes. The southern zone of arctic tundra (shrub tundra) is expected to move northward into true arctic tundra and to be replaced by boreal forest. Thus, true arctic tundra will shrink in extent and in the future will colonize only small areas of land. Only 7% of shrub tundra will occur on lands that it presently occupies; the remaining 93% is projected to move to areas presently occupied by true tundra. The extent to which shrub tundra continues to persist, therefore, depends on the extent of overlap in species composition between true tundra and shrub tundra and in the capacity of shrub tundra species to migrate poleward. Approximately 27% of the area that is now tundra is projected by IMAGE to become either agricultural lands or managed forests by the end of the twenty-first century. Much of the tundra that is converted to grassland (an additional 35% of the area that is currently tundra) will undoubtedly be managed for grazing. Thus, although management may not be the direct cause of loss of tundra to other biomes, we expect that these lands will experience increased management in the future.

Expected Changes in Drivers of Diversity

Because greenhouse-gas–induced global warming will be most pronounced at high latitudes (Kattenberg et al. 1996; Keeling et al. 1996), we expect the arctic to experience more pronounced warming than any other biome. Warming will probably occur primarily in late winter and early spring, leading to a longer growing season (Kattenberg et al. 1996; Keeling et al. 1996). The late winter–early spring warming observed in many high-latitude regions since 1980 (Chapman and Walsh 1993; Briffa et al. 1995) and the increased NDVI observed at high northern latitudes (Myneni et al. 1997) is consistent with these predictions. GCMs do not represent precipitation well at high latitudes and currently offer no consistent predictions of how high-latitude precipitation will change in the future. Most arctic regions have shown a steady increase in precipitation since 1950 (Maxwell 1997). We expect the warming that will occur in alpine regions will be less than it is at high latitudes, with a magnitude to that which occurs in nearby lowlands. As in the arctic, the expected changes in alpine precipitation are uncertain and may differ between coastal mountains (which may see increased precipitation) and continental mountains (which may have either increases or decreases in precipitation, or changes in seasonal patterns) (Barry 1992a; Barry 1992b).

The heaviest land-use changes will be in alpine ecosystems because of their proximity to population centers. Improved road access to alpine areas often results in large increases in recreational use, mining, and grazing. Such road access is often associated with increased forestry pressure at lower elevations (Miller et al. 1996). Arctic ecosystems, in contrast, are likely to experience less anthropogenic land-use change than most other ecosystems because of their distance from human population centers, their low productivity, and their harsh climatic conditions. Extensive oil, gas, and mining operations in arctic tundra since 1970 is a major new use that could potentially have additional effects on the biodiversity and functioning of these ecosystems. Although the spatial extent of these disturbances is small, their impacts can be far reaching because of the road and pipeline systems associated with them (Walker et al. 1987). Roads in particular open up previously inaccessible areas to new development, either directly related to tourism and hunting or support facilities for resource extraction. In certain habitats the direct impacts of roads on biodiversity and ecosystem function are important; additionally, roads have important direct and indirect effects on wildlife (Bäckman et al. 1979; Curatolo and Murphy 1986; McLellan and Shackleton 1988; Spencer and Port 1988; Spencer et al. 1988; Angold 1997). Because arctic diversity is often highly concentrated in "hotspots," particularly riparian areas, increased use of these areas will have greater impact than might be predicted from simple spatial extrapolation. Riparian areas are some of the only gravel sources available for the roads and pads needed for arctic energy development (Cargill and Chapin 1987; Walker et al. 1987).

Arctic ecosystems are less likely to experience heavy nitrogen deposition than many other biomes because of their distance from population centers. Nonetheless, many arctic areas are directly downwind from pollutant sources (e.g., northern Scandinavia and the Kola Peninsula). In addition, nitrogen and other atmospheric pollutants can be carried a considerable distance (e.g., from Central Europe to remote areas of the Canadian High Arctic) (Weller 1995). The relatively low precipitation of arctic ecosystems reduces the fallout of these pollutant inputs. Alpine areas will receive greater nitrogen deposition than the arctic because of their greater proximity to population centers and greater precipitation (Williams et al. 1996).

We expect relatively few novel species to be introduced to either the arctic or alpine tundra because their climates are too severe to allow most exotic introductions to survive. For example, no alien species have established at Barrow in the Alaskan arctic despite 150 years of visitation by whalers and other ships (Billings et al. 1973); all weeds in human disturbances derive from ruderal species in the local flora. In addition, both arctic and alpine ecosystems receive little import of human materials because of their inaccessibility and low level of human use. Because most of the arctic flora is circumpolar, there are relatively few climatically adapted species that could potentially be introduced from elsewhere; however, there may well be exceptions and surprises, particularly in isolated areas. For example, the introduction of house mice to many of the subantarctic islands (e.g., Marion Island), led to the subsequent introduction of cats, which became feral and had a major impact on sea birds, which are the major source of nutrients and an important cause of landscape diversity (Bloomer and Bester 1992).

In summary, in the arctic we expect climate change to be the driver of altered diversity that will change most dramatically. In contrast, alpine tundra may experience substantial changes in N deposition, climate, and land use. We expect the magnitude of change in CO_2 to be similar to that of the global average (Fung et al. 1983), and we expect the introduction of exotic species to be relatively infrequent, except where it is facilitated by climatic warming.

Changes in Diversity and Their Consequences

Arctic

The projected scenarios of climatic warming would likely increase absolute species diversity throughout the arctic, but at the cost of losing much of the existing ecosystem. A northward shift in the distributions of many species and a general increase in diversity at any particular latitude will be the major changes. The general northward shift in species distributions will result in a decrease in the extent of arctic tundra globally as the boreal forest encroaches on the tundra, and the northern limit of arctic tundra reaches the extent of land in northern latitudes. Some of this decrease in extent will be in the loss of very species-poor high arctic polar deserts, and the northward shift of more species-rich low arctic tundra vegetation types (Bliss and Matveyeva 1992). Reduced extent of polar desert regions will not result in loss of biodiversity because few species are restricted to these environments.

There are some complications to this simple northward migration model. First, controls on species distributions and diversity are not necessarily uniform from the low to the high latitudes in the arctic (Bliss and Matveyeva 1992; Chernov and Matveyeva 1997), and species will vary in their response to changing environments. Migration rates may differ at the southern and northern portions of this biome due to variation in the response of permafrost and soil development to increasing temperatures, and variation in precipitation changes across the arctic. Second, the distribution of many animal species in the arctic is tied to the distribution of plant species, and northward migration of animal species may depend on plant species migrations. Third, although many arctic species are well adapted to short- and long-term fluctuations in climate, a scenario now exists with an unprecedented combination of climate change, shifts in vegetation, and increasing land use in a region with historically very little land use or human impact. Types of land use vary spatially within the arctic. Land use is most concentrated in the low arctic and in the Russian arctic, and least concentrated in the high arctic and Canadian arctic (Chernov and Matveyeva 1997). Increases in land use that may affect biodiversity include increasing human populations, increasing resource extraction (e.g., minerals, oils, and gas), increasing extent of roads and pipelines, and increasing hunting and recreational use. Most of these types of land-use change will have strong local influences on diversity and minimal global impacts. These impacts can have regional or global consequences where these activities affect keystone species or critical reservoirs of diversity.

Changes in the Low Arctic

Changes in species composition and diversity at the boreal forest–tundra boundary will include increasing diversity with the invasion of boreal species into the tundra. The rate at which this will happen is less predictable, and conversion of tundra to forest may take centuries (Jonasson 1997; Scott et al. 1997). Even though there are records of shifts in treeline over long time periods and changes in stand density in the last century, the northern tree limit has been remarkably stable in the face of climatic variation (MacDonald et al. 1998). Northward migration of species will depend both on warming climate and on dispersal rates, colonization rates, and interactions between species (e.g., migration of boreal-adapted animal species will not likely precede migration of boreal plant species). North of the forest–tundra transition areas, the tundra is characterized by a continuous cover of vegetation dominated by woody shrubs at lower latitudes, and increasingly dominated by mixed shrub and graminoid communities to the north. Warming climate will shift the distributions of some of these species northward, but species composition and diversity will not be altered much on a global scale. The structure of the vegetation may change significantly in ways that affect permafrost conditions and wildlife.

Arctic tundra has historically had less land-use impact than other biomes, and increasing intensity of land use will place additional constraints on a variety of animal species. Although energy-related activities may have large local impacts, they become regionally important only when there is a cumulative effect of many activities in a critical area (e.g., a breeding bird nesting zone, which tend to be concentrated in certain habitats) (Walker et al. 1987; Jefferies et al. 1992). Increasing land-use intensity in the arctic will most strongly affect animal species dependent on unimpeded migration routes, or animals that serve as local resources. These include bears, wolverines, wolves, caribou, and muskoxen. Migrating birds are dependent on vegetation that will be affected by climate change, and serve as local resources, but impacts of climate and land-use change outside of the arctic may have a stronger affect on their populations than changes within the arctic.

Changes in the High Arctic

A considerable portion of the high arctic consists of polar deserts, characterized by bare ground and little vegetation (Bliss and Matveyeva 1992).
Changes to these areas will have fewer functional and practical consequences
than changes to the low arctic, where there is higher productivity and more
human interest and use. Colonization by more southerly species and subsequent increases in biodiversity can be expected with increasing summer
temperatures. Much of this colonization will radiate from polar oases, areas
of high plant, and animal diversity within the polar deserts (Svoboda and
Freedman 1988). The rate of this colonization will depend on the availability of moisture and suitable soils and substrates as well as on increasing
warmth. There are no consistent predictions of changes in precipitation, but
the general prediction of increased precipitation in maritime regions (Kattenberg et al. 1996) may mean increased precipitation for many of the high
arctic islands and coastal areas. This would tend to increase the rates of colonization and increase biodiversity in the high arctic.

Vulnerable Habitats and Critical Areas

Hotspots of diversity will change at different rates than the tundra overall,
and the effects of climate and land-use change will vary among these critical
areas. Regional diversity hotspots will be more strongly affected by land-use
change than by climate. Caribou calving and migration patterns may be most
strongly affected by shifts in climate and vegetation; however, the extant
grazing systems in the arctic are considered to be quite robust and adaptive
to change (Jefferies et al. 1992). Migratory birds will be most strongly affected
by changing vegetation in the arctic, and changing climate and land use in
their winter ranges, which are frequently to the south of treeline (Jefferies et
al. 1994; Jefferies 1997).

Climate change will likely increase diversity in oases and riparian areas. As
these areas tend to be regional diversity hotspots, they may serve as centers
of distribution along climatic gradients in the arctic. Riparian areas and other
diversity hotspots are susceptible to local diversity loss due to land-use
change (Naiman et al. 1993). Increased uses of riparian corridors in the arctic
for hydropower, recreation, and sources of gravel for road building, and
increased mining of eskers for gravel are most probable. Road building may
also affect diversity as it intersects these hotspots, increasing access and recreational and hunting pressure. Loss of diversity in these hotspots (and dispersal centers) may affect regional diversity patterns.

Caribou are an important species in the arctic, with strong links to vegetation, predator populations, and indigenous culture (Jefferies et al. 1992).
Caribou populations have responded to large shifts in climate in the past, and
population numbers show dramatic variation unrelated to climate that has yet
to be fully explained (Gunn and Skogland 1997). One possible scenario is that
the combination of changing climate and increasing land-use intensity may

significantly affect caribou and other large mammal populations in some areas. Increasing land use may decrease animals' ability to modify migration routes or recover from population declines. Caribou winter ranges along the forest–tundra border would shift northward, but would be unlikely to change substantially in character or extent. Caribou calving grounds, and spring and summer feeding grounds in many areas, are situated in a band of nonacidic tundra of mixed shrubs and sedges located between shrub tundra to the south and the Arctic Ocean to the north. Reduction of this area could affect caribou populations in Alaska, Western Canada, and Russia (Gunn and Skogland 1997; Walker et al. 2001). Loss of caribou would regionally reduce diversity of predators and insect species dependent on caribou. The response of many species important to humans to these new pressures may be strongly dependent on human wildlife management practices. There is so little known about controls on caribou populations, however, that it is very difficult to predict specific outcomes. Large increases in caribou or other important populations could also have detrimental local effects (Jefferies 1997).

Ecosystem Consequences

Northward movement of trees into arctic tundra or, to a lesser extent, the upward movement of trees into alpine tundra could cause a positive feedback to climatic warming in continuously forested areas (see Chap. 6). A dark forested surface absorbs more energy than snow-covered tundra, causing greater heating of the overlying atmosphere. For example, changes in solar input have generally been invoked to explain major changes in global climate; however, modeling simulations suggest that the northward movement of treeline during the last thermal maximum 6000 years ago accounted for half of the regional warming that occurred at that time (Foley et al. 1994). In addition, the difference in albedo between boreal forest and tundra in summer partially explains why the arctic front (creating a strong temperature gradient) often occurs at treeline (Pielke and Vidale 1996). Thus, northward movement of trees might alter the position of the arctic front.

This link between the northern forest boundary and the position of the arctic front, however, also acts as a strong negative feedback to northern migration of trees. Because the treeline serves to stabilize the position of the arctic front, temperatures just north of treeline are colder than global-scale climate models suggest (Pielke and Vidale 1996). Thus, the IMAGE model may overestimate the rate and extent of boreal forest invasion into tundra. There is also a strong negative feedback system operating on small isolated stands and individuals to the north of the continuous forest (Scott et al. 1997), so the rate of change is difficult to predict without considering interactions between these systems.

An increase in summer temperatures will likely promote an increase in the shrub cover of low arctic tundra at the expense of the moss layer. Consequences of this change include loss of insulating layers over permanently frozen ground, increased nutrient availability (possibly a temporary pulse of nutrients currently locked in permafrost), increased CO_2 release from frozen peat layers, and a decrease in high-quality forage species for arctic herbivores. Permafrost conditions affect plant growth and species composition and are maintained by both temperature and vegetation cover, so there may be feedback loops between permafrost conditions and vegetation that tend to maintain vegetation in its current state. Although the changes in composition and diversity are slight, the consequences may be far reaching because this type of tundra covers large areas of the arctic (Walker et al. 1998; Walker et al. 2001).

Alpine

Vulnerable Habitats and Critical Areas

Because of the fragmented distribution of alpine tundra and the extremely large changes in its future abundance projected by the IMAGE model, the entire global alpine ecosystem should be considered vulnerable. Most rare habitats are also associated with rare species, and thus will be critical in most cases. These particular habitats vary greatly among mountain ranges. For example, the southern Rocky Mountains glacial forelands often house rare species (Komárková 1979). Species that occur in limited habitats or near their distributional limits will generally be most at risk.

It is likely, though, that IMAGE may overestimate shifts in the alpine zone. The available data suggest that extent and overall diversity of alpine tundra have remained remarkably stable throughout past major global climate changes (Körner 1995). This is because of strong species–climate interactions at the treeline that limit upward expansion of trees into the tundra (Ives and Hansen-Bristow 1983; Armand 1992; Slatyer and Noble 1992). Nonetheless, detailed studies are already suggesting upward movement of certain species that could lead to changes in alpine diversity and function in time (Grabherr et al. 1994).

Island biogeography theory leads to the prediction that extinction rates will be higher on smaller patches; therefore, the more fragmented an alpine system is, the more vulnerable it becomes (MacArthur and Wilson 1967). Nonequilibrium conditions, however, may be maintained for long periods of time following a "shrinking" of an area, so it is difficult to predict with any precision what the levels of loss would be (Brown 1971; Lomolino et al. 1989). Because of the higher levels of endemism in alpine ecosystems, they are much more vulnerable to absolute diversity losses than are arctic ecosystems. Areas such as the European Alps, isolated and small areas of southern hemisphere alpine, and the Himalaya all are important centers of endemism.

In addition to endemic species (e.g., plants, insects, and small mammals) other sensitive alpine species include large predators and large grazers that require large home ranges or specific alpine habitats in which to survive. All tundra areas support natural grazers, although in some areas the native grazing species have become extinct (Körner 1995). Most of the large grazers use alpine areas as one part of a larger range. Like other animal groups, there are only a few bird species for which the alpine zone is a critical habitat, but major changes in extent or composition of alpine tundra could be critical for those species.

Ecosystem Consequences

The functional consequences of changes in alpine tundra extent and diversity may be far reaching because of the hydrologic importance of these areas, which often serve as watersheds for nearby population centers. Increases in nitrogen deposition may be particularly critical because these may have both direct and indirect effects. Direct effects will be changes in species composition and soil function, and indirect effects will be increased nitrogen concentrations in runoff waters as soils and vegetation cover lose their buffering capacity over time (Baron et al. 1994; Williams et al. 1996; Williams et al. 1998). Feedbacks to the climate and global system would primarily be through changes in treeline; alpine areas are some the least important zones for global carbon flux (Brooks et al. 1997; Williams et al. 1998). Alpine areas are important recreation centers in many parts of the world, and that function would be lost if these areas changed dramatically in size.

Conclusions

1. Changes in arctic tundra will be driven most strongly by climate change, and those in alpine tundra by land use (Table 5.3).
2. Differences in drivers and biodiversity changes between arctic and alpine ecosystems will be related to the different spatial configurations of the systems and their proximity to direct human use and pollution sources. Arctic

Table 5.3. Summary of changes expected in tundra biodiversity relative to other biomes

Global-change driver	Expected change in driver		Sensitivity of driver		Net effect of driver on biodiversity	
	Arctic	Alpine	Arctic	Alpine	Arctic	Alpine
Land-use change	1	2	5	5	25	50
Elevated CO$_2$	1	1	1	1	2	2
N deposition	1	3	3	3	9	27
Climate change	5	2	5	5	75	30
Biotic exchange	1	1	2	2	2	2

See Chapter 1 for explanation. Expected changes and sensitivities are ranked from low (1) to high (5). Net effects are expressed as a percentage of the maximum possible sensitivity.

tundra is spatially extensive and connected, far from pollution sources, but extremely sensitive to temperature. Alpine tundra is generally spatially fragmented, is close to pollution sources, but has been less sensitive to temperature changes in the past.

3. Arctic tundra will shift northward under a warming climate, and there will be invasions of boreal species into what is currently shrub tundra and invasions of shrubs into what is currently true tundra. Although species extinctions may be rare, the ecosystem as a whole is endangered. The IMAGE model may overestimate the rate and extent of expansion of boreal forest into tundra because of a negative feedback loop with treeline position and the arctic front, which leads to unfavorable conditions for tree establishment just to the north of the treeline. Changes in biodiversity due to land use and changes in atmospheric composition are expected to be minimal in the arctic compared with other regions.

4. The IMAGE model predicts almost complete loss of current alpine tundra on a global basis under a warming climate, but paleoecological records suggest that this may be incorrect. The upper treeline and overall diversity of tundra have remained quite constant throughout the Pleistocene; changes have primarily been localized. Nonetheless, there may well be loss of many smaller tundra fragments. Fragmentation may be aggravated by land-use change (e.g., road building and forestry). Rare species and habitats will be most sensitive to fragmentation and will be lost first. Increased nitrogen deposition is of particular concern in alpine ecosystems because it may have both direct ecosystem effects as well as secondary effects on water quality.

5. Neither the arctic nor alpine is expected to be impacted by invading species (except as a conversion to other biomes), although conditions necessary for invasions would become increasingly widespread with a warming climate.

Acknowledgments. Research leading to these generalizations was supported by NSF grants OPP-9400083, OPP-9510140, and DEB-9211776.

References

Alcamo J (ed) (1994) *IMAGE 2.0: Integrated Modeling of Global Climate Change.* Kluwer Academic Publishers, Dordrecht.

Ammann B (1995) Paleorecords of plant biodiversity in the Alps. In: Chapin FS III, Körner C (eds) *Arctic and Alpine Biodiversity: Patterns, Causes and Ecosystem Consequences*, pp. 137–149. Springer-Verlag, Berlin.

Angold PG (1997) The impact of a road upon adjacent heathland vegetation: effects on plant species composition. Journal of Applied Ecology 34:409–417.

Armand AD (1992) Sharp and gradual mountain timberlines as a result of species interaction. In: Hansen AJ, di Castri F (eds) *Landscape Boundaries*, pp. 360–378. Springer-Verlag, New York.

Bäckman L, Knutsson G, Ruhling A (1979) *Influence of Roads on the Surrounding Nature: Vegetation, Soil and Groundwater.* VTI, Fack, S-58101 Linkoping, Sweden.

Baron JS, Ojima DS, Parton WJ (1994) Analysis of nitrogen saturation potential in Rocky Mountain tundra and forest: implications for aquatic systems. Biogeochemistry 27:61.

Barry RG (1992a) Climate change in the mountains. In: Stone PB (ed) *The State of the World's Mountains*. Zed Books, London.

Barry RG (1992b) Mountain climatology and past and potential future climatic changes in mountain regions: a review. Mountain Research and Development 12:71–86.

Bartlein PJ, Prentice IC (1989) Orbital variations, climate and paleoecology. Trends in Ecology and Evolution 4:195–199.

Bee JW, Hall ER (1956) The mammals of northern Alaska. Museum of Natural History, Lawrence, KS.

Billings WD, Shaver GR, Trent AW (1973) Temperature effects on growth and respiration of roots and rhizomes in tundra graminoids. In: Bliss LC, Wielgolaski FE (eds) Proceedings of the conference, *Primary Production and Production Processes*, pp. 57–63. Tundra Biome, Dublin, Ireland.

Bliss LC, Matveyeva NV (1992) Circumpolar arctic vegetation. In: Chapin FS III, Jefferies RL, Reynolds JF, Shaver GR, Svoboda J, Chu EW (eds) *Arctic Ecosystems in a Changing Climate: an Ecophysiological Perspective*, pp. 59–89. Academic Press, San Diego.

Bloomer JP, Bester MN (1992) Control of feral cats on sub-Antarctic Marion Island, Indian Ocean. Biological Conservation 60:211–219.

Briffa KR, Jones PD, Schweingruber FH, Shiyatov SG, Cook ER (1995) Unusual twentieth-century summer warmth in a 1000-year temperature record from Siberia. Nature 376:156–160.

Brooks PD, Schmidt SK, Williams MW (1997) Winter production of CO_2 and N_2O from alpine tundra: environmental controls and relationship to inter-system C and N fluxes. Oecologia 110:403–413.

Brown JH (1971) Mammals on mountaintops: nonequilibrium insular biogeography. American Naturalist 105:467–478.

Brubaker LB, Anderson PM, Hu FS (1995) Arctic tundra biodiversity: a temporal perspective from late Quaternary pollen records. In: Chapin FSI, Körner C (eds) *Arctic and Alpine Biodiversity*, pp. 111–125. Springer-Verlag, Berlin.

Cargill SM, Chapin FS III (1987) Application of successional theory to tundra restoration: a review. Arctic and Alpine Research 19:366–372.

Carrarra PE, Trimble DA, Rubin M (1991) Holocene treeline fluctuations in the northern San Juan Mountains, Colorado, U.S.A., as indicated by radiocarbon-dated conifer wood. Arctic and Alpine Research 23:233–246.

Chabot B, Billings WD (1972) The origin and ecology of the Sierran alpine flora and vegetation. Ecological Monographs 42:163–199.

Chapin FS III, Körner C (1995) Patterns, causes, changes and consequences of biodiversity in arctic and alpine ecosystems. In: Chapin FS III, Körner C (eds) *Arctic and Alpine Biodiversity: Patterns, Causes and Ecosystem Consequences*, pp. 313–320. Springer-Verlag, Berlin.

Chapman WL, Walsh JE (1993) Recent variations of sea ice and air temperature in high latitudes. Bulletin American Meteorological Society 74:33–47.

Chernov YI (1995) Diversity of the arctic terrestrial fauna. In: Chapin FS III, Körner C (eds) *Arctic and Alpine Biodiversity: Patterns, Causes and Ecosystem Consequences*, pp. 81–95. Springer-Verlag, Berlin.

Chernov YI, Matveyeva NV (1997) Arctic ecosystems in Russia. In: Wielgolaski FE (eds) *Polar and Alpine Tundra*, pp. 361–507. Elvesier, Amsterdam.

Curatolo JA, Murphy SM (1986) The effects of pipelines, roads, and traffic on the movements of caribou, Rangifer tarandus. Canadian Field-Naturalist 100:218–224.

Elias SA (1991) Insects and climate change: fossil evidence from the Rocky Mountains. Bioscience 41:552–559.

Foley JA, Kutzbach JE, Coe MT, Levis S (1994) Feedbacks between climate and boreal forests during the Holocene epoch. Ecology 371:52–54.

Fox JF (1992) Responses of diversity and growth-form dominance to fertility in Alaskan tundra fellfield communities. Arctic and Alpine Research 24:233–237.

Fung I, Prentice K, Matthews E, Lerner J, Russell G (1983) Three-dimensional tracer model study of atmospheric CO_2: response to seasonal exchanges with the terrestrial biosphere. Journal of Geophysical Research 88:1281–1294.

Gabriel HW, Talbot SS (1984) Glossary of landscape and vegetation ecology for Alaska. Bureau of Land Management-Alaska, Technical Report 10, Alaska.

Grabherr G, Gottfried M, Pauli H (1994) Climate effects on mountain plants. Nature 369:448.

Grabherr G, Gottfried M, Gruber A, Pauli H (1995) Patterns and current changes in alpine plant diversity. In: Chapin FS III, Körner C (eds) *Arctic and Alpine Biodiversity: Patterns, Causes, and Ecosystem Consequences*, pp. 167–181. Springer-Verlag, Berlin.

Gunn A, Skogland T (1997) Responses of caribou and reindeer to global warming. In: Oechel WC, Callaghan T, Gilmanov T, Holten JI, Maxwell B, Molau U, et al. (eds) *Global Change and Arctic Terrestrial Ecosystems*, pp. 189–200. Springer-Verlag, New York.

Guthrie RD, Stoker S (1990) Paleoecological significance of mummified remains of Pleistocene horses from the North Slope of the Brooks Range, Alaska. Arctic 43:267–274.

Hafner DJ, Sullivan RM (1995) Historical and ecological biogeography of Nearctic pikas (Lagomorpha: Ochotonidae). Journal of Mammalogy 76:302–321.

Hobbie SE, Jensen DB, Chapin FS (1993) Resource supply and disturbance as controls over present and future plant diversity. In: Schultze E, Mooney HA (eds) *Biodiversity and Ecosystem Function*, pp. 385–408. Springer-Verlag, Heidelberg.

Holtmeier F-K, Broll G (1992) The influence of tree islands and microtopography on pedoecological conditions in the forest-alpine tundra ecotone on Niwot Ridge, Colorado Front Range, U.S.A. Arctic and Alpine Research 24:216–228.

Hu FS, Brubaker LB, Anderson PM (1993) A 12000 year record of vegetation change and soil development from Wien Lake, central Alaska. Candadian Journal of Botany 71:1133–1142.

Ives JD, Hansen-Bristow KJ (1983) Stability and instability of natural and modified upper timberline landscapes in the Colorado Rocky Mountains, USA. Mountain Research and Development 3:149–155.

Jefferies RL (1997) Long-term damage to sub-arctic coastal ecosystems by geese: ecological indicators and measures of ecosystem dysfunction. In: Crawford RMM (ed) *Disturbance and Recovery in Arctic Lands: An Ecological Perspective*, pp. 151–165. Kluwer Academic Publishers, Dordrecht.

Jefferies RL, Klein DR, Shaver GR (1994) Vertebrate herbivores and northern communities: reciprocal influences and responses. Oikos 71:193.

Jefferies RL, Svoboda J, Henry G, Raillard M, Ruess R (1992) Tundra grazing systems and climatic change. In: Chapin III FS, Jefferies RL, Reynolds J, Shaver GA, Svoboda J (eds) *Arctic Ecosystems in a Changing Climate: an Ecophysiological Perspective*, pp. 391–412. Academic Press, San Diego.

Jonasson S (1997) Buffering of arctic plant responses in a changing climate. In: Oechel WC, Callaghan T, Gilmanov T, Holten JI, Maxwell B, Molau U, et al. (eds) *Global Change and Arctic Terrestrial Ecosystems*, pp. 365–380. Springer-Verlag, New York.

Kattenberg A, Giorgi F, Grassl H, Meehl GA, Mitchell JFB, Stouffer RJ, et al. (1996) Climate models—projections of future climate. In: Houghton JT, Meira Filho LG,

Callander BA, Harris N, Kattenberg A, Maskell K (eds) *Climate Change 1995. The Science of Climate Change*, pp. 285–357. Cambridge University Press, Cambridge, U.K.

Keeling CD, Chin JFS, Whorf TP (1996) Increased activity of northern vegetation inferred from atmospheric CO_2 measurements. Nature 382:146–149.

Komárková V (1979) *Alpine Vegetation of the Indian Peaks Area, Front Range, Colorado Rocky Mountains*. J. Cramer, Vaduz.

Körner C (1995) Alpine plant diversity: a global survey and functional interpretations. In: Chapin FS III, Körner C (eds) *Arctic and Alpine Biodiversity: Patterns, Causes, and Ecosystem Consequences*, pp. 45–62. Springer-Verlag, Berlin.

Lomolino MV, Brown JH, Davis R (1989) Island biogeography of montane forest mammals in the American Southwest. Ecology 70:180–194.

MacArthur RH, Wilson EO (1967) *The Theory of Island Biogeography*. Princeton University Press, Princeton.

MacDonald GM (1989) Postglacial paleoecology of the subalpine forest-grassland ecotone of southwestern Alberta: new insights on vegetation and climate change in the Canadian Rocky Mountains and adjacent foothills. Palaeogeography, Palaeoclimatology, and Palaeoecology 73:155–173.

MacDonald GM, Edwards TWD, Moser KA, Pienitz R, Smol JP (1993) Rapid response of treeline vegetation and lakes to past climate warming. Nature 361:243–246.

MacDonald GM, Szeicz JM, Claricoates J, Dale KD (1998) Response of the Central Canadian treeline to recent climatic changes. Annals of the Association of American Geographers 88:183–208.

Maxwell B (1997) Recent climate patterns in the Arctic. In: Oechel WC, Callaghan T, Gilmanov T, Holten JI, Maxwell B, Molau U, et al. (eds) *Global Change and Arctic Terrestrial Ecosystems*, pp. 21–46. Springer, New York.

McLellan BN, Shackleton DM (1988) Grizzly bears and resource-extraction industries: effects of roads on behaviour, habitat use and demography. Journal of Applied Ecology 25:451–460.

Miller JR, Joyce LA, King RM (1996) Forest roads and landscape structure in the southern Rocky Mountains. Landscape Ecology 11:115.

Murray DF (1995) Causes of arctic plant diversity: origin and evolution. In: Chapin FS III, Körner C (eds) *Arctic and Alpine Biodiversity: Patterns, Causes and Ecosystem Consequences*, pp. 21–32. Springer-Verlag, Berlin.

Myneni RB, Keeling CD, Tucker CJ, Asrar G, Menani RR (1997) Increased plant growth in the northern high latitudes from 1981 to 1991. Nature 386:698–702.

Naiman RJ, Decamps H, Pollock M (1993) The role of riparian corridors in maintaining regional biodiversity. Ecological Applications 3:209–212.

Pielke RA, Vidale PL (1996) The boreal forest and the polar front. Journal of Geophysical Research-Atmospheres 100D:25755–25758.

Polunin N (1959) *Circumpolar Arctic Flora*. Clarendon Press, Oxford, U.K.

Rannie WF (1986) Summer air temperature and number of vascular species in arctic Canada. Arctic 39:133–137.

Schimel J (1995) Ecosystem consequences of microbial diversity and community structure. In: Chapin FS III, Körner C (eds) *Arctic and Alpine Biodiversity: Patterns, Causes and Ecosystem Consequences*, pp. 239–254. Springer-Verlag, Berlin.

Schweger CE, Matthews JVJ, Hopkins DM, Young SB (1982) Paleoecology of Beringia—a synthesis. In: Hopkins JVMJ DM, Schweger CE, Young SB (eds) *Paleoecology of Beringia*, pp. 425–444. Academic Press, New York.

Scott PA, Lavoie C, MacDonald GM, Sveinbjörnsson B, Mein RW (1997) Climate change and future position of arctic treeline. In: Oechel WC, Callaghan T, Gilmanov T, Holten JI, Maxwell B, Molau U, et al. (eds) *Global Change and Arctic Terrestrial Ecosystems*, pp. 245–265. Springer, New York.

Shankman D, Daly C (1988) Forest regeneration above tree limit depressed by fire in the Colorado Front Range. Bulletin of the Torrey Botanical Club 115:272–279.

Sholes ODV (1994) Plant survival. Nature 371:661.

Slatyer RO, Noble IR (1992) Dynamics of montane treelines. In: Hansen AJ, di Castri F (eds) *Landscape Boundaries*, pp. 346–359. Springer-Verlag.

Spencer HJ, Port GR (1988) Effects of roadside conditions on plants and insects. II. Soil conditions. Journal of Applied Ecology 25:709–715.

Spencer HJ, Scott NE, Port GR, Davison AW (1988) Effects of roadside conditions on plants and insects. I. Atmospheric conditions. Journal of Applied Ecology 25:699–707.

Svoboda J, Freedman B (eds) (1988) Ecology of a High Arctic lowland oasis, Alexandra Fiord (78 53′N, 75 55′W), Ellesmere Island, N.W.T., Canada. University of Toronto, Toronto.

Vartanyan SL, Garutt VE, Sher AV (1993) Holocene dwarf mammoths from Wrangel Island in the Siberian Arctic. Nature 362:337–340.

Walker DA, Webber PJ, Binnian EF, Everett KR, Lederer ND, Nordstrand EA, et al. (1987) Cumulative impacts of oil fields on northern Alaskan landscapes. Science 238:757–761.

Walker DA, Auerbach NA, Bockheim JG, Chapin FSI, Eugster W, King JY, et al. (1998) Energy and trace-gas fluxes across a soil pH boundary in the Arctic. Nature 394:469–472.

Walker DA, Bockheim JG, Chapin FSI, Eugster W, Nelson FE, Ping CL (2001) Calcium-rich tundra, wildlife, and the "Mammoth Steppe." Quaternary Science Review.

Walker MD (1995) Patterns and causes of arctic plant community diversity. In: Chapin FS III, Körner C (eds) *Arctic and Alpine Biodiversity*, pp. 3–20. Springer-Verlag, Berlin.

Weller G (1995) Global pollution and its effect on the climate of the arctic. Science of the Total Environment 160–161:19–24.

Williams MW, Brooks PD, Seastedt T (1998) Nitrogen and carbon soil synamics in response to climate change in a high-elevation ecosystem in the Rocky Mountains, U.S.A. Arctic and Alpine Research 30:26–30.

Williams MW, Baron JS, Caine N, Sommerfeld R, Sanford R (1996) Nitrogen saturation in the Rocky Mountains. Environmental Science and Technology 30:640–646.

Young SB (1971) The vascular flora of St. Lawrence Island with special reference to floristic zonation in the arctic regions. Contributions from the Gray Herbarium 201:11–115.

Zimov SA, Chuprynin VI, Oreshko AP, Chapin FS III, Chapin MC, Reynolds JF (1995) Effects of mammals on ecosystem change at the Pleistocene-Holocene boundary. In: Chapin FS III, Körner C (eds) *Arctic and Alpine Biodiversity: Patterns, Causes and Ecosystem Consequences*, pp. 127–135. Springer-Verlag, Berlin.

6. Boreal Forest

F. Stuart Chapin III and Kjell Danell

The boreal forest biome occupies $13 \times 10^6\,km^2$. It comprises approximately 25% of the world's forest land (Olson et al. 1983; Apps et al. 1993) and includes $2.6 \times 10^6\,km^2$ of peatlands (Gorham 1991). Changes in the extent or functioning of the boreal forest could substantially modify global climate through (1) release of its large stocks of soil carbon (Post et al. 1982; Kurz and Apps 1995), (2) changes in methane fluxes from peatlands (Reeburgh and Whalen 1992; Roulet and Ash 1992), or (3) changes in winter albedo and regional energy exchange (Bonan et al. 1992; Thomas and Rowntree 1992). Boreal forests have also been implicated as the "missing sink" for atmospheric carbon dioxide (Ciais et al. 1995). For these reasons, there has been considerable interest and speculation about future changes in the extent, structure, and functioning of the boreal forest.

The functioning of the boreal forest is more tightly tied to its species composition than to climate. For example, in Alaskan white spruce forests there is virtually no change in soil carbon across the entire latitudinal range of boreal forest (July mean temperature from 10 to 15°C) (Michaelson et al. 1996), whereas there is a greater than twofold range in soil carbon between conifer and hardwood stands on north and south aspects of the same hill (Van Cleve et al. 1983), due largely to differences in the effects of different plant species on soil temperature and nutrient cycling (Flanagan and Van Cleve 1983; Van Cleve et al. 1991; Pastor and Mladenoff 1992; Trofymow et al. 1995). Given these large effects of plant traits on the functioning of boreal ecosystems, it is crucial to know how the species composition and diversity of these ecosystems may change in the future.

Patterns of Boreal Diversity and Their Functional Consequences

Plants

The diversity of dominant vascular plants in the boreal forest is lower than it is in biomes to the north (tundra) or south (prairie and temperate forest). There are generally one to three tree species per stand, with single-species stands being quite common. Most boreal tree species have broad geographic distributions. For example, *Larix gmelinii* dominates most Asian boreal forests, and *Picea mariana* dominates 40% of Alaskan boreal forests. Boreal tree distributions are usually circumpolar at the genus level, with most forests dominated by *Larix, Picea, Pinus, Betula,* or *Populus* (Pastor et al. 1996). Boreal tree species differ strikingly in their rates of growth and nutrient uptake and therefore in the productivity that can be supported at a given rate of resource supply (Chapin et al. 1983; Vedrova 1995). They also differ strongly in litter quality (Chapin and Kedrowski 1983; Flanagan and Van Cleve 1983; Van Cleve et al. 1983; Trofymow et al. 1995), transpiration rates, flammability, and effects on soil temperature (Oechel and Van Cleve 1986). As a result, because boreal tree diversity is low, and the existing species differ strongly in their ecosystem impacts (i.e., functional diversity is high), loss or altered abundance of a single tree species could have profound regional consequences (Pastor et al. 1996).

Both the diversity and types of species change through succession after disturbance. Species richness of vascular plants increases until stands are 10–30 years old. It then generally declines with increasing age, as a mixture of pioneer shrubs, forbs, and grasses are displaced, sometimes by an intermediate deciduous tree phase in relatively warm sites. There is even lower vascular plant diversity in late succession, which in northern and continental regions of the boreal forest is usually coniferous with an understory consisting of a few species of dwarf shrubs (Viereck 1973; Viereck et al. 1983; Van Cleve et al. 1991; Tonteri 1998). By contrast, nonvascular plants generally increase in abundance and diversity through succession (McCune 1993), with mosses comprising up to half the production in late-successional coniferous stands (Oechel and Van Cleve 1986). Early successional boreal species generally have higher rates of nutrient uptake and growth, are more palatable to herbivores, and have higher litter quality than do late successional species (Van Cleve et al. 1991; Pastor et al. 1996). Thus, any change in disturbance regime and resulting changes in stand age structure have large effects on regional patterns of diversity, productivity, and nutrient cycling.

In general vascular plant diversity decreases to the north in boreal forest (Pastor and Mladenoff 1992), just as in tundra (see Chap. 8), whereas lichen diversity increases to the north. In the north the species composition of boreal forest is quite similar to that of tundra, differing primarily in the addition of trees. At the southern margin of boreal forest, there is similarly either a gradual transition to temperate forests in moist areas or a relatively sharp transition to prairies in continental areas (Pastor and Mladenoff 1992).

Animals

The diversity of folivorous insects in the boreal forest is generally low. Many of these species exhibit strong population fluctuations and can cause widespread defoliation and tree mortality (Kallio and Lehtonen 1973; Holling 1992). Beetles are a species-rich invertebrate group in the boreal forest. The number of beetle species declines strongly to the north in Sweden (Stokland 1994), as in many insect taxa (Väisänen and Heliövaara 1994), although there are exceptions. For example, sawfly species richness increases to the north (Kouki et al. 1994). About 70% of the threatened invertebrate species of Swedish forests are beetles (Berg et al. 1994). For the threatened forest invertebrate species of Finland, more than 60% were associated with decaying wood; this fauna is threatened in those areas of Scandinavia with extensive forest harvest (Kaila et al. 1994).

The species richness of insects increases with forest age (Niemelä 1997) and differs with the species of host tree. For example, in Sweden there are about 45 lepidopteran species associated with living trees of both Scots pine (*Pinus sylvestris*) and Norway spruce (*Picea abies*), but about four times as many associated with birches (*Betula* spp.) and willows (*Salix* spp.) (Bernes 1994). This implies that changes in age structure and species composition of forests greatly affect insect species richness.

As with plants and insects, there is a low diversity of vertebrates in the boreal forest, with this diversity increasing to the south in parallel with the longer growing season and increased proportion of hardwood species in the vegetation. As with plants, the vertebrate species are widespread, with most genera showing a circumpolar distribution (Pastor et al. 1996). Among the mammalian herbivores, there are 31 genera in the boreal forests of the northern hemisphere, of which 10 occur in both North America and Eurasia (Danell et al. submitted). No mammalian herbivore species has its distribution limited to the boreal forest region. Some species also occur in tundra, whereas others also occur in temperate forests. Because the majority of the mammalian herbivores have a broad spectrum of food plants, we expect this group of species to be relatively insensitive to small changes in plant species composition.

There are important longitudinal gradients in mammalian diversity, with midcontinents generally showing the greatest number of vertebrate herbivores (Danell et al. 1996), reflecting the greater fire frequency (Payette 1992) and therefore a greater abundance of palatable early successional species in these areas (Danell et al. 1996; Pastor et al. 1996).

Large browsing and grazing mammals (e.g., moose and deer) can at high densities change the abundance of preferred food plant species and greatly influence the rate of plant succession and ecosystem processes. Browsing mammals have a strong effect on plant diversity within stands because they are generalist herbivores that take a small quantity of many plant species (Bryant and Chapin 1986). This prevents the secondary metabolites of any single plant species from reaching toxic thresholds. They therefore tend to take a larger proportion of rare species than of common species, thus eliminating rare taxa and reducing overall plant diversity. The presence of beavers in boreal forest leads to a dramatic increase in the wetland component and creates a diversified forest landscape (Johnston 1994).

There are about 50 bird species in any one region of the boreal forest (Haila and Järvinen 1990). The diversity of forest birds in Finland decreases toward the north (Järvinen and Väisänen 1973), and species numbers decrease from west to east in both the Palaearctic and Nearctic (Haila and Järvinen 1990). In North America, about half of these are tropical migrants whose population dynamics are often influenced more by land use outside the boreal zone than by boreal processes. As expected from patterns of insect abundance and diversity, foliage insectivores are concentrated in hardwood stands and ground-feeding insectivores in conifer forests.

Microbes

The types and diversity of microbial enzymes is closely associated with forest type, reflecting differences in litter quality. In general, this enzymatic diversity is greatest in late-successional forests where there is an abundance of low-quality litter with a diverse array of chemical defenses against herbivores and pathogens. This high microbial diversity in late succession contributes to the large soil fauna and the high diversity of ground-feeding birds.

Summary

There are relatively few species unique to the boreal forest, despite its broad aerial extent. Species diversity commonly decreases to the north, although there are exceptions. Analyses of specific groups often identify regions of high diversity (Väisänen and Heliövaara 1994), but these patterns often differ among taxa. Specific groups exhibit "hot spots" of diversity (e.g., decaying logs for beetles or riparian floodplains for migratory tropical birds), but these hot spots differ among taxa, so there are a wide variety of boreal habitats that are important reservoirs of diversity.

Past Changes in Boreal Diversity

The boreal forest is a young biome, which was absent from most of its current range 6,000–8,000 years ago (Ritchie 1987; Pastor et al. 1996). Most of boreal Canada and Europe and mountainous regions of Alaska and boreal Asia were ice covered during the Pleistocene, and ice-free regions were occupied by a steppe–tundra mosaic rather than forest or peatlands (Ritchie 1987; Brubaker et al. 1995). Spruce appears to have migrated into Scandinavia in approximate equilibrium with climate (Kullman and Engelman 1997). In North America, the westward migration of white spruce preceded that of black spruce, which preceded that of pine (Ritchie and MacDonald 1986; Brubaker et al. 1995). Pine is widespread in western Canada, but it only extends into the easternmost part of Alaska, even though it grows effectively well beyond its current range, which suggests that in North America boreal taxa are still changing ranges in response to past changes in climate. In North America, boreal taxa derive primarily from midlatitudes in eastern North America and from the Yakutian Plateau of Asia, where they coexisted with many nonboreal genera during the Pleistocene (Pastor et al. 1996). Because of the short history of the boreal biome and relatively recent coexistence of boreal taxa with nonboreal taxa, we expect that highly specific coevolutionary links among the current boreal biota would be uncommon except among those interactions that are highly specific (e.g., host–parasite) and subject to strong current selection. For this reason, we expect that loss of a single species would seldom lead to a cascade of extinction of coevolved species.

Boreal forest has been subject to less-intense human impacts than southern biomes, and these impacts are largely restricted to the southern fringe. Nonetheless, these human impacts have substantially reduced biological diversity, particularly as a result of forestry activities (Sjöberg 1995). In Finland and Sweden (which are primarily boreal forests), recent extinctions are estimated to be about 3–5% for mammals, 1–3% for birds, 0% for amphibians and reptiles, 1–2% for coleoptera, lichens, and vascular plants, 2% for bryophytes, and 0.5–1% for macrofungi (Nilsson and Ericson 1997) (Table 6.1). In Finland (largely boreal forest), the number of threatened (red-listed) species includes 16 mammals (16% of total), 33 birds (26%), 2 amphibians

Table 6.1. Boreal species that have become extinct in Finland and Sweden or which are red-listed in Finland

Taxon	Recent extinctions	Red-listed species
	(% of species in taxon)	
Mammals	4	16
Birds	2	26
Amphibians	0	22
Reptiles	2	—
Coleoptera	2	9
Vascular plants	2	15
Bryophytes	2	20
Lichens	2	9
Macrofungi	1	22
Average	1.8	17.4

(22%), 326 coleoptera (9%), 232 vascular plants (15%), 168 bryophytes (20%), 135 lichens (9%), and 660 macrofungi (22%) (Nilsson and Ericson 1997). This is probably at least as high a proportion of biodiversity threatened as is that found in species-rich biomes (e.g., tropical forests). Given that there are so few species present in boreal forest, this high proportion of species loss and endangerment could significantly affect the functioning of these ecosystems. In more remote areas of boreal forest such as North America and Russia, human activities may have caused less species loss.

Expected Changes in Boreal Diversity

Distribution of the Boreal Forest

Climate warming is expected to cause a northward shift of boreal species (Pastor and Post 1988; Prentice and Fung 1990; Melillo et al. 1996). The IMAGE model projects relatively little change in the total area of boreal forest in response to reasonable scenarios of change in climate and land use (Fig. 6.1) (Alcamo 1994). This does not mean, however, that boreal forest distribution will be static. IMAGE projects that boreal forest will expand by 20%, largely due to tree colonization of tundra (77% of the total change) and of those cold steppes that receive increased precipitation (Fig. 6.1, Table 6.2).

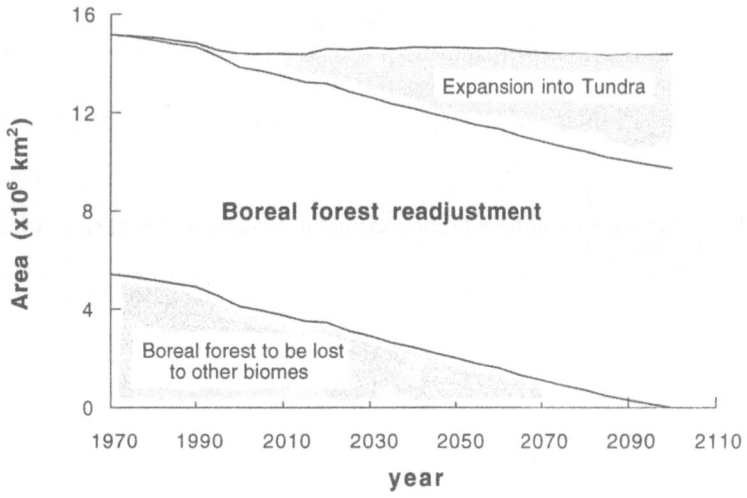

Figure 6.1. Changes in area of boreal forest projected by the IMAGE model in response to scenarios of future change in climate and land use. Although the total area of boreal forest is projected to remain relatively constant, about one third of the present boreal forest is projected to be converted to other biomes, and an additional third will be added as trees advance into tundra (Table 6.1).

Table 6.2. Conversion of other biomes to boreal forest and from boreal forest to other biomes between 1990 and 2100, as simulated by IMAGE

Biome	From boreal to other biomes	From other biomes to boreal forest
	(% of total change)	
Climate-induced changes		
Grassland	4	9
Tundra	0	77
Temperate forest	25	2
Land-use–induced changes		
Agriculture	12	8
Managed forests	59	4
Total change	100	100

By contrast, the 23% loss of boreal forest projected by IMAGE would result from an interaction of climate and land-use change. The largest initial changes are projected to be conversion to grassland and agriculture at the southern margins of continental areas, and conversion to temperate forests in areas with a maritime influence. As human population rises, however, IMAGE projects that most boreally derived grassland is converted to agriculture except in poor unproductive soils, and much of the temperate and boreal forests become managed for wood products. Thus, the model projects by 2100 that land-use change accounts for most (71%) of the loss of boreal forest. Only 57% of the current area of boreal forest is projected to persist in its current location over the next century, and this will likely undergo changes in dominant forest type in ways that depend on regional variations in climate and soil type (Pastor and Post 1988).

Regions differ in the expected changes in distribution of boreal forest. Canada and Russia, which currently have the largest expanses of tundra north of boreal forest, are projected by IMAGE to increase in total area of boreal forest (Fig. 6.2), whereas Alaska, Scandinavia, and other countries are projected by IMAGE to show a 60% reduction in area of boreal forest. These projections should be treated cautiously as scenarios of *possible* change rather than as predictions because of the many uncertainties in predicting future climate and land use and their consequences for vegetation. For example, IMAGE ignores time-lags associated with colonization and migration, so the rates of northern migration of boreal forest into tundra and of temperate trees into boreal forest are probably overestimated by the model (Chapin and Starfield 1997), particularly where there are dispersal barriers. Moreover, warming may occur more rapidly than some species can migrate (Davis and Zabinski 1992). Nonetheless, the projected patterns of change provide a reasonable basis for speculations about the consequences of global change for patterns of boreal biodiversity and are consistent with current high rates of forest harvest in China (Tian et al. 1995) and Scandinavia (Sjöberg 1995).

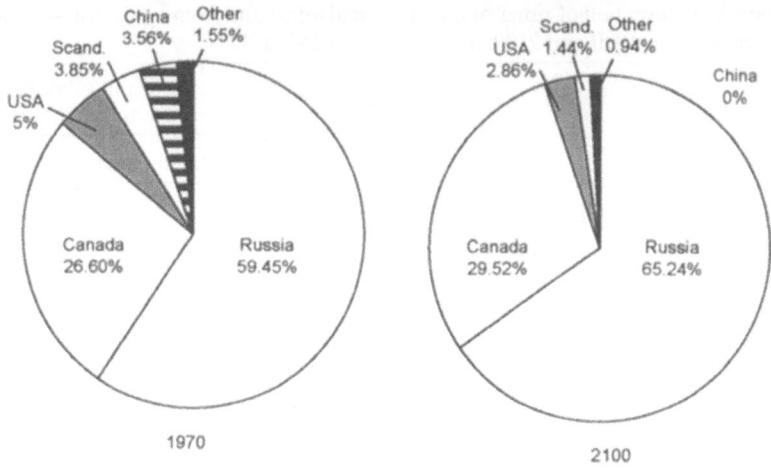

Figure 6.2. Proportion of the boreal forest that was located in different countries and regions in 1970 and the distribution of boreal forest among these regions projected in 2100 by the IMAGE model.

Causes of Altered Diversity

Species diversity in the boreal forest will undoubtedly change in complex ways due to both gains and losses of species. Moreover, the causes of altered diversity will vary temporally and spatially. Despite the complexities and uncertainties, certain predictions are possible (Table 6.3).

Expected Changes in Drivers of Diversity

There is strong agreement among general circulation models (GCMs) that greenhouse-gas–induced global warming will be most pronounced at high

Table 6.3. Summary of changes expected in boreal biodiversity relative to other biomes

Global-change drivers	Exected change in driver	Sensitivity of diversity to driver	Net effect of driver for biodiversity
Land-use change	2	5	50
Elevated CO_2	1	1	2
N deposition	2	3	18
Climate change	4	4	48
Biotic exchange	2	2	4

See Chapter 1 for explanation. Expected changes and sensitivities are ranked from low (1) to high (5). Net effects are expressed as a percentage of the maximum possible sensitivity.

latitudes (Table 6.3); warming will probably occur primarily in late winter and early spring, leading to a longer growing season (Kattenberg et al. 1996; Keeling et al. 1996). The late winter–early spring warming that has been observed in many high-latitude regions since 1980 (Chapman and Walsh 1993; Briffa et al. 1995) is consistent with these predictions. Synoptic circulation patterns guided by the Rocky and Himalayan Mountains suggest that the geographic pattern of future boreal warming will parallel current trends, with warming being greatest in Siberia, Alaska, and western Canada, and least (currently cooling) over eastern Canada. GCMs do not represent precipitation well at high latitudes and currently offer no consistent predictions of how high-latitude precipitation will change in the future; however, the general predictions that continental areas should experience greater drought and maritime areas experience greater precipitation in the future (Kattenberg et al. 1996) should apply to high latitudes.

Increased fire frequency in continental portions of the boreal forest is projected to accompany increased temperature and drought and may account for the 40% increase in area burned in the Canadian boreal forest since 1980 (Flannigan and Van Wagner 1991). This prediction seems quite robust in continental areas of low human population density (e.g., Siberia and central Canada), but it is uncertain in areas with large human populations due to possible changes in fire-control policy, as a result of changes in extent of human habitation (increasing pressure for fire suppression) and available funding (likely reducing the level of fire suppression) (Stocks 1991). The decline in fire number and extent in eastern Canada in the twentieth century is expected from the increased precipitation observed there—a pattern that is likely to continue in maritime areas in the future (Bergeron and Flannigan 1995). The contrasting trajectories in precipitation and fire frequency between continental and maritime areas should magnify the current longitudinal gradients in ecosystem type and diversity (Danell et al. 1996; Pastor et al. 1996).

Anthropogenic land-use conversion to agriculture is quite likely in those continental areas of boreal forest with suitable soils that are north of major grain belts. These areas will become suitable for dryland agriculture where regional warming increases the length of the growing season and reduces probability of frost. Most agricultural models project that these grain belts will move northward with climatic warming (Alcamo 1994). Expansion of managed forests and/or agriculture in southern boreal zones near population centers also seems likely in the future. For example, in Scandinavia, only 0.5% of productive forest land is currently protected from cutting (Sjöberg 1995). The geographic extent of this anthropogenic land-use change depends on transportation networks and world prices for agricultural and wood products. Wood harvest is currently 70% of annual wood production in Scandinavia (i.e., close to the maximum biological potential and probably exceeding a harvest rate that is sustainable) (Sjöberg 1995). Wood prices are expected to increase as the global demand for fiber exceeds global forest production early in the twenty-first century (Melillo et al. 1996). IMAGE projects that most of the loss of boreal forest will be to managed forests (Table 6.2)—primarily to monocultures of low diversity. Thus, we can develop reasonable scenarios of the *pattern* of conversion of boreal to managed forests, but the extent and rate that this will occur are quite uncertain.

Hunting is currently the major use of unmodified boreal forest. These pressures are intensive near population centers (e.g., Scandinavia) or where there are few regulations (e.g., northeast Russia). In the past intensive trapping of furbearers (e.g., beavers) had major effects on landscape structure and vegetation composition (Bridgham et al. 1995). Future impacts of game management are difficult to predict because they depend on the relative impact of humans on game species, their predators, and their habitat (Pastor et al. 1996). Thus, high human populations can increase or decrease the abundance and impact of mammalian herbivores. Scandinavian game management has increased moose numbers above their naturally regulated population size (Cederlund and Bergström 1996), whereas intensive hunting in Siberia has reduced moose numbers.

Atmospheric CO_2, nitrogen (N) deposition, and the concentrations of growth-reducing pollutants will be likely at high latitudes as part of the global trend. CO_2 is globally well mixed, so the increase in concentration will be relatively uniform across the boreal forest, with somewhat higher concentrations at high latitudes (Fung et al. 1983; Keeling et al. 1996). Future levels of atmospheric nitrous oxides, ammonia, and growth-reducing pollutants (e.g., heavy metals and sulfur dioxide) are relatively localized downwind of point sources. Large pollutant impacts will continue to be important downwind of major population centers (e.g., the southern boreal forest of Scandinavia, western Russia, and eastern North America). Additional sources occur in remote areas with economically viable ore deposits. The location and future magnitude of these remote pollutant sources depend so strongly on national policies, economics, and locations of ore deposits that they should probably be treated as stochastic elements.

The invasion of alien plant species into the boreal zone may be less likely than in more temperate biomes because of the broad circumpolar distribution of the current boreal flora (i.e., the taxa most likely to successfully establish in the boreal climate zone). As climate warms, however, we may expect invasion of alien taxa in the southern boreal zone. It is more important that invasion of insects or pathogens that disperse readily and whose distribution is currently temperature limited could cause dramatic ecological changes (Fleming and Volney 1995). Insects currently account for greater loss of tree volume than wildfire and amount to one third of the annual harvest volume in Canada (Fleming and Volney 1995). Their impact is likely to increase in the future.

Biodiversity Scenarios at the Tundra–Boreal Ecotone

At the northern limit of the boreal forest, climatic warming is highly likely to cause an expansion of trees into tundra (Starfield and Chapin 1996) and an increase in shrub density (Brubaker et al. 1995; Chapin et al. 1995), with an associated expansion of the northern limit of browsing mammals. Current and past episodes of treeline advance coincide with warm intervals that increase the rates of production of viable seeds, seed dispersal, and seedling establishment at treeline (Payette et al. 1985; Payette et al. 1989; Starfield and Chapin 1996), although there is a substantial time lag between seedling establishment and conversion to forest (Chapin and Starfield 1997). Other components of plant diversity are unlikely to be strongly affected by warming at the northern limit of boreal forest because the species currently found in the understory of northern forests are similar to those of tundra beyond the forest limit (Alexandrova 1980). The major changes in plant species diversity near the northern limits of boreal forest will probably be associated with increases in fire (Payette 1992), in response to increases in drought in continental areas and to increased fuel load as forests expand. The resulting increase in relative abundance of early successional vegetation will increase the abundance of folivorous insects and mammals, judging from their currently observed correlations with climate. In some cases, however, the northern limit of folivorous insects is directly constrained by the temperature required to complete a life cycle rather than by food plant distribution (Ayres and MacLean 1987). In these cases we expect a direct response of species range to temperature.

Some components of global change (e.g., agriculture, forestry, and invasion of temperate species) are unlikely to be important at the northern extreme of boreal forest. Other components of global change (e.g., elevated CO_2) will likely have minimal impact on north-boreal vegetation because high-latitude vegetation is strongly N-limited (Shaver and Chapin 1980; Van Cleve and Alexander 1981), which reduces its responsiveness to CO_2 (Field et al. 1992; Oechel et al. 1994). Because of the distance of most population centers from the northern treeline, pollutants and N deposition are likely to have less impact than they do in the mid- or southern boreal zone, except where there are strong prevailing southerly winds, as in Scandinavia, or local pollution sources. Lichens are quite sensitive to pollutants and N deposition and are most diverse in northern boreal forest, so even small changes in these inputs could have large impacts on abundance and diversity of these plants (Richardson and Nieboer 1981) and on animals such as reindeer that depend on them (Helle 1981).

In summary, at the northern limit of boreal forest, increased temperature will promote tree invasions (following a substantial lag time) and northward movements of temperature-limited invertebrates. Other global changes are unlikely to affect diversity strongly at the northern treeline, except where localized pollution reduces the diversity of lichens and associated herbivores.

Although future changes in diversity at the northern limit of boreal forest may be modest, those changes that occur will have dramatic functional consequences. Northward movement of trees and shrub cover would reduce winter range available for caribou and reindeer, which in turn could reduce the viability of subsistence economies of native peoples. Conversion from snow-covered tundra to a dark forest surface in winter increases the energy absorbed by the land surface and acts as a strong positive feedback to regional climate warming (Bonan et al. 1992; Thomas and Rowntree 1992; Bonan et al. 1995). Northward movement of treeline could also cause the summer position of the arctic front to move northward, greatly changing the regional temperature and synoptic weather patterns (Pielke and Vidale 1995). The increases in shrub and tree cover in the northern boreal forest seem certain to occur in areas that experience warming, but the timescale of these changes is uncertain. Fires may prevent forest expansion where the climate becomes warmer and drier, causing conversion of tundra to shrubland or grassland rather than forest (Chapin and Starfield 1997).

Biodiversity Scenarios at the Southern Boreal Ecotone

At the southern limit of boreal forest, climatic warming and increased fire frequency will substantially increase both species and landscape diversity (Baker 1993; Pastor et al. 1996), causing replacement of extensive areas of species-poor boreal forest by a forest–grassland mosaic in continental interiors (Hogg and Hurdle 1995). In maritime regions, increased temperature will promote northward migration of temperate forest species (Pastor and Post 1988), whereas increased precipitation will enhance palludification of coastal areas, reducing forest cover. In both situations, changes in host–pathogen or host–insect relationships could be important causes of plant mortality (Mattson and Haack 1987; Fleming and Volney 1995). For example, the conversion of large areas of spruce forest to shrubland and grassland in southern Alaska is associated with a temperature increase sufficient to allow spruce bark beetles to complete their life cycle in 1 year rather than 2 years (Holsten 1990). Fire frequency and prairie expansion at the southern margin would likely increase (Hogg and Hurdle 1995). The distribution of many fish populations is similarly temperature limited, such that a small temperature increase might allow invasion of new fish species, which could precipitate trophic changes that alter many ecosystem processes (Carpenter and Kitchell 1993). Any major increase in tree mortality enhances the probability of fire and opens sites for invasion of new species. In contrast to the changes at the northern treeline, increases in tree mortality and conversion of boreal forest to grassland in the south could happen quickly with minimal time lags (Smith and Shugart 1993).

Land conversion for agriculture in the southern boreal zone would cause a complete shift in plant and animal species composition on these areas, and open new habitats for invasion of weedy herbaceous species, possibly including exotic weeds.

Expansion of forestry has reduced the abundance and diversity of breeding birds (Järvinen et al. 1977; Virkkala 1987) and will probably continue to do so. Logging might reduce use by neotropical migrants in Eurasia, where they inhabit old-growth stands but increase use by neotropical migrants in North America, where they inhabit younger stands (Helle and Järvinen 1986;

Helle and Niemi 1994; Mönkkönen and Welsh 1994; Pastor et al. 1996). Lichens, which are more abundant and diverse in old-growth than in managed forests, support a high density and diversity of tree-dwelling invertebrates, which could account for the high diversity of nonmigratory passerine birds in old-growth forests of Scandinavia (Pettersson et al. 1995). Dead wood of late-successional forests also contributes to diversity of fungi and the insects that depend on them (Heliövaara and Väisänen 1984; Väisänen et al. 1993). Given the current low diversity of the southern boreal zone, invasion of exotic species has a reasonably high probability of introducing new functional groups that could substantially alter ecosystem processes, creating a positive feedback that would favor additional invasion of new species (Hobbie et al. 1993; Chapin et al. 1997).

N deposition and atmospheric pollutants might also have strong effects at the southern limit of boreal forest due to proximity of human population centers. Both drought and N deposition might make plant growth more responsive to CO_2 (Field et al. 1992) at the southern than at the northern boreal limit (Kauppi et al. 1992), leading to possible additional shifts in species composition. Excessive N deposition and pollution, however, can cause forest dieback (Schulze 1989) and major changes in community composition and reductions in diversity. For example, the reduction in lichens in polluted forests can reduce populations of birds that depend on lichen-associated insects (Zang 1990; Pettersson et al. 1995).

In summary, the southern limit of boreal forest will probably exhibit substantial increases in species and habitat diversity, due to increases in fire, conversion to agriculture, and increased insect and pathogen outbreaks. These factors will increase habitat fragmentation and open sites for colonization by new species. Other agents of global change (e.g., N deposition, pollutants, increased atmospheric CO_2, and invasion of alien species), will act synergistically to magnify these changes in diversity.

The changes in diversity at the southern limit of boreal forest will be functionally important at several respects. The increased fire and replacement of boreal forest by grassland or temperate forest will cause a net release of CO_2 to the atmosphere, contributing to global warming (Kurz and Apps 1995). The shift from boreal forest to grassland would also modify surface energy exchange with correspondingly large feedbacks to regional climate. At the stand level, the changes in structure, productivity, nutrient cycling, and trophic dynamics caused by altered species and growth-form composition would be much greater than the direct effects of altered temperature and moisture on these processes. In other words, global change would exert its greatest effects on the southern boreal forest *indirectly* through changes in species composition and diversity.

Biodiversity Scenarios in the Central Boreal Forest

In some respects, the expected changes in the central portion of the boreal forest will be intermediate between those in the north and the south, with increased temperature allowing modest increases in diversity due to north-ward migration of temperate species and increased drought and fire (where they occur), causing substantial changes in landscape structure as well as stand composition and diversity. In other respects the expected changes in the central boreal forest may be less pronounced and involve changes in the relative abundance of stand types and species already present. For example, increased fire frequency would probably not cause any major gain or loss of boreal species at the regional scale, but instead cause the shift toward more early successional stands, which would enhance the diversity of vascular plants, folivorous birds, and mammalian herbivores, but reduce the diversity of nonvascular plants, beetles, and ground-feeding birds.

Vulnerable Habitats

As described earlier, different components of diversity are concentrated in different habitats. For example, diversity of nonvascular plants, nonfolivo-rous insects, and the birds that depend on this trophic chain are concentrated in late-successional forests, which are likely to become less common in the future due to increased logging and fire. Protection of old-growth forests, par-ticularly in the southern boreal forest, may be critical to the maintenance of these components of diversity. Even protected areas will occasionally burn, however, so there must be enough areas set aside to allow some semblance of the natural fire regime to occur, even in areas protected to maintain old-growth forests.

Other components of diversity are maximized in early or midsuccessional habitats. These components of diversity are likely to increase in the future and may not require an active management program, except at those parts of the southern boreal forest where conversion of boreal forest to managed forests or agriculture becomes a major land-use change.

Conclusions

Boreal forest has lower plant diversity than biomes to the north or south. Most forests are dominated by a single tree species. Community and ecosys-tem processes are more sensitive to landscape-scale variation in dominant tree than to broad latitudinal patterns of climate. Gain or loss of even a single tree species could therefore have profound regional consequences. Specific groups exhibit "hot spots" of diversity (e.g., decaying logs for beetles or ripar-ian floodplains for migratory tropical birds), but these hot spots differ among taxa so there are a wide variety of boreal habitats that are important reser-voirs of diversity. In Finland 15–25% of the species in most taxa are currently threatened with extinction—an extent of potential species loss equivalent to that of many other biomes. Less is known about potential extinction rates in less-populated portions of the boreal forest.

The area of boreal forest is projected by the IMAGE model to remain relatively constant as boreal forest expands northward in response to climate warming and as southern boreal forest changes into agricultural and managed temperate forests. Thus, climate is likely to be largely responsible for expansion of boreal forest and land-use change for regions of loss. At the northern limit of boreal forest, increased temperature will promote tree invasions (following a substantial lag time) and northward movements of temperature-limited invertebrates. The southern limit of boreal forest will probably exhibit substantial increases in species and habitat diversity due to increases in fire, conversion to agriculture, and increased insect and pathogen outbreaks. These factors will increase habitat fragmentation and open sites for colonization by nonboreal species. These changes in diversity, although perhaps small in terms of the number of species involved, are likely to have profound effects. They will likely reduce regional albedo and act as a positive feedback to warming at the tundra–boreal boundary, but will reduce albedo and slow regional warming at the southern regional boundary. In addition to the large functional changes likely to occur within the boreal biome, the potential impacts on tropical migrant birds and on climate could affect diversity and associated ecosystem and community processes far beyond the boreal forest.

Acknowledgment. Research leading to these generalizations was supported by the Bonanza Creek Long-Term Ecological Research program.

References

Alcamo J (ed) (1994) IMAGE 2.0: integrated modeling of global climate change. Kluwer Academic Publishers, Dordrecht.

Alexandrova VD (1980) *The Arctic and Antarctic: Their Division into Geobotanical Areas.* Cambridge University Press, Cambridge, UK.

Apps MJ, Kurz WA, Luxmoore RJ, Nilsson LO, Sedjo RA, Schmidt R, et al. (1993) Boreal forests and tundra. Water Air and Soil Pollution 70:39–53.

Ayres MP, MacLean SF, Jr (1987) Development of birch leaves and the growth energetics of *Epirrita autumnata* (Geometridae). Ecology 68:558–568.

Baker WL (1993) Spatially heterogeneous multi-scale response of landscapes to fire suppression. Oikos 66:66–71.

Berg Å, Ehnström B, Gustafsson L, Hallingbäck T, Jonsell M, Weslien J (1994) Threatened plant, animal, and fungus species in Swedish forests: distribution and habitat associations. Conservation Biology 8:718–731.

Bergeron Y, Flannigan MD (1995) Predicting the effects of climate change on fire frequency in the southeastern Canadian boreal forest. Water Air and Soil Pollution 82:437–444.

Bernes C (1994) Biologisk mångfald i Sverige.

Bonan GB, Pollard D, Thompson SL (1992) Effects of boreal forest vegetation on global climate. Nature 359:716–718.

Bonan GB, Chapin FS III, Thompson SL (1995) Boreal forest and tundra ecosystems as components of the climate system. Climatic Change 29:145–167.

Bridgham SD, Johnston CA, Pastor J, Updegraff K (1995) Potential feedbacks of northern wetlands on climate change. BioScience 45:262–274.

Briffa KR, Jones PD, Schweingruber FH, Shiyatov SG, Cook ER (1995) Unusual twentieth-century summer warmth in a 1000-year temperature record from Siberia. Nature 376:156–159.

Brubaker LB, Anderson PM, Hu FS (1995) Arctic tundra biodiversity: a temporal perspective from late Quaternary pollen records. In: Chapin FS III, Körner C (eds) *Arctic and Alpine Biodiversity: Patterns, Causes and Ecosystem Consequences*, pp. 111–125. Springer-Verlag, Berlin.

Bryant JP, Chapin FS III (1986) Browsing-woody plant interactions during boreal forest plant succession. In: Van Cleve K, Chapin FS, III, Flanagan PW, Viereck LA, Dyrness CT (eds) *Forest Ecosystems in the Alaskan Taiga: A Synthesis of Structure and Function*, pp. 213–225. Springer-Verlag, New York.

Carpenter SR, Kitchell JF (1993) The trophic cascade in lakes. Cambridge University Press, Cambridge UK.

Cederlund G, Bergström R (1996) Trends in the moose-forest system in Fennoscandia, with special reference to Sweden. In: DeGraaf RM, Miller RI (eds) *Conservation of Faunal Diversity in Forested Landscapes*, pp. 265–281. Chapman and Hall, London.

Chapin FS III, Kedrowski RA (1983) Seasonal changes in nitrogen and phosphorus fractions and autumn retranslocation in evergreen and deciduous taiga trees. Ecology 64:376–391.

Chapin FS III, Starfield AM (1997) Time lags and novel ecosystems in response to transient climatic change in arctic Alaska. Climatic Change 35:449–461.

Chapin FS III, Van Cleve K, Tryon PR (1983) Influence of phosphorus on the growth and biomass allocation of Alaskan taiga tree seedlings. Canadian Journal of Forest Research 13:1092–1098.

Chapin FS III, Shaver GR, Giblin AE, Nadelhoffer KG, Laundre JA (1995) Response of arctic tundra to experimental and observed changes in climate. Ecology 76:694–711.

Chapin FS III, Walker BH, Hobbs RJ, Hooper DU, Lawton JH, Sala OE, et al. (1997) Biotic control over the functioning of ecosystems. Science 277:500–504.

Chapman WL, Walsh JE (1993) Recent variations of sea ice and air temperature in high latitudes. Bulletin of the American Meteorological Society 74:33–47.

Ciais P, Tans PP, Trolier M, White JWC, Francey RJ (1995) A large northern hemisphere terrestrial CO_2 sink indicated by the $^{13}C/^{12}C$ ratio of atmospheric CO_2. Nature 269:1098–1102.

Danell K, Lundberg P, Niemelä P (1996) Species richness in mammalian herbivores: patterns in the boreal zone. Ecography 19:404–409.

Danell K, Willebrand T, Baskin L (1998) Mammalian herbivores in the boreal forests: their numerical fluctuations and use by man. Conservation Ecology 2 ⟨www.eonsecol.org⟩.

Davis MB, Zabinski C (1992) Changes in geographical range resulting from greenhouse warming: effects on biodiversity in forests. In: Peters RL, Lovejoy T (eds) Global warming and biodiversity, pp. 297–308. Yale University Press, New Haven.

Field C, Chapin FS III, Matson PA, Mooney HA (1992) Responses of terrestrial ecosystems to the changing atmosphere: a resource-based approach. Annual Review of Ecology and Systematics 23:201–235.

Flanagan PW, Van Cleve K (1983) Nutrient cycling in relation to decomposition and organic matter quality in taiga ecosystems. Canadian Journal of Forest Research 13:795–817.

Flannigan MDF, Van Wagner CE (1991) Climate change and wildfire in Canada. Canadian Journal of Forest Research 21:66–72.

Fleming RA, Volney JA (1995) Effects of climate change on insect defoliator population processes in Canada's boreal forest: some plausible scenarios. Water Air and Soil Pollution 82:445–454.

Fung I, Prentice K, Matthews E, Lerner J, Russell G (1983) Three-dimensional tracer model study of atmospheric CO_2: response to seasonal exchanges with the terrestrial biosphere. Journal of Geophysical Research 88:1281–1294.

Gorham E (1991) Northern peatlands: role in the carbon cycle and probable responses to climatic warming. Ecological Applications 1:182–195.

Haila Y, Järvinen O (1990) Northern conifer forests and their bird species assemblages. In: Keast A (eds) Biogeography and ecology of forest bird communities, pp. 61–85. SPB Academic Publishing, The Hague.

Heliövaara K, Väisänen R (1984) Effects of modern forestry on northwestern European forest invertebrates: a synthesis. Acta Forestalia Fennica 189:1–32.

Helle P, Järvinen O (1986) Population trends of North Finnish land birds in relation to their habitat selection and changes in forest structure. Oikos 46:107–115.

Helle P, Niemi GJ (1994) Bird community dynamics in boreal forests. In: DeGraaf RM (eds) *Wildlife Conservation in Forested Landscapes*. Elsevier, The Hague.

Helle T (1981) Studies on wild forest reindeer (*Rangifer tarandus fennicus* Lönn.) and semi-domestic reindeer (*Rangifer tarandus tarandus* L.) in Finland. Acta Universitatis, Ouluensis Series A, Biologica 12:1–283.

Hobbie SE, Jensen DB, Chapin FS III (1993) Resource supply and disturbance as controls over present and future plant diversity. In: Schulze E-D, Mooney HA (eds) *Ecosystem Function of Biodiversity*, pp. 385–407. Springer-Verlag, Berlin.

Hogg EH, Hurdle PA (1995) The aspen parkland in western Canada: a dry-climate analogue for the future boreal forest? In: Apps MJ, Price DT (eds) *Boreal Forests and Global Change*, pp. 391–400. Kluwer Academic Publishers, Dordrecht.

Holling CS (1992) The role of forest insects in structuring the boreal landscape. In: Shugart HH, Leemans R, Bonan GB (eds) A systems analysis of the global boreal forest, pp. 170–191. Cambridge University Press, Cambridge, UK.

Holsten EH (1990) Spruce beetle activity in Alaska: 1920–1989. U.S. Department of Agriculture, Forest Service, Forest Pest Management, Alaska Region, Anchorage.

Järvinen O, Väisänen RA (1973) Species diversity of Finnish birds: I. Zoogeographical zonation based on land birds. Ornis Fennici 50:93–125.

Järvinen O, Väisänen R, Kuusela K (1977) Effects of modern forestry on the numbers of breeding birds in Finland in 1945–1975. Silva Fennica 11:284–294.

Johnston CA (1994) Ecological engineering of wetlands by beavers. In: Mitsch WJ (ed) *Global Wetlands: Old World and New*, pp. 379–384. Elsevier, Amsterdam.

Kaila L, Martikainen P, Punttila P, Yakovlov E (1994) Saproxylic beetles (Coleoptera) on dead birch trunks decayed by different polypore species. Annales Zoologici Fennici 31:97–108.

Kallio P, Lehtonen J (1973) Birch forest damage caused by *Oporinia automna* (Bkh.) in 1965–66 in Utsjoki, N. Finland. Report of the Kevo Subarctic Research Station 10:55–69.

Kattenberg A, Giorgi F, Grassl H, Meehl GA, Mitchell JFB, Stouffer RJ, et al. (1996) Climate models—projections of future climate. In: Houghton JT, Meira Filho LG, Callander BA, Harris N, Kattenberg A, Maskell K (eds) *Climate change 1995. The science of climate change*, pp. 285–357. Cambridge University Press, Cambridge, UK.

Kauppi PE, Mielikäinen K, Kuusela K (1992) Biomass and carbon budget of European forests, 1971 to 1990. Science 256:70–74.

Keeling CD, Chin JFS, Whorf TP (1996) Increased activity of northern vegetation inferred from atmospheric CO_2 measurements. Nature 382:146–149.

Kouki J, Niemelä P, Viitasaari M (1994) Reversed latitudinal gradient in species richness of sawflies (Hymenoptera, Symphyta). Annales Zoologici Fennici 31:83–88.

Kullman L, Engelman OE (1997) Neoglacial climatic control of subarctic *Picea abies* stand dynamics and range limit in northern Sweden. Arctic and Alpine Research 29:315–326.

Kurz WA, Apps MJ (1995) An analysis of future carbon budgets of Canadian boreal forests. Water Air and Soil Pollution 82:321–331.

Mattson WJ, Haack RA (1987) The role of drought in outbreaks of plant-eating insects. BioScience 37:110–118.

McCune B (1993) Gradients in epiphyte biomass in three *Pseudotsuga-Tsuga* forests of different ages in western Oregon and Washington. Bryologist 96:405–411.

Melillo JM, Prentice IC, Farquhar GD, Schulze E-D, Sala OE (1996) Terrestrial biotic responses to environmental change and feedbacks to climate. In: Houghton JT, Meira Filho LG, Callander BA, Harris N, Kattenberg A, Maskell K (eds) *Climate Change 1995. The Science of Climate Change*, pp. 445–481. Cambridge University Press, Cambridge, UK.

Michaelson GL, Ping C-L, Kimble JM (1996) Carbon storage and distribution in tundra soils of arctic Alaska, U.S.A. Arctic and Alpine Research 28:414–424.

Mönkkönen M, Welsh DA (1994) A biogeographical hypothesis on the effects of human caused landscape changes on the forest bird communities of Europe and North America. Annales Zoologici Fennici 31:61–70.

Niemelä J (1997) Invertebrates and boreal forest management. Conservation Biology 11:601–610.

Nilsson SG, Ericson L (1997) Conservation of plant and animal populations in theory and practice. Ecological Bulletin 46:117–139.

Oechel WC, Van Cleve K (1986) The role of bryophytes in nutrient cycling in the taiga. In: Van Cleve K, Chapin FS III, Flanagan PW, Viereck LA, Dyrness CT (eds) Forest ecosystems in the Alaskan taiga, pp. 121–137. Springer-Verlag, New York.

Oechel WC, Cowles S, Grulke N, Hastings SJ, Lawrence W, Prudhomme T, et al. (1994) Transient nature of CO_2 fertilization in arctic tundra. Nature 371:500–503.

Olson JS, Watts JA, Allison AJ (1983) *Carbon in Live Vegetation of Major World Ecosystems*. Oak Ridge National Laboratory, Oak Ridge, Tennessee.

Pastor J, Mladenoff DJ (1992) The southern boreal-northern hardwood forest border. In: Shugart HH, Leemans R, Bonan GB (eds) A systems analysis of the global boreal forest, pp. 216–240. Cambridge University Press, Cambridge, UK.

Pastor J, Post WM (1988) Responses of northern forests to CO_2-induced climate change. Nature 334:55–58.

Pastor J, Mladenoff DJ, Haila Y, Bryant J, Payette S (1996) Biodiversity and ecosystem processes in boreal regions. In: Mooney HA, Cushman JH, Medina E, Sala OE, Schulze E-D (eds) *Functional Roles of Biodiversity: A Global Perspective*, pp. 33–69. John Wiley and Sons, New York.

Payette S (1992) Fire as a controlling process in the North American boreal forest. In: Shugart HH, Leemans R, Bonan GB (eds) *A Systems Analysis of the Global Boreal Forest*, pp. 144–169. Cambridge University Press, Cambridge, UK.

Payette S, Filion L, Gauthier L, Boutin Y (1985) Secular climate change in old-growth tree-line vegetation of northern Quebec. Nature 315:135–138.

Payette S, Filion L, Delwaide A, Begin C (1989) Reconstruction of tree-line vegetation response to long-term climate change. Nature 341:429–432.

Pettersson RB, Ball JP, Renhorn K-E, Esseen P-A, Sjöberg S (1995) Invertebrate communities in boreal forest canopies as influenced by forestry and lichens with implications for passerine birds. Biological Conservation 74:57–63.

Pielke RA, Vidale PL (1995) The boreal forest and the polar front. Journal of Geophysical Research 100D:25755–25758.

Post WM, Emanuel WR, Zinke PJ, Stangenberger AG (1982) Soil carbon pools and world life zones. Nature 298:156–159.

Prentice KC, Fung IY (1990) The sensitivity of terrestrial carbon storage to climate change. Nature 346:48–51.

Reeburgh WS, Whalen SC (1992) High latitude ecosystems as CH_4 sources. Ecological Bulletin 42:62–70.

Richardson DH, Nieboer E (1981) Lichens and pollution monitoring. Endeavour 5.

Ritchie JC (1987) *Postglacial Vegetation of Canada.* Cambridge University Press, Cambridge, UK.

Ritchie JC, MacDonald GM (1986) The patterns of post-glacial spread of white spruce. Journal of Biogeography 13:527–540.

Roulet NT, Ash R (1992) Low boreal wetlands as a source of atmospheric methane. Journal of Geophysical Research 97:3739–3749.

Schulze E-D (1989) Air pollution and forest decline in a spruce (*Picea abies*) forest. Science 244:776–783.

Shaver GR, Chapin FS III (1980) Response to fertilization by various plant growth forms in an Alaskan tundra: nutrient accumulation and growth. Ecology 61:662–675.

Sjöberg K (1995) Fauna and flora management in forestry. In: Hytönen M (ed) *Multiple-Use Forestry in the Nordic Countries*, pp. 191–243. Gummerus Printing, Jyväskylä, Finland.

Smith TM, Shugart HH (1993) The transient response of terrestrial carbon storage to a perturbed climate. Nature 361:523–526.

Starfield AM, Chapin FS III (1996) Model of transient changes in arctic and boreal vegetation in response to climate and land use change. Ecological Applications 6:842–864.

Stocks BJ (1991) The extent and impact of forest fires in northern circumpolar countries. In: Levine JL (ed) *Global Biomass Burning: Atmospheric, Climatic and Biospheric Implications*, pp. 197–202. The Massachusetts Institute of Technology Press, Cambridge, MA.

Stokland JN (1994) Biological diversity and conservation strategies in Scandinavian boreal forests.

Thomas G, Rowntree PR (1992) The boreal forests and climate. Quarterly Journal of the Royal Meteorological Society 118:469–497.

Tian H, Xu H, Hall CAS (1995) Pattern and change of a boreal forest landscape in Northeastern China. Water Air and Soil Pollution 82:465–476.

Tonteri T (1998) Species richness of boreal understory forest vegetation in relation to site type and successional factors. Annales Zoologici Fennici 31:53–60.

Trofymow JA, Preston CM, Prescott CE (1995) Litter quality and its potential effect on decay rates of materials from Canadian forests. Water Air and Soil Pollution 82:215–226.

Väisänen R, Heliövaara K (1994) Hot spots of insect diversity in northern Europe. Annales Zoologici Fennici 31:71–81.

Väisänen R, Biström O, Heliövaara K (1993) Sub-cortical Coleoptera in dead pines and spruces: is primeval species composition maintained in managed forests? Biodiversity Conservation 2:95–113.

Van Cleve K, Alexander V (1981) Nitrogen cycling in tundra and boreal ecosystems. Ecological Bulletin 33:375–404.

Van Cleve K, Oliver L, Schlentner R, Viereck LA, Dyrness CT (1983) Productivity and nutrient cycling in taiga forest ecosystems. Canadian Journal of Forest Research 13:747–766.

Van Cleve K, Chapin FS, III, Dryness CT, Viereck LA (1991) Element cycling in taiga forest: state-factor control. BioScience 41:78–88.

Vedrova EF (1995) Carbon pools and fluxes of 25-year old coniferous and deciduous stands in middle Siberia. Water Air and Soil Pollution 82:239–246.

Viereck LA (1973) Wildfire in the taiga of Alaska. Quaternary Research 3:465–495.

Viereck LA, Dyrness CT, Van Cleve K, Foote MJ (1983) Vegetation, soils, and forest productivity in selected forest types in interior Alaska. Canadian Journal of Forest Research 13:703–720.

Virkkala R (1987) Effects of forest management on birds breeding in northern Finland. Annales Zoologici Fennici 24:281–294.

Zang H (1990) Population decrease of coal tit *Parus ater* in the Harz mountains due to forest damage ('Waldsterben'). Vogelwelt 111:18–28.

7. Temperate Grasslands

Osvaldo E. Sala

The temperate-grassland biome occupies a large portion of the planet. Temperate grasslands represent the potential natural vegetation of an area of 49 $\times$ 10^6 km^2, which is equivalent to 36% of the earth's surface (Shantz 1954). This estimate of the area occupied by grasslands excludes savannas, but it does include grass and shrub deserts. The area covered exclusively by grasslands is 15×10^6 km^2, which accounts for 11% of the earth's surface. The grassland biome occurs in almost all continents (Singh et al. 1983) from the Americas to Asia and from Europe to Australia (Fig. 7.1). In North America, most of the Great Plains is dominated by grasslands from the boundary with subtropical biomes in the south of the United States to the boundary with the Canadian temperate forest in the north. In South America, the pampas and large expanses of Patagonia are covered by grasslands. In Asia, vast areas are occupied by grasslands from Ukraine to China.

Grasslands are water-limited ecosystems, and their existence is associated with droughts occurring at least once during the year. Grasslands occur in areas with annual precipitation ranging between 150 and 1200 mm/year and mean annual temperatures ranging between 0 and 25°C (Lieth and Whittaker 1975). Along a precipitation gradient, temperate grasslands occur between deserts and forests, with forests taking over sites with more than 1200 mm/year of precipitation and deserts dominating regions with less than 150 mm of precipitation. A clear example of these vegetation shifts along a precipitation gradient occurs in western Patagonia, where precipitation declines

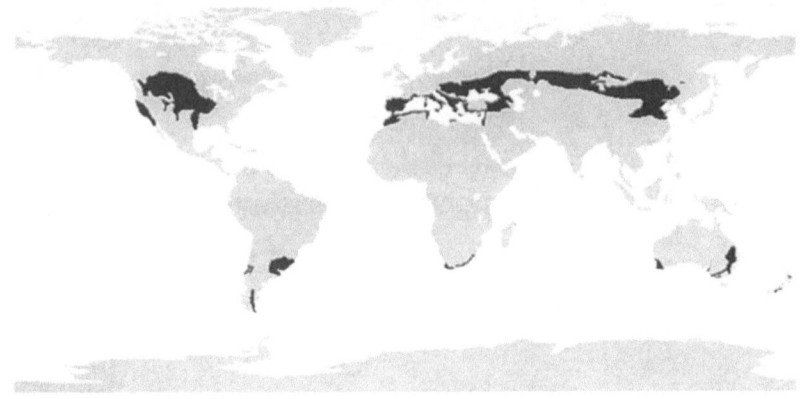

Figure 7.1. Global distribution of the grassland biome. Adapted from Bailey (1998). This figure includes Mediterranean grasslands, which will be addressed in Chapter 9.

drastically and vegetation shifts from closed forest to grassland, steppe, and desert in less than 200 km (Schulze et al. 1996).

Temperature interacts with precipitation to determine the distribution of grasslands. As temperature increases, the forest–grassland and desert–grassland boundaries occur at higher values of precipitation as a result of the increasing evaporative demand and the negative effect on the availability of water. For example, in North American Great Plains where isohyets run in a north–south direction and temperature isopleths in an east–west direction, the grassland–forest boundary has a clear SE–NW direction (Bailey 1998).

The concept of biodiversity is quite broad and encompasses different levels, from diversity within species or genetic diversity and species diversity, to ecological diversity that includes landscape diversity and diversity of functional groups (Mooney et al. 1995). In this chapter, I will focus on species and functional group diversity because most information about biodiversity has been collected at these levels.

This chapter will describe succinctly the natural patterns of biodiversity in grasslands and develop scenarios of changes in biodiversity that might occur in this biome for the year 2100. In order to construct the scenarios, I will follow the approach used throughout this book. I will independently assess the patterns of the drivers of changes in biodiversity and the sensitivity of grasslands to changes in each driver. Finally, I will combine the expected changes in drivers with the sensitivity patterns to assess total expected changes in biodiversity for the biome.

Patterns of Biodiversity in Grasslands

Biodiversity within the grassland biome varies enormously, with many native grasslands having levels of diversity as high as those characteristic of tropical forests (Sala et al. 2001). Grasslands of the Pampa, in southern South America, are an example of a high-diversity system with more than 400 species of grasses alone (Cabrera 1970). The tallgrass prairie in North America similarly has more than 250 species of plant species, most of them perennial grasses (Freeman 1998). In contrast, some grasslands of Patagonia, for example, have approximately 30 species of plants (Golluscio and Sala 1993). Diversity of vascular plants may be only a poor indicator of diversity within other taxa, however, as has been shown clearly for tropical forests (Lawton et al. 1998).

The high diversity of grassland-plant species has been simplified by grouping species into functional types. Species within a functional type share morphological and physiological characteristics that result in a common ecological role (Chapin et al. 1995). There is not a unique grouping of plant species into functional types, but groupings vary according to the purpose of the study and the availability of data. One classification divides grassland-plant species into grasses, shrubs, succulents, and herbs (Sala et al. 1997). In addition to the obvious morphological differences among these groups, species within each group share many other characteristics with important ecological meaning. For example, shrub species tend to have deeper roots than do grasses (Jackson et al. 1996); consequently, they dominate in sites with coarse-textured soils, whereas grasses dominate in finer-textured soils (Sala et al. 1997). Similarly, shrubs are more abundant in regions where precipitation occurs during the cold season because this seasonal pattern favors penetration of water into deep layers of the soil. Another classification of grassland species is based on photosynthetic pathways and divides plant species into C_3 and C_4 species. These two groups of species have differences in the anatomy of leaves and in the biochemistry of photosynthesis that result in differences in the distribution of species along latitudinal and elevation gradients. C_4 species tend to be distributed at low elevation and low latitudes, whereas C_3 have the opposite patterns (Cavagnaro 1988, Paruelo et al. 1998).

Faunal diversity is lower than it is in other biomes when we take into account the large area that grasslands occupy. Grasslands have a total of 477 species of birds and 245 species of mammals, that is only 5% of the total number of species for each group (Groombridge 1992). Specific grasslands and taxonomic groups represent important exceptions. For example, birds are quite diverse in North American tallgrass-prairie with 208 species representing 16 orders (Kauffman et al. 1998). The abundance and diversity of large mammals characterize the grassland biome. The patterns of diversity of large mammals are quite different among climatically similar grasslands. For example, the total number of species of large mammals in Africa and South

America is approximately 200 in both regions; however, the contribution of different orders is strikingly different. In Africa, the order *Artiodactyla* (bovidae) has 80 species, whereas in South America has less than 20 (Vrba 1993). In contrast, in South America, *Marsupiala* has 80 species and *Edentata* 30 species, but those orders are not present in Africa.

Diversity of below-ground organisms is known in much less detail than is diversity of above-ground organisms. There is a strong indication, however, that below-ground organisms represent a very important component of the grassland biota. Grasslands, in general terms, have more biomass and higher production below ground than they do above ground (Sims and Singh 1978). The concentration of organic matter below ground provides the substrate for diverse bacteria, fungal, and nematode groups. For example, all groups are present in the shortgrass steppe, although bacteria are the most abundant (in terms of biomass), followed by fungi, nematodes, protozoa, macroarthropods, and microarthropods (Lauenroth and Milchunas 1992). In terms of species numbers, more than 100 species of fungi and more than 200 species of nematodes have been reported for the shortgrass steppe and tallgrass prairie, respectively (Christensen and Scarborough 1969; Ransom et al. 1998).

Scenarios of Changes in Biodiversity

The Effects of Land-Use Change on Biodiversity

Changes in land use are expected to be large in grasslands (Table 7.1). They result from a broad range of human activities that culminate either in changes in land cover or simply in changes in land use. One of the most drastic

Table 7.1. Biodiversity changes in grasslands and its determinants for the year 2100

	Land use	Climate	N deposition	Biotic exchange	Atoms CO_2
Expected change in drivers	3	2	3	3	2.5
Sensitivity	5	3	2	2	3
Predicted biodiversity change	0.6	0.24	0.24	0.24	0.3

The first row presents the expected changes in the five major drivers of biodiversity change, for the grassland biome, in an arbitrary scale from 1 to 5. We used a business-as-usual scenario to estimate climate change (Haxeltine and Prentice 1996), scenario A1 of the IMAGE 2 model (Alcamo 1994) to estimate changes in land use, and the MOGUNTIA model of nitrogen deposition (Holland et al. 1999). Row 2 describes the sensitivity of grasslands to a unit change of each one of the drivers in an arbitrary scale of 1 to 5. Row 3 is the expected relative change in diversity as a result of each driver in a scale from 0 to 1. It was calculated by multiplying rows 1 and 2 and dividing by 25.

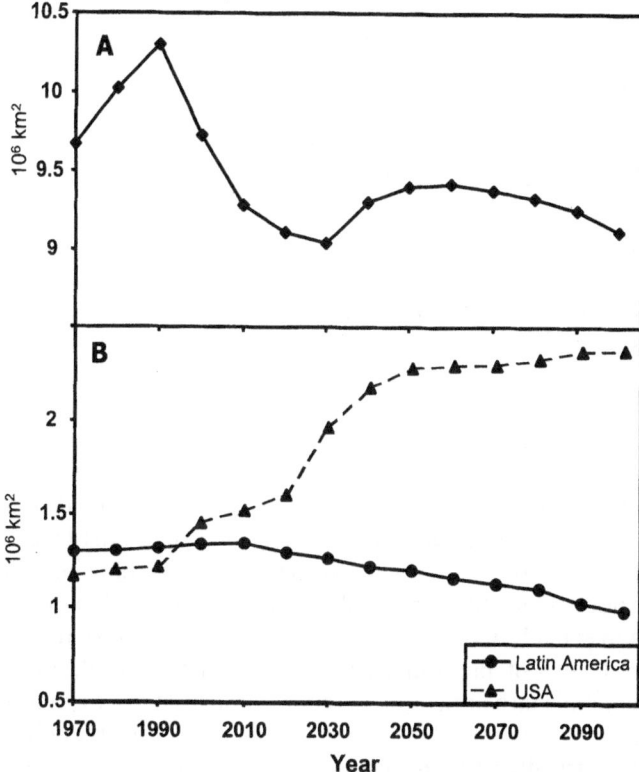

Figure 7.2. (A) Trajectory of the total area covered by grasslands in the world from 1970 until 2100 as predicted by the IMAGE 2 model (Alcamo 1994). (B) Trajectories of the area covered by grasslands in the United States and Latin America that are two contrasting regions of the 13 regions in which IMAGE 2 divides the world.

changes that grasslands may experience is the conversion into agricultural land. Temperate grasslands have optimal conditions for agriculture, with mild temperatures and moderate precipitation (Lieth and Whittaker 1975); consequently, a large fraction of native grasslands has already been converted into cropland. For example, the original tallgrass prairie in North America supports most of the current corn-belt of the United States. The vast grasslands of the Pampa are similarly now the basis for the large-scale grain exports from Argentina.

The IMAGE 2 model (Alcamo 1994), used in this exercise to estimate changes in land use, indicates that the total area covered by grasslands will remain approximately constant in the next century (Fig. 7.2A). This single datum, however, can be misleading. The total area of grasslands has losses and gains. The losses represent native grasslands that are plowed and converted into urban or agricultural land. The gains come from two sources: (1)

agricultural land that is being abandoned and returns to its potential natural vegetation; and (2) forests that are being cut down and replaced by grasslands with the purpose of feeding domestic animals. The United States and Latin America are regions that are expected to undergo opposite trajectories during the next century as a result of different social and economic forces (Fig. 7.2B). The IMAGE 2 model (Alcamo 1994) indicates that agricultural land will be reduced in the United States as a result of intensification, and that the abandoned land will revert to grassland, which was the potential natural vegetation of the region. In contrast, the grassland area in Latin America will be reduced as a result of agricultural expansion. The boreal forest is expected to undergo a similar phenomenon, although one driven by climate change instead of land-use change (see Chap. 6). The total area will remain constant, although the boreal forest will shift north with the consequent loss of boreal forest to temperate forest and grassland in the southern boundary that will be compensated by gains from areas that are currently tundra.

Habitat loss resulting from the transformation of grasslands into agricultural land is the major driver of biodiversity change in grasslands. The conversion of grasslands into agricultural land includes plowing and seeding with an exotic species that drives plant species into local extinction. If the phenomenon extends in time and space, local extinctions may turn into global extinctions. Losses of biodiversity in one part of the world, due to conversion of grasslands into agriculture, are not compensated by abandoned fields in another part of the world because of differences in the biota being gained or lost. The agricultural land that reverts to grassland in the United States has a different biota than the area of grassland that is plowed in Latin America (Fig. 7.2). Although both grasslands have similar climates, their flora and fauna are partially unrelated: North America belongs to the Holarctic realm; South America to the Neotropical realm (Udvardy 1975), and species lost in South America are lost forever. Changes in plant-species diversity drastically alter the habitat for microorganisms and above-ground animals, resulting in changes in animal species diversity. Even in a single region the weedy species that colonize an early successional grassland are quite distinct from those that occupy late successional grasslands that might be lost to agriculture or urbanization.

Another more subtle but pervasive change in land use with consequences for grassland biodiversity is overgrazing. Grazing with domestic animals is the major way of utilizing grasslands, and a large fraction of grasslands in the world is being overgrazed. Species diversity has the hump-shaped curve in response to grazing intensity, with maximum diversity occurring at intermediate intensities of grazing (Fig. 7.3) (Milchunas et al. 1988; Milchunas and Lauenroth 1993). The effect of grazing on diversity is modulated by water availability and the evolutionary grazing history of the area. The conceptual model, which was developed by Milchunas and colleagues, was corroborated by a thorough literature review that included 97 articles representing 276 data sets (Milchunas et al. 1988; Milchunas and Lauenroth

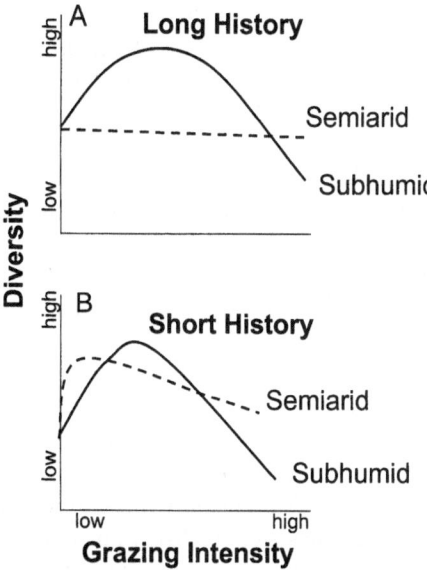

Figure 7.3. A conceptual model of the effect of grazing intensity on plant species diversity as modulated by moisture availability and evolutionary grazing history. As grazing intensity increases grasslands are invaded by species that have adaptations to herbivory increasing total diversity, further increases in grazing, decreases diversity because species that are less tolerant to grazing are driven to local extinction. Adaptations that provide resistance to drought also provide resistance to herbivory; consequently, grasslands from semiarid regions are less vulnerable to invasions and show a smaller response to grazing. Evolutionary grazing history either long (A) or short (B) further modulates the effect of grazing on diversity. Redrawn from Milchunas et al. 1988 and Milchunas and Lauenroth 1993.

1993). Grazing results in a partial or total loss of plant organs, and particular plant characteristics confer adaptive advantages under those circumstances. For example, low stature, high root–shoot ratios, location of basal meristems close to the ground and protected by basal sheaths are plant adaptations to grazing. At low levels of grazing, increases in grazing intensity favor the introduction of plant species that possess adaptations to grazing resulting in a moderate increase in species diversity. At very high levels of grazing, only the most resistant species remain in the community with the less-tolerant or intolerant species being locally driven to extinction.

Water availability modulates the effect of grazing on diversity because of the convergent selection for drought and grazing resistance. Both drought and herbivory result in the loss of plant organs, and both provide similar selection pressures. Characteristics that provide adaptations to drought resistance (e.g., high root–shoot ratio) also confer resistance to herbivory (Coughenour 1985). Grassland plant communities from semiarid regions, although subjected to low-grazing intensity, consequently, possess some adaptations (exaptations) to herbivory. As a result, the effect of grazing in semiarid regions is moderate because selection pressure does not change significantly along a grazing intensity gradient (Fig. 7.3A). In contrast, in a subhumid grassland, selection pressures shift along a gradient in grazing intensity. Low-grazing sites are dominated by plants with characteristics that allow effective competition for light (e.g., high stature and low root–shoot ratio), whereas at high-grazing intensity the opposite characteristics become most adaptive because they represent adaptations to grazing. The shift in the direction of selection pressures results in drastic changes in community composition and diversity.

The grazing history in evolutionary time further modulates the effects of grazing and moisture on diversity. Grasslands that did not have large-hooved and congregating animals during the Holocene tend to lack species with adaptations to herbivory. For example, the Inter-Mountain West in North America lacked the large herds of bison (Mack and Thompson 1982) that shaped the adjacent shortgrass steppe (Milchunas et al. 1988). Grasslands with short evolutionary-grazing history are consequently more vulnerable to invasions by species with adaptations to herbivory and shift their diversity at lower levels of grazing intensity (Fig. 7.3B).

The Effects of Climate Change on Biodiversity

The expected climate change in the grassland biome will be moderate in comparison with other biomes (Table 7.1 and Chap. 15). Global circulation models agree in predicting larger increases in temperature at higher latitudes (Kattenberg et al. 1996). Temperate grasslands are located at midlatitudes (Fig. 7.1) and will consequently experience a moderate-to-low increase in temperature as a result of the increase in greenhouse gases in the atmosphere. Changes in precipitation are more idiosyncratic and do not have clear latitudinal patterns (e.g., those described for temperature). Although generalizations about changes in precipitation for the entire biome are difficult to assemble, observations of GCM outputs indicate that most of the grassland biome will be located in areas with either no change or with a small increment depending on the season under consideration (Kattenberg et al. 1996). Soil moisture is the variable that is most closely related to the functioning of grassland ecosystems, and its status results from the combined effects of changes in precipitation and temperature. In general terms, grasslands will experience a decrease in soil moisture at the time of doubling CO_2, indicating that the warming effect will overshadow the neutral-to-slightly positive changes in precipitation (Kattenberg et al. 1996).

The functioning of grasslands is most sensitive to changes in water availability. Annual precipitation accounts for most of the year-to-year variability in primary production, whereas other environmental variables (e.g., temperature) are not related to the interannual production pattern (Lauenroth and

Sala 1992; Knapp et al. 1998). Species or groups of species respond quite differently to changes in total water availability and its seasonal distribution. Years when precipitation patterns yield high grass production consequently do not result in high forb production in North American tallgrass prairie (Knapp et al. 1998). In the Patagonian steppe, conditions leading to high production of shrubs also do not coincide with those yielding high production of grasses (Jobbágy and Sala 2000).

Directional changes in water availability (i.e., those predicted under climate change scenarios) will affect certain grassland species in detriment of others; consequently, they will have major detrimental effects on grassland biodiversity (Table 7.1). The overall effect of climate change on grassland biodiversity will be high, more as a result of the high sensitivity of grasslands than of the expected changes in climate. Prolonged changes in water availability will certainly alter the competitive balance among species that, in turn, will lead to local extinctions. If changes are sustained through time and encompass a large area, local extinctions may result in global extinctions. Water availability may decrease as a result of a direct decrease in precipitation or as a consequence of an increase in temperature that increases evaporative demand and negatively affects the ecosystem–water balance. The more drought-resistant plant species of a community will have a competitive advantage over the more mesic species in regions where water availability is expected to decrease. On the contrary, in regions where water availability will increase, plant species with high relative growth rate and high shoot–root ratio will outcompete and ultimately drive to local extinction other species with characteristics that made them more drought resistant and less capable of using higher water availability. Plant characteristics associated with drought resistance and its components of tolerance and avoidance include morphological and physiological features (e.g., rooting depth, capacity for osmotic regulation, stomatal control of transpiration water losses, or duration of life cycles) (Kramer 1969). Enormous variability has been reported among plant species in the characteristics associated with drought resistance, highlighting that directional changes in water availability will inevitably result in large changes in biodiversity.

Soil microorganisms show differential responses to soil water content in a way that is similar to the differential responses reported for plant species (Freckman et al. 1987). Soil–water thresholds at which microorganisms turn into drought-induced quiescence vary among species. As a result, it appears that changes in the frequency of wet-soil conditions may result in changes in the composition of the community of microorganisms and ultimately in their biodiversity.

The Effects of Changes in Nitrogen Deposition on Biodiversity

Nitrogen deposition has increased over background levels as a direct result of human activity (see Chap. 3). Nitrogen deposition occurs mostly downwind of regional pollution sources and not far from them (Holland et al. 1999). Areas of industrial concentration and urbanization are located mostly in regions where the potential natural vegetation is temperate forests (Headrick 1990). Eastern North America, northern Europe, and Japan currently concentrate most of the industrial activity and show the highest levels of NO_3 and NH_4 deposition (Holland et al. 1999). In the twenty-first century, industrialization and the associated pollution will expand into other regions (e.g., China and countries of Southeast Asia). Temperate grasslands are and will be adjacent to these high-industrial concentration areas and, in most cases, are downwind from them. The vicinity of temperate grasslands to sources of pollution and specifically to regions where fossil fuel combustion is concentrated indicates that they will be subjected to heavy nitrogen deposition (Table 7.1).

Increases in nitrogen availability have a negative effect on plant biodiversity in grasslands (Berendse and Elberse 1990; Huenneke et al. 1990; Tilman 1993). For example, Tilman (1993) performed a long-term experiment where he added nitrogen fertilizer from 0 to $27 g/m^2/year$. After 11 years of treatment, the species richness had declined significantly such that plots that received the most nitrogen had less than 50% of the original plant biodiversity (Fig. 7.4). The loss did not occur uniformly across different functional groups, with forbs losing many more species than perennial grasses. The loss of species within the forb group, which included the nitrogen-fixing legumes, suggest that there might be associated losses in the nitrogen-fixing bacteria. This is an example about how changes in plant-species composition driven

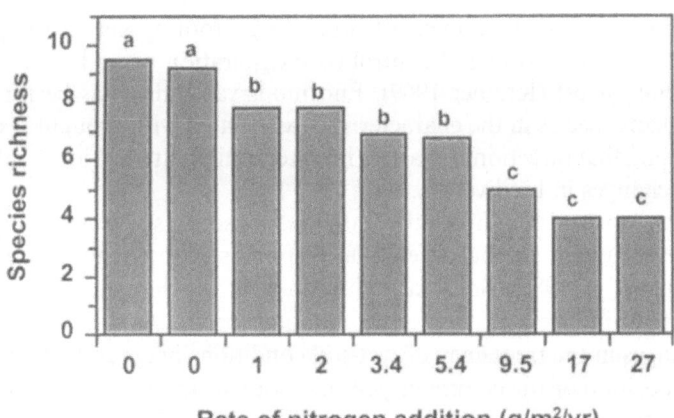

Figure 7.4. The effect of increasing N availability on plant species richness in the North American tallgrass prairie. Nitrogen fertilizer was added annually during an 11-year period, and results show species richness at the end of the experiment. Redrawn from Tilman 1993.

by increased nitrogen availability may affect microbial composition. Studies with forest soils indicate losses of ectomycorrhizal fungi diversity as a result of nitrogen deposition (Arnolds 1991).

The Effects of Biotic Exchange on Biodiversity

Introductions of exotic species into grassland ecosystems first increase local diversity, but, later, exotic species may outcompete native species, driving them to local extinction and reducing diversity. The magnitude of the successful invasions and their consequences for biodiversity depend on the amounts of exotic propagules (seeds and individuals) that reach a given grass-land site and its sensitivity. Prediction of "invasibility" of grasslands is difficult, however, because some systems that become seriously invaded contrast with other apparently similar grasslands that do not become invaded even when receiving similar loads of invading species (Mack 1989).

Seeds of exotic plant species or exotic animals arrive in grassland ecosystems largely as a result of human activities. In some cases, humans have introduced plant species because of their superior forage quality that consequently escaped from cultivated paddocks and invaded an entire grassland region (Mack 1989). In other cases, introductions occur accidentally when agriculture tools or animals are exchanged among regions, bringing plant seeds and animals. The amount of propagules received by ecosystems depends directly on the level of human activity. Grasslands are expected to receive high levels of propagules (Table 7.1) because they are among the biomes with the highest human impact. Temperate grasslands have been and will continue being prime locations for agriculture. A large fraction of grasslands has consequently been transformed into croplands, which is an activity that brought along many alien plant species and their associated animals. The alien-plant group ranges from cultivated varieties to weedy species and the associated pests and parasites.

Grassland types vary dramatically in their invasibility. North America provides two of the most striking examples of grasslands that have been invaded to a degree that has drastically transformed their functioning (Mack 1989). The Inter-Mountain West has been invaded by *Bromus tectorum* after the introduction of cattle by European settlers. This annual grass species outcompeted the native perennial grasses, resulting in severe changes in the fire regime, in the biogeochemistry of the region, and, finally, in the animal-carrying capacity. California's Central Valley was similarly originally dominated by perennial grass species of the genera *Stipa*, *Aristida*, or *Festuca*. After the arrival of European settlers and their cattle, the annual grass species *Avena barbata* replaced perennial grasses so completely that few remnant fragments are currently left in their pristine condition. Almost adjacent to some of the most vulnerable grasslands are other grasslands that have received many alien plants but were not invaded. For example, the North American shortgrass steppe, which is adjacent to the Inter-Mountain West although separated by the massive Rocky Mountains, has proven to be quite resistant to invasions. Mack (1989) suggests that the major characteristics that distinguish between the two types of grasslands and which determine their invasibility is the presence of large-hooved, congregating mammals in the Holocene and dominance by cespitose grasses.

It appears that grasslands have moderate vulnerability to invasions, or sensitivity to invasions, in comparison to other biomes (Table 7.1). The factors that make one ecosystem more vulnerable to invasions than another is a question of active debate and research. Environmental severity is one factor that has been proposed to control invasibility. More mesic environments may be more vulnerable to invasions than are xeric ecosystems (Rejmánek 1989). Vulnerability to invasions could also increase with the frequency and intensity of disturbances (Rejmánek 1989). Finally, original biodiversity levels could be associated with vulnerability to invasions, with the more diverse ecosystems being less vulnerable (Rejmánek 1996). Experimental studies confirm the effect of biodiversity on determining vulnerability to invasions, but they highlight the role of other factors that often overshadow the biodiversity effect (Levine 2000).

In synthesis, grasslands, as a biome, receive abundant propagules of exotic species because they are located in regions that have been severely transformed by agriculture, and they are close to major heavily industrialized zones. Grasslands as a whole are moderately vulnerable to invasions, although there is large variability in invasibility within the grassland biome.

The Effects of Increased CO_2 Concentration on Biodiversity

The change in the concentration of CO_2 experienced by grasslands will be similar to that experienced by all other biomes because atmospheric CO_2 mixes globally in short periods of time (Fung et al. 1987). Nonetheless, grasslands are expected to be among the most responsive ecosystems to increased CO_2 concentrations. A synthesis of all the elevated CO_2 experiments up to 1995 showed that grasslands were the ecosystems in which productivity responded most strongly to a doubling CO_2 (Koch and Mooney 1995). The Californian annual grassland showed a 40–60% of increase in productivity relative to productivity under ambient CO_2; this was surpassed only by Kansas tallgrass prairie. Grasslands were already hypothesized to be among the most responsive ecosystems before scientists started whole-ecosystem elevated CO_2 experiments (Mooney et al. 1991). Perhaps the most consistent response to elevated CO_2 observed in different kinds of experiments has been a reduction in stomatal conductance and transpiration that result in an increase in water-use efficiency. For example, in a Californian annual grassland, doubling CO_2 resulted in a 50% reduction in stomatal conductance and transpiration, and a doubling of water-use efficiency (Jackson et al. 1994). An alteration of the water-use efficiency will have the largest impact on ecosystems that are frequently limited by water availability. Grasslands, which are shaped by water limitation, are consequently among the ecosystems with greatest potential response to elevated CO_2.

A CO_2-induced change in water-use efficiency can affect the entire ecosystem–water balance, modifying the seasonal pattern of water availability as well as the total amount of water available for plants (Jackson et al. 1998). Changes in water availability and the distribution pattern in the soil profile and throughout the growing season will alter the competitive balance of the plant species in the community. For example, plants that grow faster and have higher transpiration rate will outcompete the slower growing and more drought-resistant species. Field et al. (1995) found that increasing CO_2 in an annual grassland resulted in a large increase in soil water content predominantly at the end of the growing season. This increase will benefit a small group of species that are vegetative during late season, but it will not benefit the dominant species *Avena barbata* that is senescent at that time. This is an example of how changes in stomatal conductance may result in changes in species composition. It is expected that alterations of the competitive balance may result in local extinctions. If the phenomenon is extended in space and time, local extinctions may result in global extinctions.

Changes in CO_2 may affect animals by altering the physical and chemical environments. Elevated CO_2 increased the carbon–nitrogen ratio of leaf tissue in several grassland ecosystems (Field et al. 1995; Owensby et al. 1996). Increases in the carbon–nitrogen ratio negatively affect the quality of the diet of animals feeding on those plants. In an experiment using large open-top chambers in tallgrass prairie, Owensby et al. (1996) found that forage quality, estimated by the proportion of acid detergent fiber and in vitro dry-matter digestibility, sharply decreased under elevated CO_2 conditions relative to ambient conditions. Moreover, sheep grazing inside the chambers showed a lower intake than did control sheep. Owensby et al. (1996) speculated that the CO_2 impact will be greater for wild ruminants than it is for domestic ones. Results with ruminants in grasslands seem to be different than results obtained with insects and trees. Insects fed with leaves of trees grown under elevated CO_2 showed increased consumption rate, although the insect growth rate was decreased (Lindroth 1996).

Conclusions

Our exercise suggests that biodiversity in temperate grassland ecosystems will be severely affected by the year 2100. The five drivers of biodiversity change (i.e, land use, climate, nitrogen deposition, biotic exchange, and atmospheric CO_2) will have moderate-to-high effects on grassland biodiversity (Table 7.1). The mesic climatic conditions, which are characteristic of grasslands, make them ideal for agriculture; consequently, a large fraction of the grassland area will be transformed into cropland. The geographical location of grasslands, adjacent to urban and highly industrialized areas, results in grasslands receiving high human impact including high nitrogen deposition and large amounts of propagules of exotic plant and animal species. The changes of drivers will be large in grasslands, and their sensitivity to those changes will be very high (Table 7.1). Grasslands host species with a wide variety of morphological and physiological characteristics. For example, within the plants, grassland diversity ranges from shrubs to grasses, from plants with C_3 to C_4 photosynthetic pathways, and from deep- to shallow-rooted species. Changes in environmental conditions will alter the competitive balance of grassland species rapidly. Changes in the competitive balance will yield changes in abundance that will result in local extinctions in the short-to-medium term. When changes in drivers persist in time or affect large expanses of grasslands, local extinctions will turn into global extinctions.

The scenario for changes in biodiversity in grasslands contrasts with the scenarios for other biomes that show equally large impacts (e.g., tropical forests or southern temperate forests) (see Chaps. 11 and 12). Those two forest biomes will undergo large changes in biodiversity mostly as a result of very large changes in a single driver (land use). In contrast, grasslands will show moderate-to-high effects that result from all five drivers. From this point of view, the grassland scenario is similar to the scenario for Mediterranean ecosystems (see Chap. 9).

One of the limitations of this exercise has been the scale at which I was able to work. Scenarios at the scale of the entire biome mask specific differences among grassland types that will be large for some drivers. For example, even though some grassland types will continue to be plowed and transformed into agricultural land, others will experience the opposite pattern with cropland being abandoned and reverting slowly to the grassland condition. The scale limitation affects both the magnitude of the error associated with this exercise and constrains its applicability because most management decisions occur at more-detailed scales than the one used here.

Another limitation of this exercise is that it has not considered the interactions among drivers of biodiversity change. It is likely that there will be important synergistic interactions among drivers. The combined effects on biodiversity of nitrogen and CO_2 enrichment will probably be much larger than the sum of the independent effects of those two drivers. The effect of habitat fragmentation on biodiversity could similarly be enlarged by the presence of abundant propagules of exotic species. Exotic species may not succeed in invading a particular grassland type if the environment had not been modified by climate change or elevated CO_2. I can also envision an antagonistic interaction among drivers. For example, when alteration of the environment is very large (e.g., in the case of plowing that destroys all native vegetation) other drivers may not cause further losses of diversity. Our understanding of interactions is so limited that I have chosen not to include them in this chapter.

Mitigation of the expected changes in grassland biodiversity will result from reduction in the rate of change of drivers at the global scale and development of detailed management plans tailored for the particular ecological, social, and economic conditions of each grassland type. The development of management plans will require a better understanding of interactions among drivers as well as trade-offs among the goods and services that grasslands yield.

Acknowledgments. I thank A.T. Austin, M. Oesterheld, and F.S. Chapin III for valuable suggestions for this chapter and insightful discussions throughout the entire exercise of developing the biodiversity scenarios, and J.P. Guerschman and L. Vivanco for assistance with the figures. This work was supported by the Inter-American Institute for Global Change Research, the National Center for Ecological Analysis and Synthesis at the University of California Santa Barbara, Consejo Nacional de Investigaciones Científicas y Técnicas, Fondo para la Investigacíon Científica y Tecnológica, and University of Buenos Aires.

References

Alcamo J (1994) *Image 2: Integrated Modeling of Global Climate Change.* Kluwer Academic Publishers, Dordrecht.

Arnolds E (1991) Decline of ectomycorrhizal fungi in Europe. Agriculture, Ecosystems, and Environment 35:209–244.

Bailey RG (1998) *Ecoregions: The Ecosystem Geography of the Oceans and Continents.* Springer-Verlag, New York.

Berendse F, Elberse WT (1990) Competition and nutrient availability in heathland and grassland ecosystems. In: Grace JB, Tilman D (eds) *Perspectives on Plant Competition*, pp. 93–116. Academic Press, San Diego.

Cabrera A (1970) *Flora de la Provincia de Buenos Aires: Gramineas.* Instituto Nacional de Tecnología Agropecuaria, Buenos Aires.

Cavagnaro JB (1988) Distribution of C_3 and C_4 grasses at different altitudes in a temperate arid region of Argentina. Oecologia 76:273–277.

Chapin FS, Lubchenco J, Reynolds HL (1995) Biodiversity effects on patterns and processes of communities and ecosystems. In: Mooney HA, Lubchenco J, Dirzo R, Sala OE (eds) *Biodiversity and Ecosystem Functioning: Basic Principles*, pp. 289–300. Cambridge University Press, Cambridge, UK.

Christensen M, Scarborough AM (1969) Soil microfaunal investigations, Pawnee site. Rep. 23, U.S. IBP Grassland Biome, Colorado State University, Fort Collins.

Coughenour MB (1985) Graminoid responses to grazing by large herbivores: adaptations, exaptations, and interacting processes. Annals of the Missouri Botanical Garden 72:852–863.

Field C, Chapin F, Chiariello N, Holland E, Mooney H (1995) The Jasper Ridge CO_2 experiment: design and motivation. In: Koch G, Mooney H (eds) *Carbon Dioxide and Terrestrial Ecosystems*, pp. 121–145. Academic Press, New York.

Freckman DW, Whitford WG, Steiberger Y (1987) Effect of irrigation on nematode population dynamics and activity in desert soils. Biology and Fertility of Soils 3:3–10.

Freeman CC (1998) The flora of Konza prairie. A historical review and contemporary patterns. In: Knapp A, Briggs J, Hartnett D, Collins S (eds) Grassland Dynamics: Long-Term Ecological Research in Tallgrass Prairie, pp. 69–80. Oxford University Press, New York.

Fung IY, Tucker CJ, Prentice KC (1987) Application of advanced very high resolution radiometer vegetation index to study atmosphere-biosphere exchange of CO_2. Journal of Geophysical Research 92D:2999–3015.

Golluscio RA, Sala OE (1993) Plant functional types and ecological strategies in Patagonian forbs. Journal of Vegetation Science 4:839–846.

Groombridge B (ed) (1992) *Global Biodiversity: Status of the Earth's Living Resources.* Chapman and Hall, London.

Headrick DR (1990) Technological change. In: Turner BL, Clark WC, Kates RW, Richards JF, Mathews JT, Meyer WB (eds) *Earth as transformed by human action*, pp. 55–67. Cambridge University Press, Cambridge, UK.

Holland EA, Dentener FJ, Braswell BH, Sulzman JM (1999) Contemporary and preindustrial global reactive nitrogen budgets. Biogeochemistry 46:7–43.

Huenneke LF, Hamburg SP, Koide R, Mooney HA, Vitousek PM (1990) Effects of soil resources on plant invasion and community structure in California serpentine grassland. Ecology 71:478–491.

Jackson RB, Sala OE, Field CB, Mooney HA (1994) CO_2 alters water use, carbon gain, and yield for the dominant species in a natural grassland. Oecologia 98:257–262.

Jackson RB, Canadell J, Ehleringer JR, Mooney HA, Sala OE, Schulze ED (1996) A global analysis of root distributions for terrestrial biomes. Oecologia 108:389–411.

Jackson RB, Sala OE, Paruelo JM, Mooney HA (1998) Ecosystem water fluxes for two grasslands in elevated CO_2: a modeling analysis. Oecologia 113:537–546.

Jobbágy E, Sala O (2000) Controls of grass and shrub aboveground production in the Patagonian steppe. Ecological Applications 10:541–549.

Kattenberg A, Giorgi F, Grassl H, Meehl GA, Mitchell JFB, Stouffer RJ, et al. (1996) Climate Models—Projections of future climate. In: *Climate Change: The IPCC Scientific Assessment*, pp. 285–358. Cambridge University Press, Cambridge, UK.

Kauffman DW, Fay PA, Kaufman G, Zimmerman JL (1998) Diversity of terrestrial macrofauna. In: Knapp A, Briggs J, Hartnett D, Collins S (eds) Grassland dynamics: Long-term ecological research in tallgrass prairie, pp. 101–112. Oxford University Press, New York.

Knapp A, Briggs J, Blair J, Turner C (1998) Patterns and controls of aboveground net primary production in tallgrass prairie. In: Knapp A, Briggs J, Hartnett D, Collins S (eds) *Grassland Dynamics: Long-Term Ecological Research in Tallgrass Prairie*, pp. 193–221. Oxford University Press, New York.

Koch G, Mooney H (1995) Responses of terrestrial ecosystems to elevated CO_2: a synthesis and summary. In: Koch G, Mooney H (eds) *Carbon Dioxide and Terrestrial Ecosystems*, pp. 415–429. Academic Press, New York.

Kramer PJ (1969) *Plant and Soil Water Relationships: A Modern Synthesis.* McGraw-Hill Book Company, New York.

Lauenroth WK, Milchunas DG (1992) Short-grass steppe. In: Coupland RT (ed) *Natural Grasslands: Introduction and Western Hemisphere*, pp. 183–226. Elsevier, Amsterdam.

Lauenroth WK, Sala OE (1992) Long-term forage production of North American shortgrass steppe. Ecological Applications 2:397–403.

Lawton J, Bignelli D, Bolton B, Bloemers G, Eggleton P, Hammond P, et al. (1998) Biodiversity inventories, indicator taxa and effects of habitat modification in tropical forest. Nature 391:72–76.

Levine J (2000) Species diversity and biological invasions: relating local process to community pattern. Science 288:852–854.

Lieth H, Whittaker R (1975) Primary productivity of the biosphere. Springer-Verlag, New York.

Lindroth RL (1996) CO_2-mediated changes in tree chemistry and tree-lepidoptera interactions. In: Koch GW, Mooney HA (eds) *Carbon Dioxide and Terrestrial Ecosystems*, pp. 102–120. Academic Press, San Diego.

Mack RN (1989) Temperate grasslands vulnerable to plant invasions: characteristics and consequences. In: Drake JA, Mooney HA, diCastri F, Groves RH, Kruger FJ,

Rejmánek M, et al. (eds) *Biological Invasions: A Global Perspective*, pp. 155–179. John Wiley and Sons, New York.

Mack RN, Thompson JN (1982) Evolution in steppe with few large, hooved mammals. The American Naturalist 119:757–773.

Milchunas DG, Lauenroth WK (1993) Quantitative effects of grazing on vegetation and soils over a global range of environments. Ecological Monographs 63:327–366.

Milchunas DG, Sala OE, Lauenroth WK (1988) A generalized model of the effects of grazing by large herbivores on grassland community structure. The American Naturalist 132:87–106.

Mooney HA, Drake BG, Luxmoore RJ, Oechel WC, Pitelka LF (1991) Predicting ecosystem responses to elevated CO_2 concentrations. BioScience 41:96–104.

Mooney HA, Lubchenco J, Dirzo R, Sala OE (1995) *Biodiversity and Ecosystem Functioning: Basic Principles*. Cambridge University Press, Cambridge, UK.

Owensby C, Ham J, Knapp A, Rice C, Coyne P, Auen L (1996) Ecosystem-level responses of tallgrass prairie to elevated CO_2. In Koch G, Mooney H (eds) *Carbon Dioxide and Terrestrial Ecosystems*, pp. 147–162. Academic Press, New York.

Paruelo J, Jobbágy E, Sala O, Lauenroth W, Burke I (1998) Functional and structural convergence of temperate grassland and shrubland ecosystems. Ecological Applications 8:194–206.

Ransom MD, Rice CW, Todd TC, Wehmueller WA (1998) Soils and soil biota. In: Knapp A, Briggs J, Hartnett D, Collins S (eds) *Grassland Dynamics: Long-Term Ecological Research in Tallgrass Prairie*, pp. 48–66. Oxford University Press, New York.

Rejmánek M (1989) Invasibility of plant communities. In: Drake JA, Mooney HA, diCastri F, Groves RH, Kruger FJ, Rejmánek M, et al. (eds) *Biological Invasions: A Global Perspective*, pp. 155–179. John Wiley and Sons, New York.

Rejmánek, M (1996) Species richness and resistance to invasions. In: Orians GH, Dirzo R, Cushman JH (eds) *Biodiversity and Ecosystem Processes in Tropical Forests*, pp. 153–172. Springer-Verlag, Berlin.

Sala OE, Austin AT, Vivanco L (2001) Temperate grassland and shrubland Ecosystems. In: Levin S (ed) *Encyclopedia of Biodiversity* 7, 5:627–635. Academic Press, San Diego.

Sala OE, Lauenroth WK, Golluscio RA (1997) Plant functional types in temperate semi-arid regions. In: Smith TM, Shugart HH, Woodward FI (eds) *Plant Functional Types*, pp. 217–233. Cambridge University Press, Cambridge, UK.

Schulze ED, Mooney HA, Sala OE, Jobbágy E, Buchmann N, Bauer G, et al. (1996) Rooting depth, water availability, and vegetation cover along an aridity gradient in Patagonia. Oecologia 108:503–511.

Shantz H (1954) The place of grasslands in the earth's cover of vegetation. Ecology 35:142–145.

Sims PL, Singh JS (1978) The structure and function of ten western North American grasslands. III. Net primary production, turnover and efficiencies of energy capture and water use. Journal of Ecology 66:573–597.

Singh JS, Lauenroth WK, Milchunas DG (1983) Geography of grassland ecosystems. Progress in Physical Geography 7:46–80.

Tilman D (1993) Species richness of experimental productivity gradients: how important is colonization limitation? Ecology 74:2179–2191.

Udvardy M (1975) A classification of the biogeographical provinces of the world. Report 18, IUCN, Morges.

Vrba ES (1993) Mammal evolution in the African Neogene and a new look at the Great American Interchange. In: Goldblatt P (ed) *Biological Relationships between Africa and South America*, pp. 391–432. Yale University Press, New Haven.

8. Tropical Savanna

Brian Walker

The savanna biome is defined here, following Huntley and Walker (1982), as those regions of the world characterized by: (1) a strongly seasonal climate with hot or warm wet summers and cool dry winters, and (2) a mixed life-form vegetation in which both woody and herbaceous plants play a significant ecological role. In tropical and subtropical savannas the herbaceous layer is dominated by C_4 grasses, whereas the woody plants are C_3 trees and shrubs. In temperate savannas the grasses are generally also C_3. Because of their greater extent this chapter will focus on the tropical/subtropical savannas.

Patterns of Diversity

Savannas have a very high diversity, both alpha (because of the mixed life form and a highly variable climate) and beta (marked spatial differences). There are large numbers of all species types and a large tail of minor species. Some of these minor species are "passengers" (Walker 1992), vulnerable to any changes in savanna function, but some may have important functions at particular times.

The savanna biota has a very long evolutionary history, and community composition has developed without the periodic destruction associated with glaciation events in higher latitudes. As a result savanna species have differentiated (niche packing) to make maximum use of the environment. Thus,

Silva (1987) describes the phenological spread between co-dominant grass species in the Barinas savannas of Venezuela, which gives rise to a higher number of species than would persist if they all had the same growth phenology response to the timing of rainfall. The marked interannual variability in the seasonal pattern of rainfall allows all the phenological types to persist over time. The species also exhibit marked differences in terms of a range of other functional attributes (e.g., seed dormancy, root–shoot ratio, capacity for vegetative spread, and seed predation).

The term *biodiversity* implies the full spectrum of species, and it is as well to note that we know very little about most of the species in the savannas. The statement that "as a first approximation all species are insects" applies strongly in the savannas, but there are also a host of mites, nematodes, protozoa, fungi, and other groups of species about which we are virtually ignorant. It is therefore not possible to make predictions about their future other than in terms of likely associated changes with changes in the vegetation and larger fauna.

Numbers of Species and Biogeographic Patterns

Knowledge of species in the savanna biome as a whole is very poor, and there is no good estimate for total species numbers; however, diversity is very high in all cases because of the multiple life form structure of the savannas and their very variable climate. Some idea of diversity can be obtained from a few examples:

- The Nylsvley savanna in South Africa (cf Scholes and Walker 1993)
  80–100 plant species per 0.1 ha
  319 plant species in one savanna community
  325 species of birds
  67 species of mammals
- Northern Australia (Walker and Gillison, 1982)
  >80 plant species per hectare
  >80% of the 503 *Eucalyptus* species are in the savannas
- Brazil (Eiten, 1982)
  300–330 vascular plant species per hectare

Functional Group Diversity

Comparisons of functional groups need to be made on the basis of two different kinds of functional diversity (Walker 1996):

1. The diversity of different kinds of species in terms of the ecosystem function they perform (e.g., herbivory, nitrogen fixation, etc.).
2. The diversity of species' environmental responses within the different functional groups (i.e., in terms of their responses to different kinds of environmental stresses and disturbances).

The first kind of functional diversity is important in terms of the maintenance of ecosystem function. The second kind is important in terms of the ability of the ecosystem functional group concerned to continue performing its particular function in the face of changes in the environment. This functional response diversity is a measure of the resilience of the ecosystem.

Both kinds of functional diversity are high in the savannas, although the first in particular varies among savanna types. For example, present day African savannas have a larger range of functionally different herbivores and predators than do their counterparts in South America or Australia. These latter two continents suffered a Pleistocene loss of the equivalent diversity of large mammals. There is no fossil record of the associated changes in the plant species diversity that results from the loss of these herbivores. In addition, in Australia, at least, effects of domestic livestock have probably obscured any differences in plant species diversity between Africa and Australia (Walker et al. in press). Silva's (1987) account of phenological diversity within Venezuelan grasses is an example of the second kind of diversity.

A minimum set of characteristic plant functional groups in the savannas could be: trees, shrubs, C_4 grasses, and C_3 forbs. Different response types occur within these groups to cope with one or more environmental stresses, as follows:

trees—drought resistance, frost resistance, high temperature survival, fire resistance, palatibility to herbivores, defense against herbivory (chemical and physical), frost resistance, high temperatures, droughts, and so on. The life-history attributes that contribute to these environmental responses would, in turn, include such features as rooting depth, deciduosity, and physiological properties (e.g., cell osmotic potential, bark thickness, sclerophylly, spininess, etc.).
shrubs—as for the trees, and shade tolerance.
grasses—longevity, palatibility and resistance to grazing, drought resistance, fire tolerance, phenological differences in response to variability in rainfall seasonality, temperature response, light response (shade tolerance), rooting patterns (including carbohydrate exudation in low fertility soils), and seed dormancy.
forbs—longevity, seed dormancy, dispersal, and palatibility.

Determinants of Savannas and Local Patterns

Savanna vegetation has two primary determinants—plant available moisture (PAM) and available nutrients (PAN)—and two secondary determinants— fire and herbivory (Tothill and Mott 1985, Walker 1987). The biomass of vegetation is primarily determined by PAM (Walker and Langridge, in press). There is little difference between the equilibrium ("climax") biomass on a nutrient-poor soil and a nutrient-rich soil. The rate at which the biomass increases (one aspect of ecosystem function), however, is much higher on nutrient-rich soils (Scholes and Walker 1993). The species of plants are determined by both PAM and PAN; differences in soil nutrient status, in particular, lead to very different species compositions.

Fire has three main effects: (1) It reduces the overall amount of woody vegetation (reduces the woody:grass ratio), (2) it changes the structure of the woody vegetation by reducing the shrub layer, thereby creating more of an open parklike vegetation and (3) it changes the species composition of both the herbaceous and woody layers, to fire tolerant species. Very few animal species in the savannas are directly impacted by fire. Changes in composition of the fauna associated with fire are due in the main to fire-induced changes in the vegetation.

Herbivory in savannas is a major process involving an array of herbivores from large, grazing herd species through solitary browsing mammals to grazing and browsing insects (e.g., grasshoppers and caterpillars). The long-term evolutionary consequence of this pressure is the existence of quite different sets of species on fertile and infertile soils. On low-fertility soils nitrogen levels are low and plants have evolved a variety of chemically based deterrents against herbivores, both insect and mammalian. High C–N ratios have given rise to concentrations of phenolics that are toxic to animals. On the high-fertility soils plants have evolved a range of physical deterrents (e.g., thorns), at least against the mammalian herbivores.

With respect to determining savanna structure and composition, however, the herbivores can be treated as two main groups, the large grazing mammals (including domestic livestock) and the eruptive insects. This statement is an oversimplification, but it highlights the dominant effects: In particular, under changing environmental conditions, there may well be other functional types of herbivores that prove to have a significant effect.

Any particular savanna type does not consist of a single combination of mixed woody–grass vegetation. In addition to the spatial heterogeneity associated with catena effects, there are a number of other communities within the broadscale matrix of woody–grass combinations that are determined by special conditions [e.g., wetlands, seasonally hydromorphic grasslands (vleis, dambos, flooded savannas)] rocky outcrops and riverine fringes. Although these communities generally occupy only a small fraction of the biome they play a disproportionately important role in terms of their contribution to the biodiversity of the savanna, in two ways; they have their own unique species, and they serve as crucial refucia, providing food and/or cover at critical times of the year for the broadscale savanna species (Stafford-Smith and Morton 1990). What happens to these small areas in the future will have a major effect on the savanna biome's biodiversity.

Past Losses of Diversity

The history of savanna use differs markedly according to continent. Africa has the longest record, given the current understanding of hominid and human evolution, and Africa and India have the two savanna regions that did not experience a loss of large mammalian herbivores in the Pleistocene.

South American and Australian savannas were subjected to intense grazing pressure by livestock herds from around the middle of the last century, following thousands of years without the influence of grazing herds. Because botanists followed some decades behind the livestock, early losses of particularly vulnerable plants (e.g., early growing, palatable species) would not have been recorded. Extinctions of vertebrate species in the savannas over the past few thousand years has been negligible, even where introduced vertebrate species have had a significant impact (e.g., in Northern Australia) (Woinarski and Braithwaite 1990).

Knowledge of invertebrates and microbes is too deficient to say anything about losses in their diversity.

The Future Environment and Its Influence on Biodiversity

Scenarios of Altered Diversity

The IMAGE scenario suggests that the aerial extent of savannas over the world as a whole (in $10^3 km^2$) will decrease by more than two thirds from 1990 to 2050 (from 8335 to 2746), and will then increase a little by 2100 (up to 3278). The biggest changes, by far, to 2050 are due to conversion of savanna into agricultural land (4767), predominantly in Africa. Small areas of scrubland (340) mixed forest (20) and grassland (3) become savanna. By 2100 some $863 \times 10^3 km^2$ of agricultural land are converted back into savannas, mostly in South America, in response to improved economic conditions.

Drivers of Change

In his analysis of species extinctions Diamond (1989) identified what he called the "evil quartet"—overkill, habitat destruction and fragmentation, introduced alien species, and chains of extinction.

These four causes of extinction certainly all still apply in the savannas, but it is helpful in terms of a prospective analysis to recast them a little with the added cause of direct changes in the atmosphere and climate. In terms of impact on biodiversity the order of future changes in the savannas is:

land use (e.g., habitat loss, degradation, and fragmentation)
alien species
CO_2
climate

As a major extinction process, overkill in the future savannas will be limited
to the larger vertebrates. Diamond's final evil factor (i.e., chains of extinction
is a pervasive influence that has no special savanna characteristics.

Land Use

Four forms of land use predominate:

1. Conversion—loss of native savanna vegetation to arable land
                —loss of riverine habitat through livestock and removal of
                trees, and altered river flows
                —loss of wetlands (permanent and seasonal) through grazing,
                drainage and ploughing.
2. Change in fire regime.
3. Change in grazing.
4. Direct harvesting of particular species for food and for medicines (a form
   of overkill).

The effects of land use on biodiversity vary, and will continue to vary, around
the world. The IMAGE scenarios suggest some intensification in Australia
and Latin America, but by far the biggest changes in terms of loss of savanna
to agriculture, are forecast to be in Africa, in the more humid regions. Nevertheless, in Africa, India and Southeast Asia the bulk of successful *primary*
land conversion changes have already taken place. Intensity of use may
increase as populations rise, but there are few potentially arable areas left,
outside the tsetse fly infested areas of Africa, that have not already been converted to agriculture at some time in the past. The enormous changes in the
IMAGE scenario stem from the assumption that all the people will continue
to be able to feed themselves. The development of new short-season varieties
of sorghum or other crops through new plant-breeding techniques may
enable the encroachment of agriculture further into the semi-arid savannas,
but the effects will be small relative to the changes that have already taken
place.

In the Sahel savannas (J-C Menaut, personal communication) virtually
every piece of potentially arable land has been used for agriculture at some
time or another. The vegetation therefore exists as a mosaic of fallow patches
of varying ages (from a few years to a decade or two) since the time of last
cultivation. With increasing population pressure the "postcultural recovery
time" gets shorter. Menaut (personal communication) depicts the plant diversity of the savanna in relation to time since cultivation. The early unstable
mix of pioneer annuals is replaced by a low-diversity community of small
perennial grasses that gradually give way to a high-diversity community of
perennial grasses and dicotyledonous plants. As rotation time declines, so
does the proportion of the area in the low-diversity phase increases. The
effects of future land use change will obviously depend on where along the
successional axis the region currently exists. An important consequence of a

shift toward the annual grass phase is that there is a concomitant decline in C and N input to the soil via perennial roots, and a decline in soil fertility (e.g., Whitford et al. 1987).

In South America and Australia the majority of the savanna region is still under extensive forms of use. Livestock grazing is the predominant effect, although there is an increasing trend to agriculture, for example, soy-bean production in South America. Conversion to agriculture in these regions is limited to the humid savannas.

Changes in fire and grazing regimes are strongly (negatively) correlated. Increasing grazing pressure leads to (1) a decrease in the diversity of perennial grasses and (2) a decrease in fire frequency (reduced fuel loads) accompanied by an increase in the abundance, but not necessarily the diversity, of woody plant species.

The IMAGE scenario is based on coarse soils data (the best available), and no topography. The forecasted extensive conversions will in reality be highly fragmented at a fine landscape scale. Rocky slopes, hill tops, drainage lines and otherwise unsuitable areas for cultivation will be excluded.

In terms of vulnerability and likely losses of species the wetland and riverine habitats rank highest. Large reductions in river flows are likely as demand for water increases. The loss of riverine vegetation is already significant. Wetlands will be subjected to increasing pressure as potentially productive systems. And as described earlier, wetlands and riverine habitats are biodiversity hot spots in the savannas.

Alien Species

The impact of alien species is a general problem in all biomes, and it needs to be recognized as a serious threat to biodiversity in the savannas. There are three categories of aliens, identified by the kinds of effect they have:

1. Displacement of ecologically equivalent native species:
   Examples are African grasses in the cerrados of Brazil, the llanos of Venezuela and in the humid savannas of northern Australia, and the shrub *Mimosa pigra* in northern Australia. The African grasses have led to local disappearance of native grasses, although there is no evidence that complete extinction has occurred.
2. Impact on other species:
   Introduced predators [e.g., the cane toad (*Bufo marinus*) in Australia] are classic examples. Apart from poisoning native predators (e.g., the goanna) that eat it, the cane toad has a depressing effect on prey species of local amphibians. The greatest impacts of introduced mammals, however, have been in the temperate regions of the world. Introduced vertebrates have had little success in establishing themselves in the complex and competitive trophic web of African savannas, and are therefore of little significance. The exception to this, of course, are domestic livestock (which would not survive in Africa without their human carers).

3. Change in ecosystem structure and processes:
   The Mitchell grass plains of tropical Australia are naturally treeless
   because the deep cracking clay soils prevent establishment of native trees.
   In the 1930s *Acacia nilotica* was introduced from Pakistan to provide
   shade for sheep. It was able to grow on these soils, and the plains are now
   being invaded by the trees. Through their deep roots and taller canopies
   they have completely altered the structure and the water balance of the
   ecosystem. The effects on the full complement of native species are difficult to unravel from the effects of sheep and cattle grazing.

The rate of movement of species around the world is increasing, and the
impact of alien species is a global change phenomenon that ranks high in the
savannas as a threatening process to biodiversity, although it is difficult to
predict for any particular locality.

Elevated CO_2

Elevated CO_2 may not have a particularly significant direct effect on ecosystem metabolism in savannas, beyond an expected influence on water use efficiency. Even here, though, the effect at the ecosystem level is likely to be
considerably less than it is at the individual plant level, through a number of
negative feedback mechanisms that lead to down regulation (for example, see
Korner 1996). In terms of biodiversity, however, of all biomes the savannas
are likely to respond most to CO_2. The expectation is that an increase in C_3
woody plants, both shrubs and trees, at the expense of C_4 grasses, although
evidence suggests that this may not be a consistent difference (Mooney et al.
1999). Such changes ("bush encroachment," "woody weed invasion") have
already occurred in savannas on all continents as a consequence of heavy and
continuous grazing by domestic livestock, not because of CO_2. The effects
on biodiversity are confounded by the effects of grazing, but the change to
a woody dominated community is accompanied by very significant changes
in other species. The herbaceous layer species under a tree canopy are different to those out in the open, due to changes in microclimate and soil nutrient status (Scholes and Walker 1993). The invertebrates and avifauna in open
and wooded savannas are also different. Mixed savanna communities have
both complements. As one or other vegetation component declines (woody
plants or open grassland), the likelihood of the loss of particular associated
species increases. For example, populations of the lepidopteran larvae (*Cirina
forda*), which feed principally on *Burkea africana*, which is the dominant tree
of a widespread type of savanna in southern Africa, can only maintain themselves above some minimum density of trees.

Field evidence regarding changes in C–N ratios in plants under elevated
CO_2 is variable (Mooney et al. in prep.), and even where the ratio does
increase in live leaf tissue, it seems not to carry through to the litter derived
from such plants. Nevertheless, any such increase in C–N in grasses would
likely lead to an increase in perennials over annuals, through the continued

effects of increased water use efficiency and decreased grazing pressure (induced by the higher C–N).

Elevated CO_2 will not act on its own. The associated climatic changes are likely to bring about changes in production and fuel load that will lead to changes in fire regimes. The effects will be different along the gradient from the very humid to the arid savannas. At the very humid end fires are virtually annual, so little change in fire frequency is likely. The biggest impacts are likely to be in the intermediate, subhumid savannas, where any changes in fire frequency led to significant changes in the woody–grass balance.

According to the predicted changes in savanna biome vegetation from BIOME 3, all of the savanna regions of the world will become woodier due, mostly, to the physiological effects included in the model that favor C_3 plants at high CO_2. They will also become more productive. Whether such changes to markedly woodier vegetation actually occur will be very much dictated by what happens to the fire regimes. In the extensive savanna regions with little management, it will depend on whether the potential increase in fire frequency is offset by continued high grazing. If C–N ratios in the grass layer increase, digestibility and therefore total consumption of grass will decrease, leaving a greater amount for fuel build up. In managed savannas fire is determined by the manager. In any event, the predicted vegetation changes in the BIOME 3 scenario, in particular for the savanna biome, will be significantly modified by the combined effects of grazing and fire.

Climate

Rainfall

The coefficient of variation in rainfall in the savannas is very high, and the biota is accordingly well adapted to large fluctuations. Williams et al. (1995) calculate that in northern Australia a 10% change in rainfall would require between 70 and 300 years to be detected statistically. They conclude that there will be no big changes in the biota in the next 100 years as a result of response to changes in the mean annual rainfall. Changes in the seasonal distribution of rainfall or in the frequency of extremes, however, could have very marked effects. In their area a 1-in-60 "wet" year becomes a 1-in-20 "wet" year for a 10% increase in mean annual rainfall. They conclude that changes in within-season patterns of rainfall will have the biggest effects, especially for the grasses and herbs.

In general, changes in mean climatic conditions lead to changes in ecosystem function (e.g., productivity), whereas changes in the variance and pattern of climate will lead to changes in species. Any losses in species that result from such changes in climatic variance reduce the resilience of the savanna to subsequent climatic extremes. This is because the resilience of savannas derives from having a number of different species in each functional type of species, capable of performing that function under different conditions. The loss of species therefore acts in a positive feedback manner to further reduce both ecosystem function and resilience. From the perspective of biodiversity scenarios, therefore, the emphasis should be on getting better forecasts of changes in climate variability from the GCM community.

A general conclusion from future climate scenarios, based on BIOME 3 and particular site-specific studies of savannas, is that the various belts of savanna types will shift along present moisture gradients. The arid grass/shrub savannas will develop into mixed semi-arid savannas with emerging deciduous trees in the shrub layer. At the humid end the transition into seasonal deciduous forest will depend on the fire regime. The present ecotone between the tropical rainforest of northeastern Australia and the sclerophyll (*Eucalyptus* dominated) forest with a strong herbaceous layer, is determined by fire (Harrington and Sanderson 1994). The rainforest is continually in the process of invading the sclerophyll forest and is only kept at bay by periodic fire. In the absence of fire, rainforest species establish under the sclerophyll canopy and grow through to overtop and dominate the forest. The eucalypts are unable to establish under a rainforest canopy. The dominant plant species and some of the related fauna [e.g., the northern race of the sugar glider (*Petaurus breviceps*)] are restricted to this narrow ecotone. Whether or not the ecotone is able to shift in accordance with climate change will depend strongly on land-use management practices.

The transient nature of changes in species composition at any point along the moisture gradient will be the net result of loss of species through mortality, and gain of new species through immigration. There will be a lag effect in this, depending on the longevity of the species. The longer lived the species, the bigger the lag effect, and the more out of equilibrium with the climate the community will become. This will have the effect of episodic, sudden major changes in composition as cohorts of long-lived species reach the end of their lifespan, compared with smaller, more continuous changes in communities of short-lived species. Because the BIOME3 results suggest mostly increasing moisture conditions, invasion by new species will be a continuing pressure on the existing communities, and availability of new safe sites for establishment will likely be the rate-limiting process (assuming species' potential migration rates match or exceed the rate of climate change). In those areas where moisture conditions are likely to decline under climate change, mortality in the existing community will likely be more extensive and faster than invasion by new species, and there will be a transient decline in diversity. This effect would be magnified where land use has already changed the "natural" existing community.

Menaut (personal communication) considers that a major effect of rainfall change in the Sahel is likely to result from the combination of an increase in total rainfall, giving rise to increased net primary production, and an increase in the length of the dry season. The net effect will be an increase in fires and therefore an increase in the grass–woody ratio.

BIOME 3 predicts that most of the moist savannas of the world will be replaced by tropical deciduous forest. The cerrados of Brazil will virtually disappear, as will the moist savanna belt in Africa that surrounds the central tropical rainforest, north and south of the equator, and Australia's northern and northeastern tall grass savannas. Again, it is necessary to note that these scenarios are due to the combined effects of changes in climate and the physiological effects of increased CO_2. They do not include the effects of fire or grazing, which would slow or prevent this transition, or conversion to cultivated lands, which would accelerate the loss of savanna.

The T-106 precipitation anomaly scenario suggests that the South American, the West and Southern African, and the Indian savannas will get substantially wetter. The northern Australian savannas will get slightly wetter, or experience no change. This scenario supports the BIOME 3 prediction of increased woody vegetation in the savannas.

Temperature

The effects of increases in mean temperatures have mostly been taken into account in the changes in water balance. Changes in the extremes of temperature, however, will have particular and significant effects on species.

Increases in the occurrence of extreme high temperatures will lead to reduction in establishment of sensitive plants. Kennard and Walker (1973) showed that the establishment of the grass *Panicum maximum* in SE Zimbabwe was prevented by very high soil surface temperatures during the first 9 days after germination. Grasses differ in their sensitivities to high temperatures.

Westoby (personal communication) has examined the distribution of eucalypt species in Australia in relation to temperature and found that just over half of them have temperature envelopes narrower than 3°C. This temperature tolerance is therefore less than the forecast 3°C shift poleward for a double-CO_2 climate in the temperate regions. The magnitude of the effect may be less in the tropics, but it is still there. Westoby points out that in terms of biodiversity it is not so much the loss of eucalypts, but rather the fate of the very large number of associated insects, that is important. If the plant species are able to migrate with the shifting temperature envelope, all is well. The insects will move with them, but if the plant species are unable to move fast enough, the insects (which could move on their own) will need to dissociate themselves from their present host species in order to survive. Westoby's crude calculation is that 20–30% of the total temperate insect fauna may be confronted with this dilemma (conservatively, around 200 species of insects for every bird and mammal species). We do not have even the basis for calculating what percentage of insects so affected might survive this dissociation between plant host and temperature envelope. We know next to nothing about the insect faunas of savanna plants and their host specificity.

Low temperatures are also important determinants of the species composition of savannas. Rushworth (1975) has described the influence of occasional frosts on the species composition of the vegetation on the Kalahari sands in western Zimbabwe. The subtropical and tropical interior plateau of southern Africa experiences frosts in the dry, winter months. Frost hollows occur in the extensive, flat Kalahari sands, and a "black" frost (down to −10°C) occurs every few years. Saplings of the dominant tree of the region, *Baikiaea plurijuga*, are killed by such frosts, and the result is a patchy vegetation with higher overall species diversity. We lack knowledge about the minimum temperature relations of savanna species in general, but a decrease in the frequency of very low temperatures will likely lead to significant changes in species composition wherever frosts currently occur.

Interactive Effects

As alluded to several times already, the individual effects of the various global changes cannot be considered in isolation. It is their interactive effects that will determine the outcome on biodiversity. In their summary of potential climate change effects in northen Australia, Williams et al. (1995) illustrate this with examples from the local fauna. The bilby (*Macrotis lagotis*), which is a small, rabbit-sized marsupial, has contracted from a very wide range to a small northern part of its essentially arid distribution because of land use and alien, introduced species. It is now likely to be caught between climate change and environmental degradation. The same problem occurs for the tropical golden bandicoot, which has contracted to its northern fringe. Climate change will increase the risk of local (and eventually total) extinction.

The biggest single factor in the decline of savanna biodiversity is land use. In particular, the destruction of habitat and habitat fragmentation. Land use is highly selective, with the most fertile, flat areas going first. An overall 20–30% conversion to arable land can mean the complete loss of particular communities. For the remaining biota, the interactive effects of fragmentation, the presence of alien species, and climate change greatly increase the risks of extinction. The land-use change has already occurred in much of Africa and Southeast Asia, making these savannas especially vulnerable to climate change and the introduction of alien species.

Summary of Biodiversity Changes

Table 8.1 is a summary of five regional savanna scenarios. The participants were asked to rank the likely impact of a significant change in each driver on each potential change, on a score of 1–5, and also to rank the expected change in the drivers and the time lag involved in ecosystem responses to such changes. The savanna regions, and the scientists who developed the

Table 8.1. Average of expected changes (1 = very small; 5 = very large) in the biodiversity of savannas in response to various drivers of change, based on five regional assessments (see text for explanation)

| | Δ Land use | | | | | | | | | Δ Atmosphere | | | | | | | | | Alien species | | | Δ Climate | | | | | | | | |
| | Habitat loss | | | Fragmentation | | | Intensification | | | Nitrogen | | | Pollution | | | CO$_2$ | | | | | | Δ Precip. | | | Δ Temp. | | | Δ Seasonality | | |
Expected changes in biodiversity	avg	max	min	avg	max	min	avg	max	min	avg	max	min	avg	max	min	avg	max	min	avg	max	min	avg	max	min	avg	max	min	avg	max	min
Δ Species diversity																														
loss of species	3.8	5.0	2.0	3.4	5.0	2.0	2.8	4.0	2.0	1.7	3.0	1.0	1.6	3.0	1.0	0.8	1.0	0.0	2.3	4.0	1.0	2.0	3.0	1.0	1.0	2.0	0.0	2.0	3.0	1.0
global extinctions	2.5	5.0	1.0	2.0	3.0	1.0	2.5	4.0	1.0	1.0	1.0	1.0	1.5	3.0	1.0	1.0	1.0	1.0	0.8	1.0	0.0	1.0	1.0	1.0	0.8	1.0	0.0	1.3	2.0	1.0
addition of species	1.8	4.0	1.0	2.0	4.0	1.0	1.6	4.0	1.0	1.0	1.0	1.0	1.4	2.0	1.0	1.2	2.0	1.0	2.0	4.0	1.0	1.2	2.0	0.0	1.0	2.0	0.0	1.8	2.0	1.0
Δ rel. abundance	4.0	5.0	2.0	3.2	5.0	2.0	3.4	4.0	3.0	1.7	2.0	1.0	2.4	4.0	1.0	2.2	3.0	1.0	2.8	4.0	2.0	2.3	3.5	1.0	1.4	3.0	1.0	2.7	3.5	2.0
Δ Fct. group diversity																														
loss	3.6	5.0	2.0	3.2	5.0	2.0	2.8	4.0	2.0	1.7	2.0	1.0	2.2	4.0	1.0	1.3	3.0	0.0	2.8	4.0	0.0	2.3	3.5	1.0	1.2	2.0	1.0	2.1	3.5	1.0
addition	1.8	4.0	1.0	2.0	4.0	1.0	1.8	4.0	1.0	1.7	2.0	1.0	1.6	2.0	1.0	1.8	3.0	1.0	2.0	4.0	1.0	1.8	3.0	1.0	1.2	2.0	1.0	1.8	3.0	1.0
Δ rel. abundance	3.5	5.0	2.0	3.0	4.0	2.0	3.8	4.0	3.0	2.0	3.0	1.0	2.0	4.0	1.0	2.3	4.0	1.0	3.5	5.0	2.0	2.9	4.5	2.0	1.3	2.0	1.0	2.9	4.5	2.0
Δ Landscape diversity	3.4	5.0	2.0	3.2	5.0	2.0	3.0	4.0	2.0	1.3	2.0	1.0	1.0	1.0	1.0	1.2	2.0	1.0	3.0	5.0	2.0	2.4	4.0	2.0	1.0	2.0	0.0	2.2	4.0	1.0
Δ Functioning																														
direct effects	3.4	5.0	2.0	3.0	5.0	2.0	3.2	5.0	2.0	1.7	3.0	1.0	1.8	3.0	1.0	1.6	3.0	1.0	3.2	5.0	2.0	2.8	4.0	2.0	1.6	3.0	1.0	2.8	4.0	2.0
indirect effects	3.2	5.0	2.0	2.5	3.0	2.0	3.4	5.0	2.0	1.0	1.0	1.0	1.8	3.0	1.0	2.2	5.0	1.0	3.0	5.0	1.0	2.6	3.0	2.0	1.6	3.0	1.0	2.6	3.0	1.0
Δ ecosystem services	3.4	5.0	2.0	2.6	4.0	2.0	2.8	5.0	2.0	1.3	2.0	1.0	2.0	3.0	1.0	1.8	3.0	1.0	3.2	5.0	2.0	2.2	3.0	2.0	1.6	3.0	1.0	2.4	3.0	1.0
Expected Δ in drivers	3.0	4.5	1.5	2.5	4.0	1.5	3.1	4.0	2.5	1.7	3.0	1.0	1.9	3.0	1.0	2.8	4.0	2.0	2.0	4.0	1.0	3.0	4.0	2.0	2.6	4.0	2.0	2.6	4.0	2.0
Time lag in Δ in diversity	4.0	5.0	3.0	3.4	4.0	2.0	3.8	5.0	3.0	2.5	4.5	1.0	2.9	4.5	1.0	2.4	4.0	1.0	3.5	4.5	2.0	2.6	4.0	1.0	2.4	4.0	1.0	2.6	4.0	1.0

scenarios, are: humid northern Australian savannas in the region around Darwin (Dick Williams), Southeast Asian savannas and savanna forests (Philip Stott), seasonal nonflooded neotropical savannas in Venezuela (Zdravko Baruch), dry African savannas (Jim Ellis), and Indian savannas (Jai Singh).

Explanation and Case Studies

It is clear from the distribution of the values in Table 8.1 that the major changes in savanna biodiversity in the immediate and near future will be via habitat loss, fragmentation, and intensification of land use. Nitrogen deposition and other pollution (should they occur) are likely to have relatively minor effects except for secondary effects through changes in ecosystem function induced through increased nitrogen, but the likelihood of these drivers becoming significant in the savannas is small. Increased CO_2 is almost certain, and will have some effects, although the expected changes in savannas are not considered to be more profound than changes that would result from other atmospheric pollution. Alien species follow land use in terms of potential impact on savanna biodiversity, and the likelihood of it occurring is high at least in some savannas. As explained earlier, the C_3–C_4 interaction is expected to be stronger in the savannas than it is in most other biomes (although, again, it is important to note evidence of how variable this might be), and changes in the relative abundances of species and functional types of plants will be significant—depending on the downregulating effects of other constraints to plant growth at the ecosystem level.

In terms of the sequence of changes, the CO_2 effects will lag somewhat behind the land-use effects, although both will continue for many decades. Alien species effects are occurring now, and this process is likely to increase with increasing globalization. Climate change effects will lag behind the others both because of the response time of the physical atmospheric system and because the savannas have such a high coefficient of rainfall variation. Any changes in the seasonal distribution of rainfall, however, will have immediate effects.

Apart from introduced alien species, the arrival of which is very hard to predict, the savannas are unlikely to gain new species from neighboring biomes. Bird species distributions within the biome shift considerably in response to cycles or fluctuations in rainfall, but this seldom results in the appearance of "new" species. Losses of species, however, are predicted in response to any shift in the seasonal distribution of rain, and through displacement by introduced aliens. Total extinction of species is most likely to be brought about by selective loss of particular habitats and, again, through a shift in seasonality. The probability of total extinction as a result of direct, individual effects of changes in these drivers, however, is considered to be small. It is the interactive effects that will most likely result in total extinctions (cf. the situation for bilbies in northern Australia).

Overall, the biggest changes in savanna biodiversity will be in terms of relative abundances of species and functional types (especially the woody and grass components) and the areas of greatest uncertainty in trying to predict what will happen are in relation to changes in seasonality, and its effects, and the indirect effects associated with changes in ecosystem function.

Overview of Vulnerability

A scenario for the future of biodiversity in the savannas needs to be considered on a regional basis.

South Asia (primarily India and Pakistan) is already greatly modified. Future changes will be much less than those that have already occurred. The same applies to southeast Asia (which has, or had, only moist savannas). All the land that can be successfully cultivated, barring a few protected pockets, has already been cleared and is either under cultivation or is in some fallow state. From 1995 to 2030 IMAGE scenario A1 suggests an increase in land under crops from 1.9 million $\times 10^3$ km^2 to 2.1 m, mixed crops and pastures from 110 k to 517 k, a decline in savannas from 744 k to 484 k, and a decline in tropical woodlands from 135 k to 71 k The future will see losses of species from remaining isolated fragments like lights going out on a darkened landscape.

The humid parts of Africa's savannas are in much the same situation as Asia. The more extensive, less-populated (often tsetse fly–infested), semi-arid regions are still biologically more intact. Their future is largely dependent on technological advances in crop breeding and biological control, and current progress indicates that much of this region will be developed for agriculture before 2050. It therefore represents a region that will experience perhaps the greatest future losses and changes in savanna biodiversity. For the period 1995–2030 IMAGE suggests a complete loss of tropical woodlands, from 1.5 m $\times 10^3$ km^2 to zero, a decline in savannas from 4 million down to a mere 6000, with an increase in land under crops (for the whole of Africa, not just the savanna biome) from 452 k to 4.9 million. The IMAGE model assumes, however, that the projected human population will continue to produce the food required to meet the minimum daily intake for a healthy person. This is already not being achieved. A transient feature of the increasing pressure for agricultural land will be a shortening of the rotation time for fallow lands, and this will almost certainly lead to local and regional losses of species. Knowledge of soil types is also limited and may be more of a constraining factor to the spread of agriculture than would appear in the scale of the IMAGE analysis.

According to IMAGE South American savannas will decrease over the 1995–2030 period from 2.7 m to 2.0 m $\times$ 10E3 km^2 and tropical woodlands from 4.1 m to 3.8 m. Land under crops in Latin America will apparently decrease from 358k to 296k, but pastures will increase from 4.6m to 6.0m.

Gaps and Future Needs

Improved scenarios of future biodiversity in the savannas will require:

1. Data
   A statistically representative sample of the savannas for:
   i. Estimates of present α and β diverity, in all functional groups.
   ii. Present patterns of land use, according to an agreed, standardized set of land-use categories, at c 100m resolution.
   iii. Predictions of future land-use patterns.
   iv. Better forecasts, from GCM and climate models, of changes in the seasonal distributions and extremes of rainfall, and in extremes of temperatures.
2. Processes
   i. Free Air Carbon Enrichment (FACE) experiments with C_3 woody plants, C_3 grasses (where they occur), C_3 forbs, and C_4 grasses, with different soil fertilities and, if possible, with and without grazing. Only in this way can we get empirical evidence on the net effect of CO_2 increase on relative species composition and changes in soil organic matter. (This assumes that migration rates for different species within these categories is not limiting).
   ii. Studies on growth phenology patterns in different savannas and their sensitivity to changes in seasonal rainfall patterns.
   iii. Experimental manipulation of the seasonal pattern of available soil moisture, with a priori predictions of expected species changes.
3. Models of savanna species dynamics that include functional groups, response types, and metapopulation dynamics in relation to land-use patterns, aimed at predicting species composition.

Postscript

Whatever happens in the future, the changes in biodiversity of the savannas will be better monitored than were the changes in the past. We will have better and more timely information with which to test our hypotheses, and to develop strategies to cope with the changes. Only time will tell whether this information gets used to make a difference, or merely provides an interesting epitaph for the savanna regions concerned.

Acknowledgments. I thank Mark Westoby, Jean-Claude Menaut, Bob Scholes, Alan Andersen, Dick Braithwaite, and Steve Morton for their ideas on the future of savanna biodiversity, and Philip Stott, Jai Singh, Zdravko

Baruch, Dick Williams, and Jim Ellis for developing the five regional scenarios of savanna biodiversity futures.

References

Diamond J (1989) Overview of recent extinctions. In: Western D, Pearl M (eds) *Conservation in the Twenty-First Century*, pp. 37–43. Oxford University Press, New York.

Eiten G (1982) Brazilian savannas. In: Huntley BJ, Walker BH (1982) *Ecology of Tropical Savannas*. pp. 25–47. Springer-Verlag, New York.

Harrington GN, Sanderson KD (1994) Recent contraction of wet sclerophyll forest in the wet tropics of Queensland due to invasion by rainforest. Pacific Conservation Biology 1:319–327.

Huntley BJ, Walker BH (eds) (1982) *Ecology of Tropical Savannas*. Springer-Verlag, New York.

Kennard DG, Walker BH (1973) Relationships between tree canopy cover and *Panicum maximum* in the vicinity of Fort Victoria. Rhodesian Journal of Agricultural Research 71:145–153.

Korner C (1996) The response of complex multispecies systems to elevated CO_2. In: Walker BH, Steffen W (eds) *Global Change and Terrestrial Ecosystems*. Cambridge University Press, Cambridge, UK.

Mooney HA, Canadell J, Chapin FS III, Ehleringer J, Korner C, McMurtrie RE et al. (1999) Ecosystems physiology responses to global change. In: Walker BH, Steffen W, Canadell J, Ingram J (eds) The Terrestrial Biosphere and Global Change. Implications for Natural and Managed Ecosystems, pp. 141–189. Cambridge University Press, Cambridge, UK.

Rushworth JE (1975) The floristic, physiognomic and biomass structure of Kalahari sand shrub vegetation in relation to fire and frost in Wankie National Park, Rhodesia. M.S.c. thesis, University of Rhodesia.

Scholes RJ, Walker BH (1993) Nylsvley: The study of an African savanna. Cambridge University Press, Cambridge, UK.

Silva J (1987) Responses of savannas to stress and disturbance: species dynamics. In: Walker BH (ed) *Determinants of Tropical Savannas*, pp. 141–156. IRL Press Ltd., Oxford, UK.

Stafford-Smith MD, Morton SR (1990) A framework for the ecology of arid Australia. *Journal of Arid Environments* 18:255–278.

Tothill JC, Mott JJ (eds) (1985) *Ecology and Management of the World's Savannas*. Australian Academy of Science, Canberra.

Walker BH (ed) (1987) *Determinants of Tropical Savannas*. IRL Press Limited, Oxford.

Walker BH (1992) Biodiversity and ecological redundancy. *Conservation Biology* 6(1):18–23.

Walker BH (1997) Functional types in non-equilibrium ecosystems. In: Smith TM, Shugart HH (eds) *Plant Functional Types*, pp. 91–93. Cambridge University Press, Cambridge, UK.

Walker BH, Langridge J (1997) Predicting savanna vegetation structure on the basis of plant available moisture (PAM) and plant available nutrients (PAN): a case study from Australia. *Journal of Biogeography* (In press).

Walker BH, McFarlane FR, Langridge JL (1997) Grass growth in response to time of rainfall and season along a climate gradient in Australian rangelands. *The Rangeland Journal* (In press).

Walker J, Gillison AN (1982) Australian savannas. In: Huntley BJ, Walker BH (1982) *Ecology of Tropical Savannas*, pp. 55–24. Springer-Verlag, New York.

Whitford WG, Reynolds JF, Cunningham GL (1987) How desertification affects nitrogen limitations of primary production on Chihuahan desert watersheds. In: Alden EF, Ganzales C, Mair W (eds) Strategies for classification and management of native vegetation for food production in arid zones, pp. 143–153. *U.S. Forest Service, General Technical Rep.* RM-150.

Williams RJ, Cook GD, Braithwaite RW, Anderson AN, Corbett LK (1995) Australia's wet-dry tropics: identifying the sensitive zones. In: Pernatta J, Leemans R, Elder D, Humphrey S (eds) *The Impact of Climate Change on Ecosystems and Species: Terrestrial Biomes*, pp. 39–64. IUCN, Gland, Switzerland.

Woinarski JCZ, Braithwaite RW (1990) Conservation Foci for Australian Birds and Mammals. *Search* 21:65–68.

9. Mediterranean-Climate Ecosystems

Harold A. Mooney, Mary T. Kalin Arroyo,
William J. Bond, Josep Canadell, Richard J. Hobbs,
Sandra Lavorel, and Ronald P. Neilson

In this chapter we will review the current status of biological diversity in the mediterranean-climate regions of the world, making comparisons among them, as well as examining the threats to biological systems now and in the future. Mediterranean-type climates, which are characterized by a predominantly winter rainfall regime, exist in five regions of the world: parts of California, South Africa, Chile, southern Australia, and the Mediterranean basin. Even though all of these regions share broad climatic conditions and contain ecosystems that have similar structures and dynamics (Hobbs et al. 1995), each region has a unique history of human habitation and use. There is a long history of comparative biotic studies among these regions upon which we draw; however, there are no comparisons of the social drivers of land-use change among these areas. Our analyses are therefore less complete than we would like.

There are many limitations to the development of plausible scenarios of biodiversity change in mediterranean-climate regions. At the most general level, all mediterranean regions are relatively narrow and predominantly along the coasts of continents. For this reason they fall within general circulation model (GCM) grid cells that are mostly water. Further, the metric of a GCM output, average temperature increase, is not too helpful for biogeographic predictions, particularly in this topographically diverse region. Most serious, though, is the lack of good resolution on predictions on the amount and seasonality of rainfall, one of the most important controllers of biotic patterns in this summer-drought region. A further complication is the lack of information on what might happen to upwelling patterns of the coastal waters, an important controller of local climate. We probably have a better capability to predict what will happen to land use, with the ever-growing populations, than we have for climate.

In all mediterranean-climate regions, there are an extraordinary number of biomes: grasslands, shrublands, woodlands, and forests of varying types and hence numerous ecotones. To a large degree these ecotones are controlled by climate, both past and present, and by fire regimes and soil types. Land-use patterns have more recently changed boundaries either completely, or indirectly, by grazing, for example. Some of these boundaries are complex. For example, some boundaries between shrublands and grasslands are partly controlled by mammals (Bartholomew, 1970). Others (e.g., the boundaries between redwood forest and adjacent meadows) may be partly biotically controlled because the redwoods augment local precipitation by trapping fog (Ingraham and Matthews, 1995).

It is now well accepted that biomes will not shift en masse with climate change; rather, the distributions of species will be affected individually. Because we have so little information on the detailed ecology of most species in these regions we thus face an enormous challenge in predicting potential shifts in species abundances and distributions.

Mediterranean-climate regions have an unusual number of biomes and biome boundaries, and they are also generally rich in endemic and rare species, many with very localized distributions. It is these latter types for which the prognosis under global change is not easy. Many of these organisms are now localized in parks and preserves, or in fragments of relatively undisturbed habitat surrounded by degraded or altered landscapes, leaving little place for migration in response to climatic change, even if this were possible on biological grounds. Thus, predictions based solely on climate-controlled biogeography may not be realistic because of the importance of new land-surface alterations in determining migration potential.

Finally, the biota is changing rapidly in composition in all mediterranean-climate regions due to the increasing load of invasive species. We do not yet have a robust theory to tell us which new spcies will successfully invade in the coming years under present conditions, much less under conditions of climate change.

In the following sections we first view the richness and patterning of biodiversity in the various mediterranean-climate regions. We will then examine the potential consequences of global changes on these patterns. Our coverage is of necessity uneven because equal information is not available from all of these regions.

Biodiversity Patterns

Australia

Because of the broad range of precipitation, soils, and landforms, the mediterranean area of southwestern Australia encompasses a wide array of ecosystem types, ranging from closed forests to open savanna woodlands and shrublands (Beard and Sprenger 1984ab; Beard 1990; Hobbs 1992; Hopper 1992). Forest vegetation is generally restricted to the higher rainfall areas (>750 mm); at lower rainfalls the distribution of woodland, mallee, shrubland, and heath vegetation types is determined mainly by substrate type. There are additional types that occupy a relatively small area but are important ecologically [e.g., permanent and ephemeral-lake systems (both salt and freshwater) and granite outcrops].

Southwestern Australia has a high level of plant species diversity, both within and among communities. The Southwest Botanical Province (0.3×10^6 km^2) is a recognized hotspot for floristic biodiversity and endemism (Groombridge 1992; Hopper et al. 1996). Three decades ago Beard (1965) recorded just 3611 named taxa for the Southwest Botanical Province. Now an estimate of 8000 species in this region seems reasonable, with about 75% endemism.

The flora of the southwest is dominated by genera of woody perennials in families such as the Myrtaceae, Proteaceae, Fabaceae, and Epacridaceae. The flora has radiated greatly at the species level, whereas diversity at the family and genus levels is relatively low (Lamont et al. 1984). In the southwest, *Acacia* has at least 400+ species, *Eucalyptus* 300+, *Grevillea* 200+, *Stylidium* and *Melaleuca* 150+, and *Hakea* and *Caladenia* 100+. Such speciation has been concentrated in the Transitional Rainfall Zone for the majority of woody perennial taxa (Hopper 1979; Lamont, et al. 1984).

The remarkable species richness is the end-product of long and complex evolutionary processes, stimulated by environmental perturbations in the late Tertiary and Quaternary (Hopper 1979, 1992; Hopper et al. 1996). The high floristic diversity of the region is attributed to this long evolutionary history in isolation from the rest of Australia, the lack of glaciation, and complex historical and current disturbance patterns. The evolution and persistence of so many rare locally endemic species, particularly in the Transitional Rainfall Zone flora, has been interpreted as being in response to dynamic environments subject to recurrent and unpredictable environmental change (Main 1982; Pate and Hopper 1993).

Although southwestern Australia retains elements of the mammal fauna that are now missing from the rest of the continent (e.g., Kennedy 1990), levels of faunal diversity and endemism do not, in general, match those of the flora. Many of the species in the region are representatives of either arid or more mesic zones, and the mediterranean-climate region represents the limits of their distributions. Such species are commonly confined to the periphery of the mediterranean-climate region. This pattern reflects the transitional nature of the mediterranean-climatic zone between the drier inland areas and the more mesic environments to the east and southeast. The low level of mammalian endemism suggests that the distribution of many vertebrate species in the region may largely be a consequence of historical and biogeographic factors rather than a result of adaptations to local ecological conditions. Much of the vertebrate fauna has broad environmental tolerances that enable it to persist under a wide range of conditions, including those presented by the marked seasonal climate of mediterranean regions.

Birds also show a low level of endemism. Only nine species are endemic to the mediterranean-climate zone (Schodde 1981), although more than 190 species have been recorded in the western Australian wheatbelt region alone (Saunders and Ingram 1995). Of more than 80 species of mammals recorded from the Western Australian mediterranean region, 16 (approximately 20%) are endemic to it, whereas approximately 28% of the mediterranean zone reptiles are endemic. In contrast, 64% of the 36 frog species found in the region are endemic (data from distribution maps of Strahan 1983; Cogger 1986). Many other invertebrate groups also show a high degree of endemism that reflects the dispersal characteristics of particular taxonomic groups. More vagile groups (e.g., butterflies and grasshoppers) have a higher proportion of species that extend beyond the region, whereas many terrestrial and epigaeic arthropods (e.g., scorpions and mites) have very restricted distributions (Mark Harvey, personal communication).

Mediterranean Basin

The Mediterranean flora encompasses a total of about 25,000 plant species within an area of $2.3 \times 10^6 \text{km}^2$. Species richness averages from around 10–30 species per square meter to 30–50 per 100m^2 and more than 100 species per ha (Westman 1988). In a comparison of 0.1 ha plots from all mediterranean climatic regions, the highest species richness was found in disturbed areas in Israel (Mooney 1988).

High diversity in the Mediterranean results from a number of factors. The Mediterranean region is located at the confluence of three important biogeographic regions: paleoarctic, North African, and Caucasian. It has acquired elements from all three regions (Quézel 1985). In addition, the physical geography is very diverse, with rugged topography (mountain ranges reaching above 4000m located at short distances from the coast of the Mediterranean Sea) and a large number of islands and peninsulas. Related to this geographic diversity is a wide range of climatic conditions ranging from warm semi-desert conditions in North Africa to mild or even cold winters, including frequent frost, to the north. The general topographic and climatic complexity has been overlain by a long history of human occupation that has involved continuous disturbance for several millennia. The region has been logged, cultivated, grazed, and burned for a long enough time for evolutionary changes to have taken place in the flora in response to these disturbances.

In addition to high overall species diversity, the region is characterized by a high landscape-scale (ß) diversity. This results from the complex physical geography and the long history of human occupation with a fine-grained distribution of land uses. Many Mediterranean landscapes are mosaics of communities at different successional stages that are the result of differing disturbance regimes and histories. Examples of these highly diverse managed landscapes are the traditional intermixed patches of woodlands, shrublands, pastures, and croplands, with a high diversity of annual grasses capable of surviving multiple stresses (e.g., fire, drought, grazing, and cutting) (Naveh and Whittaker 1979; Blondel and Aronson 1995). These types of landscapes are commonly found in France, Spain (Dehesas), and Portugal (Montados) (Joffre et al. 1988; Ibañez et al. 1989).

High endemism is also a characteristic of the Mediterranean flora. Endemism is positively correlated with species richness, and it is highest in the Iberian Peninsula, Greece, Morocco, and Turkey. In general, calcareous areas tend to have higher species richness than areas on acid substrates. Of the 25,000 species, 37% are endemic to the region, representing about 10,800 taxa of the total Mediterranean vascular flora (Greuter 1995). Others have estimated the percentage of endemic species as high as 50% (Quézel 1985). This high level of endemism has been linked to the complex geography noted earlier, and the thousands of islands and islets particularly in the eastern Mediterranean Basin (Quézel 1985; Quézel and Médail 1995). Greece alone has as many as 2000 island and islets (Tzanoudakis and Panitsa 1995). The eight largest island territories of the Mediterranean have more than 1000 single-area endemics. The high level of endemism may also reflect the high number of annuals which, due to their short generation time, rapidly evolve and speciate (Quézel 1995).

Of the 3583 species endemic to individual Mediterranean countries nearly 2000 are rare or threatened (Léon et al. 1985; Cody 1986). In contrast, if nonendemic species are included, Greuter 1994) estimated that 4251 species are threatened, as a result of agriculture, over-grazing, deforestation, and urbanization. Somewhat different numbers are given by The World Conservation Monitoring Center (in Ramade 1990) that suggested that of 4777 species endemic to a single Mediterranean country (excluding Syria, Lebanon, and Turkey), 2758 are Rare, 180 Endangered, 344 Vulnerable, and 454 of Indeterminate Status.

The Mediterranean region has 70–80% of the Palearctic fauna of birds, mammals, reptiles, and amphibians (Oosterbroek 1994) and 75% of the insect orders (Balletto and Casale 1991) and a high degree of richness and endemism (Table 9.1).

Table 9.1. Estimated total number of species (and endemics, as a percentage of the total number of species) for vertebrates, and number species and subspecies for insect groups in the Mediterranean region

	Breeding birds	Mammals	Reptiles	Amphibians	Neuropteran	Rhopalocen	Tipulidae
Total spp	399	195	163	65	461	677	498
Endemics	45 (11%)	40 (21%)	77 (47%)	27 (42%)	230 (50%)	416 (61%)	361 (72%)

Data from Oosterbroek 1994.

The regions with highest florisitic species richness are West Asia Minor (Greek islands along the Turkey coast, Cyprus, part of Turkey, Lebanon, and Israel), the Balkans (former Yugoslavia, Albania, Greece, European part of Turkey), followed by the Iberian Peninsula. If East Asia Minor is considered (Eastern Turkey, Armenia, northwest Iran, southern halves of Georgia and Azerbaydzhan), species richness is among the highest of all. Italy and the Iberian Peninsula show the highest higher taxa diversity of all areas of the Mediterranean (Oosterbroek and Arntzen 1992).

For vertebrate fauna, reptiles and amphibians are the groups with the highest percentage of endemic species, whereas insects have the highest endemic percentage of all in part because both species and subspecies are taken into account.

In the northern Mediterranean, open landscape mosaics are also the cause of high bird community richness, which decreases with an increasingly closed canopy of evergreen forests (Tellería et al. 1992). A great deal of bird diversity is due to the fact that the Mediterranean region is an important stopover place for the Afro-Palearctic migration, and croplands, meadows, and wetlands are critical to that migration route (Farina 1989). Coastal and inland wetlands are the biodiversity hot spots for waterbirds (Van der Hane and Van den Berk 1994).

Another group with special value for its contribution to the total biodiversity pool is the freshwater fish. There are 229 endemic fish taxa (132 species and 97 subspecies) in the northern Mediterranean that represent 13 families (Crivelli and Maitland 1995).

South Africa

The southwestern Cape region of South Africa has a diversity of ecosystem types related to the range of mean annual rainfall, soil type and topography. Under higher rainfall conditions (ca. 800–1600 mm pa) remnants of closed evergreen forests occur in sites that are well protected from fire. Under lower rainfall conditions remnants of related evergreen shrublands, but with many more spinescent and succulent species, survive along the coast and occasionally on inland lower mountain slopes. These shrublands are also fire sensitive and restricted to areas that seldom burn. Most of the mountains and coastal lowlands of the Cape, however, are dominated by fynbos, a fire-prone shrubland characterised by the presence of stiff, sclerophyllous, reed-like Restionaceae, ericoid shrubs and broad-leaved shrubby members of the Proteaceae. Fynbos occurs on nutrient-poor soils over a wide rainfall gradient from 250 mm to 3000 mm per annum. It is replaced by renosterveld shrublands, which are also fire-prone, and dominated by Asteraceae and a grassy herbaceous layer (often dominated by C4 grasses) on clay rich soils below 700 mm rainfall. At the drier ends of fynbos distribution (<250 mm rainfall pa), succulent shrublands are found. These are not fire-prone. Unlike Californian succulent shrublands, the dominant elements are leaf succulents.

The summer-dry climate region of South Africa thus supports a rich diversity of ecosystem types with evergreen forests and shrublands that do not burn, fire-prone heathlike shrublands, fire-prone asteraceaous shrublands, and succulent shrublands. The region is strikingly different from the rest of South Africa and most other mediterranean regions, in having virtually no grasslands and savannas.

The flora of the Cape Region, which is characterized by substantial winter rainfall, is among the richest in the world for similar-sized areas. There are some 8600 plant species of which 68% are endemic (Bond and Goldblatt 1984). Seven families and 248 genera are endemic, or nearly so, to the region. In comparison to other mediterranean-type regions, the Cape has unusually large species–family ratios (i.e., very species-rich genera), but generic diversity is not unusual. Western Australia shows similar patterns of high levels of intrageneric diversity (Linder et al. 1992).

The succulent shrublands are extremely rich in species, particularly leaf-succulents in the Mesembryanthemaceae. The biome includes the Namaqualand flora of the west coast, which is extremely rich in annuals and geophytes and provides a spring floral display that attracts thousands of visitors annually. The biome has more than 5000 species, of which more than 50% are endemic, making it the richest semi-arid vegetation anywhere (Milton et al. 1997).

Southern Africa has a rich vertebrate fauna in world terms, but most of it is concentrated in summer rainfall savannas (Siegfried 1989). The two winter rainfall biomes (fynbos and succulent Karoo) have conspicuously low densities, both of larger vertebrates and insects, but measures of species richness per unit area are relatively high for amphibians, reptiles, birds, and mammals in southern Africa (Siegfried 1989). The fauna does not match the flora in endemism, as was noted for Australia. Fynbos has only seven endemic mammals and six endemic bird species. Included among the mammals were the extinct blue antelope, the bontebok, the Cape mountain zebra, the Cape dune mole rat (a large fossorial species occurring on lowland sandy soils), and several species of rodents and shrews. There are higher levels of endemism in other groups. Nine of 30 frog species and nearly half of the 30 freshwater fish species are endemic. The lizard fauna is rich with at least 50 species, but is still being explored. There appears to be a number of closely related vicariant endemics in groups such as dwarf chameleons and geckoes. There is a high diversity of tortoises (12 species), especially in the succulent Karoo, and one endemic fynbos species.

Chile

The mediterranean-climate area of central Chile is an area of high vegetation diversity and species richness. Together with the winter rainfall deserts to the north, also of mediterranean tendency, it comprises one of the world's 24 globally recognized threatened hotspots (Mittermeier et al. 1988), and has

Table 9.2. Species richness and endemism in the native mediterranean-type climate flora of central Chile based on the area between 32°S and 40°S, all vegetation types included

Ecoregion	Area ($km^2 \times 10^3$)	No. species	Species endemic to Flora of Chile	Percent species endemic to Flora of Chile	No. species endemic to area	Percent species endemic to area
Summer rainfall area	154	ca. 825				
Winter rainfall areas (desert + medit)	300	3429	1821	53.1	1602	46.7
Deserts	145	1893	1100	58.1	605	31.9
Mediterranean	155	2537	1176	46.3	593	23.4
Cool temperate area	303	ca. 1360				
Continental Chile*	737	5082	2630	51.8	2630	51.8

Data are for species and varieties; from Marticorena 1990.
From Arroyo and Cavieres 1997.
Data are also given for the winter rainfall deserts (area between 25°S and 32°S and the coastal deserts Regions to around Pisagua), the summer rainfall area (Regions north of 25°S, excluding the coastal deserts), the cool temperate area (area between 40°S to 56°S), and continental Chile.

been considered as a World Center of Plant Diversity (Davis et al. 1997). Local vegetation diversity has been enhanced by major vegetation reorganization in the Pleistocene and Holocene (Villagrán 1994). The highly dissected landscape, which results from Pleistocene uplifting the dominant Andean chain, has provided a theater for much local speciation, especially in the herbaceous flora. Because it is positioned at the crossroads of two major biogeographic regions, the Neotropics and the ancient Gondwanan province, it is not uncommon to find dominant woody taxa of widely different biogeographical origins existing side by side in the mediterranean vegetation of central Chile. Indeed, the mediterranean vegetation intergrades imperceptibly into southern temperate rainforest, which is a situation that would appear to be unique among the southern hemisphere mediterranean-type climate areas.

Central Chile (the mediterranean climate portions in a strict sense) has a rich and endemic flora (Table 9.2) with high life-form diversity (Montenegro and Ginocchi 1995; Arroyo and Cavieres 1997; Arroyo et al. 1997). Compared with the mediterranean flora of the California Floristic Province, which is its closest analog (Mooney 1977), central Chile stands out for its higher proportion of woody species and genera, and a significantly smaller annual flora (Arroyo et al. 1997). Whereas 30% of the native flora of the California

Floristic Province is annual, a mere 16% of Chilean mediterranean species are annual. Twenty-one percent of the mediterranean flora of Chile is woody, in comparison with 14% in the California Floristic Province. This difference in life-form composition in the two presently climatically similar areas has been related to an accumulated historical effect of a more equitable climate in southern South America (Arroyo et al. 1997).

For overall richness, Arroyo et al. (1997) cite a total of 2395 vascular plant species (i.e., angiosperms, gymnosperms, and ferns), in 591 genera, that occur between the northern limit of the Province of Choapa (31°S–31°30′S) to the southern limits of the Provinces of Concepción and Ñuble (ca. 37°15′S). There are 437 herbaceous genera in central Chile and 180 genera containing woody species (Arroyo et al. 1997). Thus, a very high proportion of the phylogenetic richness is found in the herbaceous flora. For the larger area between 32° and 40°S there are 2537 species (Arroyo and Cavieres 1997). Species accumulation curves for the central Chilean region approach an asymptote, which suggests that the total flora is relatively well known (Maldonado et al. 1995).

The hyperarid winter rainfall deserts, which is also a mediterranean-type climate in a wider sense, support 1893 species (Table 9.2). Together, the more typical mediterranean-type climate and the hyperarid winter rainfall deserts (i.e., winter rainfall area) support a flora of some 3429 species.

The richness of the mediterranean-climate area is best appreciated by comparing the number of species present with other major ecoregions of Chile, bearing in mind, size of land area (Table 9.2). Although only slightly larger than the drier winter-rainfall desert area, it has more than 600 more species. The cool temperate forest zone (including all vegetation types) south of 40°S, for an area a little under twice the size of central Chile (Table 9.2), has only around recorded 1360 species. Plant species-richness in central Chile is relatively low in comparison with other mediterranean-type climate areas (Cowling et al. 1996). In comparison with California, central Chile has fewer genera than are found in the California Floristic Province for a given area (Arroyo et al. 1997). Increasing isolation from potential sources of biota east of the Andean chain that ensued as of the rapid uplift of the Andes in the Pliocene and Pleistocene (Arroyo et al. 1997) appears to have resulted in fewer genera dispersing into the area of central Chile.

Endemism in the mediterranean-climate area, a relatively small geographical area (Table 9.2), is around 23%; however, an outstanding 47% of species in the larger winter rainfall area, which is the more natural biogeographic unit in Chile, are endemic. Endemism is also high in the winter rainfall deserts per se (32%). The high level of endemism considering the full winter rainfall area has been enhanced by continental Chile's islandlike nature. Bordered to the west by the Pacific Ocean, to the east by the high Andean crest and to the north by an area of extreme aridity, interchange with adjacent continental areas has been restricted over the course of evolution of Chile's mediterranean flora (Arroyo et al. 1996, 1998; Villagrán and Hinojosa 1997).

Table 9.3. Species richness (and percentage endemism where available) in some well-studied groups of animals in the mediterranean area of Chile

Group	Asilidae	Butterflies	Ants	Apodidea	Fresh water fish	Amphibians	Reptiles	Land birds	Mammals
Species	72	97	51	187	24	25	39	179	51
Endemic species (%)	—	—	—		7 (29%)	12 (48%)	13 (33%)	1 (1%)	3 (6%)

Data refer to the area between 32–40°S. Data for amphibians, reptiles, land birds, and mammals from Arroyo et al. 1998. Data for Asilidae compiled from Artigas 1970; butterflies: Peña and Ugarte 1997; ants: Snelling and Hunt 1975; Apoidea: Toro 1986; fresh water fish: Arratia 1981. See Table 9.2 for data on vascular plants.

The mediterranean-type climate area of Chile shows both high species-richness and high vegetation diversity. The main vegetation types are relict rainforest, evergreen *Nothofagus* forest, deciduous *Nothofagus* forest, sclerophyllous forest, matorral, northern coastal matorral, succculent and inland scrub, montane sclerophyllous forest, montane coniferous forest, and alpine (Arroyo et al. 1997).

Species richness data for Chilean mediterranean-climate region groups, in addition to plants, are shown in Table 9.3. Overall, species richness is higher in the mediterranean areas in comparison with other ecoregions of Chile (data not shown), particularly considering that the southern temperate region is almost twice the size of the mediterranean-climate region. Notable peaks in species richness in the mediterranean area are seen in butterflies, Asilidae, ants, reptiles, plants, and Apoidea.

California

As in other mediterranean-climate regions California has a diversity of ecosystem types. This is in part due to its complex history as well as to topographic diversity. California has relictual forests (e.g., the redwood forests) that are remnants of vegetation types that were much more widely distributed in the past, when summer rainfall was still present. These forests are now restricted to the coastal fog belt where they are able to supplement the rain with fog drip in the summer. California has extensive development of scrubland, chaparral (dominated by many species from the genera *Arctostaphylos*, *Ceanothus*, and *Adenostoma*). In moister sites forests occur dominated by evergreen *Quercus* species. At the drier ends of the chaparral distribution, coastal and succulent scrublands are found. The coastal hills and inland valleys, where not totally covered by agriculture, are dominated by deciduous *Quercus* species and alien annual grass savannas. At higher elevations species-rich montane coniferous forests occur. Thus, there are evergreen and deciduous forests, savannas, and scrublands of various types within the summer-dry climate of California.

The mediterranean climatic region covers an area of about 130,000 km² in Chile and 250,000 km² in California (Arroyo et al. 1995), with about 2500 and

4240 plant species, respectively. The Chilean flora has less within-genus diversity and a lower representation of annuals. A ful one third of the California flora is endemic, whereas the Chilean mediterranean climatic flora is about 25% endemic, reflecting an apparently wider habitat distribution of many Chilean plants in contrast to their Californian counterparts (e.g., gamma diversity for birds is greater in California than it is in Chile) (Cody et al. 1977). The Chilean flora has only about 16% annuals, whereas California has 30%. This difference may relate to the more equable climate of Chile.

Summary of Biodiversity Patterns in Mediterrean-Climate Regions

All of the mediterranean-climate regions of the world are species-rich with a high degree of endemism, particularly for plants, but less so for birds and mammals. These differences are due to the degree of natural dispersability of these latter organisms. Further, these areas have very high numbers of rare and threatened species, as well as have high biome diversity, including grasslands, shrublands, woodlands, and forests. Most of these regions are rich in species for certain key genera, although Chile seems to be an exception to this trend. There are also differences in the proportional species in various life forms (e.g., California has a larger fraction of annual herbs in the flora than does Chile).

As discussed later, the biota of Chile, California, and the Mediterranean Basin have more generalized pollinator and disperser relations than does the biota of South Africa and Australia. This has implications for community stability under global change.

In summary, these areas are similar in their high species and ecosystem richness but they differ in a number of important properties relating to life-form richness and biotic interactions that will be influenced by global change.

Climatic Patterns, Climate Change Scenarios, and Impacts on Biota

Australia

Under the current climatic regime, considerable variation in rainfall amount and distribution occurs across the mediterranean-climate zone of Western Australia (Hobbs 1992). Rainfall declines from about 1500 mm/year in the extreme southwest (Gentilli 1989) to about 250 mm at the eastern edge of the area, which could be classed as mediterranean, and the length of the summer drought increases correspondingly.

Australian and South African mediterranean climates differ from those of the other mediterranean regions (i.e., the Mediterranean Basin, California, and Chile) by having summer rainfall as a common, if unpredictable, occurrence (di Castri 1981). The current winter rainfall regime has been in effect for only the last 2.5 million years. During that period there have been numerous warm wet/cold dry fluctuations (Churchill 1968; Bowler 1982; Lamont et al. 1984).

Annual variations in rainfall amounts and distributions, coupled with relatively short historical records, make the detection of any long-term trends in rainfall in the mediterranean region of Western Australia difficult (Saunders and Hobbs 1992). Pittock (1988) indicated a possible decline in winter rainfall of 2.9–4.8% per decade between 1913 and 1986 for southwestern Australia. It is uncertain whether these changes result directly from changes in regional climatic patterns, or from an interaction with changing land cover as a result of vegetation clearance for agriculture (Smith et al. 1992; Lyons et al. 1993; Whetton et al. 1994).

Scenarios of climate change, in response to CO_2 increases, suggest temperature increases of between 0.5 and 1.6°C by 2030 for southwestern Australia (Whetton et al. 1994). Five climate models used for this study all indicated an increase in summer rainfalls and small (not necessarily significant) decreases in winter rainfall. Increases in summer rainfall ranged from 9 to 33% and decreases in winter from 2 to 14%.

The impacts of any likely climate change on biodiversity in southwestern Australia are hard to predict because of a lack of basic information on environmental tolerances of species and because the altered and fragmented nature of many ecosystems is likely to inhibit or modify any possible biotic response to climate. Any changes in current rainfall patterns are, however, likely to have profound effects on land use in the wheatbelt region, with follow-on impacts on biodiversity. Reductions in winter rainfall are liable to render large areas increasingly marginal for agricultural production, potentially leading to their abandonment. Whether this will result in the reinstatement of native vegetation without management intervention is debatable (Yates and Hobbs 1997). Increased summer rainfall may lead to increasing erosion and runoff unless extensive revegetation has occurred. Increased summer rainfall may also favor the increased spread of the disease *Phytopthora cinammomi*. Increased temperatures are likely to mean a higher incidence of extreme temperatures, with implications for the biota. For example, extreme summer temperatures have been observed to lead to tree deaths.

Mediterranean Basin

GCMs are predicting a warming of about 2°C in winter, and from 2 to 3°C in summer for the Mediterranean regions by 2100 (Palutikof and Wigley 1996). Changes in total rainfall and temporal distribution are more poorly understood, but current predictions suggests that there will be slightly increased precipitation during winter and a decrease of summer precipitation by 5–15%. There is already evidence of decreased cloudiness and precipitation in the central and western Mediterranean Basin (Maheras 1988) and increased temperatures over the last century (Piñol et al. 1998). The latter study showed an average increase of mean annual temperature of 0.10°C per decade for the period 1910–1994, and an increase of 0.3°C per decade during the last 20 years.

The effects of climate change on biodiversity in the Mediterranean Basin, and for mediterranean-climate regions in general, are expected to occur based on the following grounds. First, increased evapotranspiration will lead to increased water stress that will bring changes in community composition as species assemblages are reorganized according to the new environmental conditions. Second, an increase in winter temperatures is expected to affect biodiversity in the following ways.

First decrease or loss of relict temperate species present only in marginal habitats (north-facing mountain or cliff slopes): These species are more likely to be affected mostly by a decrease in water availability than by increased temperature directly, so predictions need to be linked to rainfall scenarios. These species might also be affected by decrease in frosts because many of them require chilling for germination. Second decrease or loss of species with late season germination because of large risk of establishment failure brought on by lack of water, and because of space preemption by early germinating species that will establish larger biomass throughout winter. Third consequences for recruitment of woody seedlings are uncertain. Woody seedlings are sensitive to competition for water (e.g., recruitment is rare and limited to "good years") (Bacilieri et al. 1993).

Fourth, warmer climate will push closed canopy evergreen-oak forests northward (Piñol et al. 1995) and ultimately, will shift forest communities in the south toward low-cover desert shrublands with the associated consequences on the water and carbon fluxes.

Changes in human activity and global warming will overall reduce plant diversity in such sensitive areas as temperate/mediterranean and mediterranean/desert ecotones. The abundance of endemic species will decrease with increasing aridity and there will be an increase in the abundance of annual species (Holzapfel et al. 1992). Studies in the southwest of the Iberian Peninsula show that species living in the dry end of the mediterranean climate have a low dispersal capacity that will prevent them from a successful northward displacement as rapid warming takes place (Merino et al. 1995). Other species, however, such as some evergreen trees of the genus *Quercus* evidently have a much faster dispersion capacity as they were found from southern Spain to Sweden during the northward migration that took place in the first half of the Holocene (Birks 1990).

South Africa

GCM scenarios for the Cape region, with a doubling of CO_2, predict a rise in average temperature of 1–2.5°C, with uncertainty on the direction of change in precipitation, in total or seasonally (Hewitson and Crane 1996, Hudson 1997).

The possible consequences of climate change for biome limits in southern Africa have been explored by correlative studies (Ellery et al. 1991), primarily in summer rainfall areas. Euston-Brown (1995) studied the causes of biome boundaries near the eastern limits of the fynbos biome using both correlative methods and translocation experiments across altitudinal and geological gradients. He showed that several functional groups, characteristic of fynbos, are at their climatic limits. Global warming or rainfall reduction would therefore lead to the retreat of fynbos elements to higher, cooler elevations. The steep mountainous terrain would provide a topographic refuge for sensitive groups on mesic pole-facing aspects but they would be eliminated from lower elevations and from north-facing slopes.

The most sensitive functional groups were members of the Restionaceae and Proteaceae. Ericoid shrubs were the least sensitive and survived and grew when transferred to lower, warmer elevations. Functional groups representative of adjacent grassland and closed thicket formations (broad–leaved, non–fire-prone shrublands) survived transfer to fynbos soils at higher altitudes, suggesting that disturbance, rather than direct climatic or edaphic controls, limit their current distribution. Euston-Brown's (1995) study therefore suggests that the fynbos biome will shrink if effective precipitation decreases due to global warming. Patches will survive at higher elevations and on mesic slopes. The vegetation-replacing fynbos would likely depend more on the interaction between disturbance regime and climate than on climate alone. If fires become more frequent, then grasslands are likely to spread from the east to occupy fynbos areas (Trollope 1973). If fires are less frequent, then fynbos will be replaced by non–fire-prone thicket vegetation (Manders et al. 1992). Correlative studies of succulent Karoo species richness and growth-form diversity with climatic parameters show that both are positively correlated with climatic heterogeneity and rainfall evenness (Cowling et al. 1994). The extent of succulent Karoo is therefore most sensitive to a change in the seasonality of rainfall. Both biomes are geographically isolated at the southwestern tip of Africa. There is no refuge lying at cooler southern latitudes comparable to, say, South America. If global warming causes significant temperature change, fynbos remnants can only retreat to higher elevations in the mountains or more mesic slopes or soils. This must place many fynbos species at risk of extinction. Succulent Karoo occupies a flatter landscape, especially along the west coast. In the absence of topographic refugia, the entire biome is severely threatened with extinction. Both biomes are intimately associated with winter rainfall climates. If global change leads to rapid change in rainfall seasonality, large-scale extinction could be expected.

In contrast to other mediterranean regions, the mediterranean shrublands of South Africa and Australia are characterized by many plant species with specialized insect or vertebrate pollinated flowers and hundreds of species dependent on ants for seed dispersal (Johnson 1992). Given the short lifespan of many plant species, and therefore their high dependence on seeds, these systems must rank as among the most vulnerable anywhere to cascading extinctions caused by loss of mutualist partners (Bond 1994). It is still poorly known whether climate or land-use change will result in the loss of such partners (Bond 1995).

California and Chile

The projections for climate change for Chile and California differ, due to differences in land–sea ratios (Trenberth 1993). The mediterranean-climate latitudes of California are expected to warm nearly twice as fast as these regions in Chile. Projections of temperature change with a CO_2 doubling will average 3°C warmer in the north and less than 2°C warmer in the south, at latitudes of around 40°C.

In an analysis of the climate change expected for California with a doubling of CO_2, Knox (1991) predicted a rise in average temperature of 2–4°C, precipitation changing between plus to minus 20% of current values, snow level rising 100 m/degree increase in temperature, increased temperatures of shallow water bodies, and a poleward shift of storm tracks.

One of the distinctive features of coastal Californian and Chile is fog. Fog results from the condensation of moisture contained in the air moving in from the Pacific as it hits the cold water surfaces along the coast caused by upwelling. Subsiding air in these regions forms a cap and results in the fog layer being close to the ground. Bakun (1990) predicts that climate warming will cause a greater differential temperature between land and water and hence increase the winds that lead to upwelling, thus increasing fog amount. McBean (1996) cites work that predicts an opposite result due to effects on subsiding air; however, with either an increase or decrease there would be important effects on the fog-requiring vegetation of these regions. There has been a trend over the past decades for an increase in wind stress off of the Californian coasts (Hsieh and Boer 1992).

The fates of the biota of California and Chile, as with other mediterranean-climate regions, are uncertain because of the many interactions and feedbacks that will occur between climate changes per se and with other global changes. Botkin et al. (1991), however, make the following predictions for California based on a doubling of CO_2 concentrations, noting the difficulties of making predictions based on the topographic diversity of this region:

Increasing temperatures and decreasing soil moisture will lead to increased fire frequency in the short run with an acceleration of vegetation change. In the longer run there will be a decrease in forest biomass. Change in forest composition is due to such factors as the loss of the winter chill conditions necessary for germination and growth of Douglas fir, as one example. Other impacts may include a change in tempo of recycling due to increased temperature, and movement of forests upslope from present locations.

Knox (1991) concluded that 20–50% of the area currently occupied by a given vegetation type would be preempted by new ones. Other analyses for California are given in Keeley and Mooney (1993), Westman and Malanson (1992), and Malanson and O'Leary (1995).

Invasive species will likely be more pervasive and numerous with the community disruption that will accompany rapid climate change. If there is a shift in the seasonality of rainfall, we could see changes in abundance of C_4 weeds that need summer rain or irrigation.

The biogeographic model MAPPS (Neilson 1995) projects vegetation response to different climatic scenarios for doubled CO_2, utilizing growth form responses to climate rather than responses of whole biomes or of individual species, as given in Malanson and O'Leary (1995). Figure 9.1A (see color insert) gives the current aggregated vegetation distribution in California according to the MAPPS model (derived from simulations explained in Neilson et al. (1998) for three different Global Circulation Models [OSU, GISS, and the Hadley Center Model (HADCM2SUL)]. The Hadley model, which incorporates sulphate aerosols into the simulations, has the greatest impact on a doubled CO_2 world, with current desert systems becoming grasslands and shrublands turning more into woodlands (i.e., an apparently more mesic California). The discrepancy in responses of a single vegetation model to different global climate-change scenarios highlights the fact that we are still in a very early stage of making predictions of what the future will bring.

Summary of Expected Climate Change and Impacts

All mediterranean-climate regions will show an increased temperature with a doubling of atmospheric CO_2 concentration. The amount of change will differ among these regions in relation to the prevailing water-to-land mass. The temperature changes will have a large impact on the hydrology, and hence biology, of these regions, including impacts on snow-melt periods, which are important biologically in these summer dry areas. Most regions will see an increased amount of precipitation; however, how these changes will play out seasonally is not agreed upon either for a given area or among areas. Understanding changes in seasonality of preciptiation in these arid regions is crucial for making realistic predictions of climate impact on the biota. In addition, how coastal fog patterns will be affected is not agreed upon, yet it has large potential importance in controlling biotic patterns. Fire climates will be affected with increasing temperatures and evapotranspiration, as discussed later, with differing biotic consequences in each region.

Direct impacts of climate change on the biota of mediterranean-climate regions are predicted to be profound. Mediterranean-climate regions are particularly vulnerable because of their high degree of endemism and localized species distributions. Temperature effects alone will have a big impact

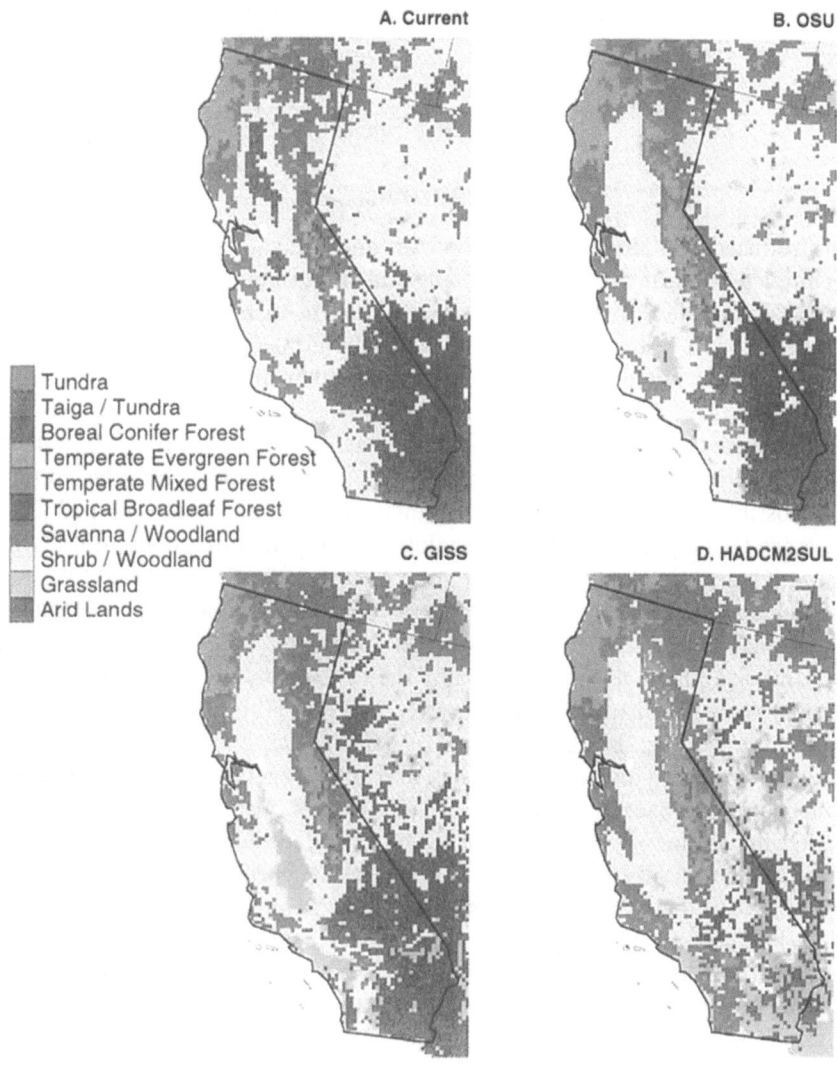

Figure 9.1. Modeled responses of California vegetation to different climate model projections for a doubled CO_2 world. (A) Present vegetation, (B) response to the OSU climate model, (C) to GISS model, and (D) to the Hadley Center model (HADCM2SUL) (see color insert).

on biotic distributions in mountainous regions, and along north–south gradients, as has already been observed for butterflies. Temperature-driven changes in evapotranspiration will have direct and indirect effects (changing fire climates) on the biota. The high numbers of narrowly distributed species in these regions will certainly be threatened by a rapidly changing climate. Shifts in the many ecotones between the diversity of biome types will be expected. These general trends will be influenced by the other global changes discussed later.

Invasive Species

Australia

Introduced predators, particularly foxes and cats, are a major cause of the complete or near extinction of many Australian marsupials (Burbidge and McKenzie 1989; Friend 1990). Indeed, Western Australian forests have provided the last refuge for a number of marsupial species, presumably because fox numbers remained lower there than in other areas. Predator control is now practiced in many areas, with obvious success indicated by increases in abundances of native mammals (Kinnear et al. 1988; Friend 1990). Poisoning of introduced predators using 1080 (sodium monofluoroacetate) poison is possible because the resident mammal populations have developed a tolerance to the poison through coevolution with plants that contain it (Twigg and King 1991). A large-scale program of fox eradication, entitled "Western Shield," promises to allow recovery or reintroduction of endangered mammals in many forest and nonforest areas.

Invasive plants are another major threat to many mediterranean-climate ecosystems in Australia (Humphries et al. 1991; Humphries 1993). Individual invaders [e.g., bridal creeper (*Myrsiphyllum asparagoides*)] have the potential to crowd out native species and alter vegetation composition and structure. Herbaceous species, especially grasses, also have the potential to alter fire regimes by changing the structure and availability of fuel (D'Antonio and Vitousek 1992). About 1500 species are currently naturalized across all of Australia, with 848 species recorded from Western Australia (Keighery 1991). There are 220 species that are recognized as noxious weeds across Australia (Parsons and Cuthbertson 1992), and many are problems in native ecosystems. Additional species, such as invasive pine species (Richardson et al. 1994), may become problems in the future (Hobbs 1993a).

The relatively high levels of invasion by mammals, plants, and fish in Western Australia undoubtedly result mainly from the deliberate introduction of these organisms for agriculture, horticulture, or sport. New species of plants continue to be introduced (Rejmánek and Randall 1994; Keighery 1995), and some rapidly extend their range (Dodd and Moore 1993). Australia continues to import plant species without due consideration for the potential threats of invasiveness. For most animal groups, the rates of introduction are now probably slower, although the invasion of a southwestern Australian river system by *Perca fluviatillis* (Hutchison and Armstrong 1993) indicates that the problem continues.

The introduced fungus *Phytopthora cinammomi* is a major factor influencing forest and shrub ecosystems in Western Australia and in other parts of the country (Dell and Malajczuk 1989). Depending on the severity of attack,

Phytopthora causes the loss of overstorey and understorey plant species. A large number of plant species are susceptible to the fungus, and it is likely that others are impacted by the changes in microclimate and local moisture relations caused by loss of overstorey cover. This fungus and other diseases are also important in other nonforest communities, including the floristically rich kwongan heath (Wills 1993; Wills and Keighery 1994).

Mediterranean Basin

Most of the invasive species in the Mediterranean have been human-facilitated or intentionally introduced, either associated with human settlement or agricultural trade. Thus, invasions will probably continue in the future as barriers among the European Union countries disappear and human migration increases, particularly from the south to the north.

In southern France about 60 species belonging to 17 different families are presently naturalized (Guillerm and Maillet 1982). Four families are strongly represented: Asteraceae, Poaceae, Amaranthaceae, and Solanaceae. These species originate from temperate, mediterranean-climate, or subtropical regions of the Americas and Asia.

An interesting trend concerns the increase in subtropical and tropical species. This increase is strongly linked to irrigated land where these species can thrive because their lifecycle is offset relative to native agricultural weeds that are predominantly winter annuals. Indeed C_4 metabolism is a common feature among exotics.

It has been suggested that abandoned fields, which are disturbed and relatively fertile, are highly invasible because of their simple food webs and trophic interactions. As an example, *Opuntia ficus-indica* has been reported to invade abandoned fields in the northeast of the Iberian Peninsula (M. Vilà, personal communication). This species may be further favored by predicted increases of atmospheric CO_2 (García de Cotaza and Nobel 1990) and water stress (Luo and Nobel 1993). Abandoned fields in southern France are invaded by two alien woody species—tree of heaven (*Ailanthus elegantissima*) and wattle (*Acacia dealbata*) (Blondel and Aronson 1995)—although revegation after field abandonment is largely controlled by the seed availability from the surrounding native woody vegetation.

In general, plant invasions seem to be more frequent and often successful, and follow more predictable patterns than do animal invasions (di Castri 1991). The introduction of freshwater fish species has been the most numerous among vertebrates, yet the effects on local populations and ecosystem functioning have been poorly studied (Crivelli 1995). To the present, 70 introduced exotic or translocated fish species and subspecies have been reported in the northern mediterranean region. All of them have been intentional human introductions (Crivelli 1995). Continuous species introductions, wetland depletion, and increased water pollution, by a more resource-demanding human population, will lead to an overall impoverishment of native populations of freshwater fish species in the next decades.

South Africa

Alien invasive plants, mostly trees or shrubs from other mediterranean regions, pose by far the greatest immediate threat to the Cape flora (MacDonald 1989, Richardson et al. 1996). Pines and Hakea species (Australian Proteaceae) have invaded 14% of the remaining areas of mountain fynbos, whereas Australian *Acacia* species have invaded up to 68% of the remaining area of lowland fynbos. The invasive species overtop native shrubland species, suppressing most of them and leading to local extinction and a marked reduction in biodiversity (Richardson et al. 1996). Among the most significant changes in ecosystem processes is the reduction in runoff from invaded catchments estimated to be between 30 and 70% lower than uninvaded fynbos (Le Maitre et al. 1996). Alien species also alter fuel conditions, causing more intense fires that in turn cause accelerated soil erosion after fire (Richardson et al. 1997). The invasive acacias accumulate nutrients in the surface soil layers to the point where it becomes very difficult to restore fynbos species adapted to nutrient-poor conditions in areas cleared of aliens (Stock and Allsopp 1992).

Alien invasion is much less significant in the succulent shrublands; however, *Prosopis* (mesquite) is a locally important invader along water courses with species of Chenopodiaceae and *Opuntia* spreading locally in dryland areas (Richardson et al. 1997).

In South Africa, the threat of invasive trees to water supply has led to extensive alien clearing using mechanical methods supplemented by release of biological control agents. About 60,000 ha of the 1.15 million ha of fynbos catchments are treated each year by felling dense stands of *Pinus* and *Hakea* species. Efforts at controlling the problem include the deployment of thousands of previously unemployed workers in an ambitious attempt at large-scale environmental restoration through alien removal. Economic analyses indicate that savings in terms of increased water run-off make the expense of alien clearing cost-effective (van Wilgen et al. 1996).

Chile

Alien invasive species are widely distributed altitudinally and latitudinally in the Chilean mediterranean-climate area today. Marticorena (1990) provides a figure of 657 introduced taxa (i.e., species, subspecies, and varieties) for all of continental Chile, but this figure is probably now closer to 700 (Marticorena, unpublished data). About two thirds of these species are in the mediterranean-climate area, with more than 400 in the metropolitan region (Santiago–Valparaiso district) where Chile's population is concentrated. As in other mediterranean-climatic areas, an overwhelming number of the introduced species are of Old World origin. Most are herbaceous, with many annuals. Important sources of invasive species in central Chile are cultivated plants in botanical gardens, escaped medicinal plants, and forage plants (Matthei 1995). Many of these intentionally introduced species today are now serious weeds of agricultural crops and common inhabitants of disturbed areas. In structurally undisturbed vegetation, the relatively low percentage of native annuals in the flora of central Chile (Arroyo et al. 1995) is a feature that could engender greater susceptibility to invasion by exotic species than in climatically similar California. A surprising number of weedy species in the far north of Chile are of subtropical origin (Matthei 1995) and could penetrate further south with global warming, especially where irrigation agriculture is practiced, as is the case in the central valley. Increased trade and traffic between the countries of the southern Pacific and the lowering of trade barriers among the countries of Mercosur are expected to increase the number of invasive species establishing in central Chile in the coming century. A number of the world's worst weeds occurring in New Zealand and Australia, Argentina, and Uruguay have not yet been reported in Chile (Holm 1997).

California

There are more than 1000 established alien plants in the Californian flora (Rejmanek and Randall, 1994) representing about 20% of the flora. The numbers of new plant invasives may be slowing in California, although this is certainly not the case for marine organisms (Cohen and Carlton 1998). There are many more invasive species than there are endangered and threatened ones in California (Hobbs and Mooney 1997), even though California has one of the highest numbers of endangered species in the United States. Many of these invasive species have had large ecosystem impacts in terrestrial, freshwater, and coastal marine ecosystems (Dukes and Mooney 1999).

Summary of Invasive Species

All of the mediterranean-climate regions are heavily impacted by invasive species with a prognosis of greater numbers of invasives becoming established in the future with land-use and climate change. These areas differ in that in some regions (i.e., South Africa) invasive shrubs and trees are doing the most damage, whereas the herbaceous plants are the greatest pests in Chile and California. Australia, has many invasive mammals also causing damage.

Land-Use and Cover Change

An overview of the impact of land use on the biodiversity of mediterranean-type ecosystems has appeared (Rundel et al. 1998). We will address potential land-use changes in each of these regions separately.

Australia

Producing future scenarios requires an understanding of the changes already underway because of trends in land use, particularly in relation to urban growth, forestry, and agricultural development. Urban development in the Perth metropolitan region is spreading rapidly as the population of the city grows. Particular threats to the remaining biodiversity come from altered fire regimes, weed invasion, and, in the case of wetlands, altered hydrology and eutrophication. Arson is a major cause of bushfires (Dixon et al. 1995; Pigott and Loneragan 1995).

Forest systems of Western Australia have been managed in a variety of ways for timber production over the past century (Dargavel 1995). The most important impacts on forest systems are forest management practices, in particular timber harvesting, fuel reduction burning, and the impacts of invasive species, including the introduced pathogen, *Phytopthora cinammomi*. The jarrah (*Eucalyptus marginata*) forest is currently subjected to widespread short-rotation fuel reduction burning, which has evolved to reduce the risk of destructive wildfires. Controversy surrounds the questions of whether such a burning regime is effective and whether it has adverse impacts on the forest ecosystem (McGrath 1985; Tingay 1985; Underwood et al. 1985). It has been claimed that the current regime mimics the regime prevailing prior to European settlement (Hallam 1975; Burrows et al. 1995), although the evidence for this is not compelling, and basic knowledge on Aboriginal fire regimes is scant (Williams and Gill 1995). Among other factors future scenarios depend on the interaction between *Phytopthora* and changes in rainfall amounts and distributions, the long-term impacts of continued fuel reduction burning, and the success of current attempts to control feral predators.

In agricultural landscapes, profound changes in nutrient, energy, and water fluxes have occurred as a result of the replacement of perennial vegetation with predominantly annual crops and pastures (Hobbs and Saunders 1993). Changed energy fluxes and wind regimes result from the change from tall evergreen vegetation to the alternating short annual vegetation and bare soil associated with agriculture (Hobbs 1993b). In addition to influencing local microclimates, these changes may have regional scale implications. Rainfall patterns in southwestern Australia have changed over the past century, such that annual amounts have declined in the agricultural areas, but increased in the adjacent inland areas (Pittock 1988). Smith et al. (1992) have suggested that these changes could be associated with the effects of changing albedo on cloud formation. Evidence for this comes from a phenomenon frequently observed in satellite images for the area, whereby clouds are found only above the large areas of native vegetation remaining in the wheatbelt and to the east of the line between cleared and uncleared land.

The landscape has changed from one in which considerable internal nutrient redistribution occurs, but little net loss is evident, to one in which large inputs of fertilizer occur, and large losses of nutrients result from wind and water erosion and harvesting (Hobbs 1993b). In the fragmented wheatbelt landscape, remnant native vegetation on low-nutrient soils are set in an agricultural matrix with elevated nutrient levels and increased mobility of topsoil. Loss of topsoil by wind and water erosion, especially during episodic storm events, is a major problem over much of the wheatbelt and can bring nutrient-rich soil into remnant areas and hence alter the nutrient regime. This has important implications for weed invasion, which is promoted by nutrient increase (Hobbs and Atkins 1988).

Rising water tables also threaten biodiversity. Within uncleared vegetation, plant architecture, soil structure, and soil biotic activity channel and redistribute rainfall, and most is utilized and transpired or evaporated, with little runoff or transfer to the water table (Nulsen et al. 1986; McFarlane et al. 1992; McFarlane et al. 1993). Removal of native vegetation and replacement with annual crops and pastures results in considerably less efficient use of rainfall, increased runoff, and greater inputs to the water table. This has led to rising water tables, with rises of 50 cm/year or more recorded in some areas (George et al. 1995) Beneath much of the wheatbelt are considerable quantities of salt stored at depth, resulting from a long history of input in ocean-derived rainfall coupled with poor drainage (Hingston and Galaitis 1976). As water tables rise, the stored salt is mobilized, and, wherever the water table reaches the surface, salinization occurs, posing a threat to a significant proportion of the wheatbelt (McFarlane et al. 1993; Nulsen 1993; George et al. 1995). Although native vegetation within remnants still uses water more efficiently than the surrounding crops and pastures, the overriding influence of the agricultural matrix means that water tables are rising even under large remnants (Salama et al. 1994). Many low-lying areas have already been affected, and most of the previously fresh water lakes have gone saline (Froend et al. 1987; Froend and McComb 1991; Hobbs et al. 1993). In some areas the water table is rising so quickly that remnant vegetation will be destroyed within the next 5–10 years, whereas the timescale is longer but the problem is no less insidious in lower rainfall areas.

The processes already established will in large part determine future trends in biodiversity. Extensive habitat removal and fragmentation over the past 50–100 years is likely to have created a situation where some species have been lost already, but others remain in populations that are not viable in the long term. The "extinction debt" is thus likely to result in continued loss of species from the region, unless restorative measures are undertaken. Future scenarios for the agricultural area thus depend in large part on the type and magnitude of human response to the changes discussed earlier. Maintenance of a "business as usual" farming system of mixed annual cropping and pastures will result in continued degradation of both agricultural and conservation lands through rising water tables, salinization, and erosion. On the other hand, extensive modification of farming practices to include considerably more perennial vegetation (e.g., alley farming, tree plantations, and perennial pastures) may allow the halting or reversal of current trends.

Mediterranean Basin

The decrease of farming and agricultural practices in the northern Mediterranean Basin, as a result of more "efficient" production systems and growing urbanization, is leading to an abandonment of farmland and subsequent encroachment of woody vegetation (Moreno et al. 1998). This phenomenon started at the beginning of the century and is now being accelerated by the new Common Agriculture Policy of the European Community. During the period 1965–1985 rangelands and croplands receded by 7% and 9%, respectively (Le Houréou 1992). This is leading to an increase in homogenization of the landscape and loss of highly diverse and species-rich systems, although increasing urbanization will counteract the vegetation expansion. This differs from the situation in Australia, where fragmentation is the cause of loss of diversity—here, diversity loss results from the reduction in management input in systems with a long history of human management. Unique cases of loss of highly diverse landscapes are the loss of managed systems (e.g., the "dehesas" in Spain and "montados" in Portugal) (García et al. 1994), and the mosaic landscapes that intermix natural and managed patches in France and Italy. All of these systems have high species and landscape diversity.

In general, faunal diversity is also higher in regions where the landscape offers some patchiness with open areas, shrublands, and forests, and biodiversity decreases in landscapes dominated by closed-canopy forests (Tellería et al. 1992). With the continuous abandonment of cropland (e.g., the wheat fields in Spain that maintain large and diverse bird populations) (Diaz and Tellería 1994), the future avifouna diversity is predicted to decline.

Other types of activities are increasingly causing major anthropogenic disturbances (e.g., road construction, urban development, and forest clear cutting). Natural disturbances by wild mammals are increasing (e.g., the case of wild boar whose populations are building up in many forests of the western Mediterranean).

In general, changes in management strategies that alter landscape units and its structure will bring the largest changes in biodiversity (birds, Tellería et al. 1992). Increases in human population and tourism will have the strongest impact on the flora and fauna of islands with a predicted overall impoverishment (Balearic Islands, Mus 1995). Finally, increasing water demand for an expanding urban population and for use in more intensive agroindustrial farming is predicted to lower water tables in the more arid regions of the Mediterranean. This lowering of the water table affects the dynamics of temporary ponds and streams that are causing a decline of amphibian populations in the central Iberian Peninsula (Rodríguez-Jiménez 1988).

Trends in the European Mediterranean contrast with those of North Africa and Near East, where arid steppes and rangelands are being cleared for grain production. During the period 1965–1985 forest and shrublands receded by 3% and cropland expanded by 5% (Le Houréou 1992). In the next decades the Mediterranean coast of North Africa will also see a fast-growing tourist industry moving on the coastal areas and replacing natural habitats and farmlands, with a consequent loss of habitat and hence of biodiversity.

Wetlands and coastal dunes are the notable exception to the stability in regional species diversity. Indeed, the Mediterranean coast is subjected to ever-increasing pressures from urban and tourism-oriented development. Wetlands and their species are under serious threats due to draining, spraying with pesticides, and repeated disturbance (e.g., by recreation vehicles) (Naveh and Liebermann 1994). As a result, numbers of declining or endangered species are higher for coastal regions (Barbéro 1989).

Many crucial habitats in the Mediterranean Basin, in relation to biodiveristy, are very vulnerable to changing land-use patterns. River deltas and salt marshes in the Mediterranean Basin are focal breeding areas, and in many cases they are internationally important stopover points in the Afro-Paleoarctic migration system. This migration route goes from as far north as Sweden, over the Mediterranean region, to East Africa and the Middle East (Bairlein 1991). These areas are hotspots for animal biodiversity, especially waterbirds (Van der Hane and Van Den Berk 1994). They are particularly vulnerable because climate change will bring sea-level rise, salinity increase, and changes in the rates of sediment deposition (Sánchez-Arcilla et al. 1996). This will be in addition to the human impacts by upstream dams, catchment erosion, and organic pollution that are expected to increase as human population increases.

South Africa

About one third of the area occupied by fynbos and one quarter occupied by succulent Karoo has been transformed (Rebelo 1997). Fynbos covers both steep mountainous terrain and lowland areas. Only 7% of mountain fynbos has been transformed, largely by afforestation, but 50% of lowland fynbos and more than 90% of renosterveld have been converted to cropland or plantation forestry. Fifteen percent of the total fynbos area is conserved, largely in the steep mountainous areas. A little more than 2% of succulent Karoo falls into protected areas.

Many areas, especially along the coast, are undergoing rapid urban development. Farmlands outside protected areas have traditionally been used for extensive livestock farming. This form of farming has generally been considered relatively benign in terms of biodiversity impacts. Innovations in farming, however, can cause extremely rapid land-use change. For example, the introduction of fixed-point irrigation from aquifers caused proliferation of crop farming in sandy west coast areas previously of very low agricultural value. Ostrich farming in the succulent Karoo has also led to very rapid habitat transformation of marginal farming land rich in native plant species. These unpredictable changes to the rural status quo indicate the critical importance of protected areas for long-term conservation.

South Africa is a rapidly urbanizing society and is one of the few countries in Africa south of the Sahara with more than half its population living
in cities. This may help reduce pressure for transformation of natural vegetation for agricultural purposes. The spread of cities, however, is already
threatening many sensitive areas. Both biomes are sensitive to such localized
development because of the many localized endemic species.

California and Chile

Habitat loss in California has been substantial, and the growing human
population insures that it will increase. Virtually all of the wetland, riparian,
annual grassland, coastal sage, and vernal pools have either been destroyed
or substantially modified (Hobbs and Mooney, 1997). Rivers have been
dammed and diverted, marshes have been filled, and the coastal plains have
been extensively developed. There does seem to be a movement to reverse the
trends of the past with efforts to restore riparian areas in particular and to
protect and restore wetlands. A massive effort to restore the biotic systems
of the San Francisco Bay is underway that involves a coalition of farmers,
conservationists, and fisherman. There are similar innovative plans to develop
the remaining coastal sage areas sustainably.

In Chile, habitat modification has been more modest, influencing primarily the central zone where the population is most concentrated. In the
mediterranean area of Chile 24% of the land area is dedicated to intensive
agriculture, 13% to plantation forestry, and 1% urbanization, such that close
to 40% of the land area is subjected to intensive use (Arroyo et al. 1998).
Perhaps as much as 50% of the remaining shrublands are subjected to
grazing, although not necesarily of an intensive kind. Pine and eucalyptus
plantations that now cover over 2 million hectares (CONAF 1997) are particularly important and are centered over the area of maximum tree species
and woody plant endemism (Armesto et al. 1988). Unlike in all other
mediterranean-type climate areas, natural lightning-ignited fires do not occur
in central Chile, and fire is probably not a natural factor that has molded
adaptive responses (Armesto et al. 1995). From the 1970s to the 1990s,
around 15,000–49,000 ha/year of natural vegetation in Chile burned (INFOR
1997), with most of the fires occurring in the mediterranean area. Such unnat-
ural fires will both affect local plant and animal biodiversity and provoke
further spread of fast-growing invasive species in central Chile. The subarid
regions to the north have been degraded considerably by overgrazing (Hajek
et al. 1990). With economic growth and the development of the south, in particular, we can expect a much accelerated habitat modification.

The exploitation of forests in California and Chile has had a very different history. For example, in California, a long period of the utilization of
coastal redwoods has left less than 5% of old growth, and this is principally
in parks. Even the Sierra Nevada has been gradually logged of old growth in
the National Forests. The Giant Sequoia was saved from exploitation princi-

pally because it was not economically feasible to harvest. Plantation forestry has been practiced in the northern part of California, but principally with native species. Early plantations of *Eucalyptus* in southern California never became commercially viable.

Agriculture is much more extensively developed in California than it is in Chile. Agriculture is very resource intensive, utilizing 80% of all of the available fresh water of California. Large water projects subsidize agricultural activities, often to the detriment of the long-term system sustainability, as happened in the irrigation of the west side of the San Joaquin valley. These water projects have altered the total hydrology of the state in a major way. Further water-intensive crops (e.g., rice) are grown in a climate that could not sustain it without the water that has been diverted from aquatic ecosystems.

Summary of Land-Use Change

Overall, land use will be by far the most important factor for the future of biodiversity in the Mediterranean. Patterns of extinction, therefore, are likely to be linked with political and socioeconomic factors that will determine coastal development, agricultural policy, and forestry. Although the mediterranean-climate regions of the world share a common set of ecosystem types and, to a certain extent, future climatic regimes, they differ greatly in present and projected population sizes, growth rates, and densities, as well as GNPs, all of which will have a large impact on land use, biotic invasions, atmospheric composition, and, hence, biotic impacts (Hobbs et al. 1995). Further, these regions differ in the amount of local resource extraction, again having differential effects on biotic systems. Even if their future climates were common, therefore, the prospects for biotic systems will be quite different. In California, for example, population growth is pushing urbanization into areas that were formerly farmland; in turn, farming is utilizing more marginal lands. In other mediterranean-climate regions we are seeing abandonment of farmland and regression to natural systems. The attractiveness of mediterranean climates will continue to make these areas grow in population and economy disproportionately; hence, the social dimensions of change will exceed those of many other regions of the world.

Other Global Changes

Increased Atmospheric CO_2

Monoliths of Mediterranean grasslands exposed to elevated CO_2 show small NPP responses because low nutrient availability limits the extent of the production response (Navas et al. 1995). Elevated CO_2, however, is expected to change species composition because of the highly variable interspecific CO_2 response that has been observed in a variety of Mediterranean species

(Roumet et al. 1996). There were differential species responses in a calcareous grassland in Italy exposed to elevated CO_2 (F. Miglietta, personal
communication).

There is still little evidence on the effects of elevated CO_2 on ecosystems
of the Mediterranean region, but a moderate increase in productivity is
expected largely due to improved water relations (Hättenschwiler et al. 1997).
The study of 14 Mediterranean species collected over the past 240 years
(Peñuelas and Matamala 1990) and the study of *Olea europea* over the last
several thousand years (Beerling and Chaloner 1993) showed a consistent
decrease in stomatal density. Potential water savings can favor late growing
season species, and, hence, drive important changes in community composition, as has been reported for a California annual grassland exposed to elevated CO_2 (Field et al. 1996). The increase in water savings in grasslands may
result in invasions by shrub species, as presently occurs in wet years (Williams
et al. 1987). Likewise, in many cases elevated CO_2 favors N_2-fixing plants, as
in the case of *Lotus* in assemblages of annuals species dominated by *Bromus*
and *Avena* (J. Canadell, unpublished). Thus, drought-tolerant N_2-fixers are
good candidates to expand in a warmer Mediterranean with high CO_2.

Nitrogen Deposition

Nitrogen and acid deposition are important regional global change issues
that affect most ecosystems. Alterations in productivity, nutrient flux, and
trophic changes are documented for the western Mediterranean and are attr-
ributed to increased nitrogen supplies, to a large extent from atmospheric
deposition (Paerl 1993, 1995). Nitrogen deposition impacts plant community
composition by favoring fast-growing species and decreasing overall biodiversity, as seen experimentally elsewhere (Wedin and Tilman 1996). In addition, a large fraction of the endangered species grow in habitats that are
N-deficient (Ellenberg 1993); therefore, increasing eutrophication from
atmospheric deposition may bring further complications to biodiversity conservation (Mooney et al. 1998).

Acid rain has been detected in regions of the Mediterranean where there
are no local sources of acid loading [e.g., the Pyrenees (Spain–France),
Sardinia (Italy), and Patras (Greece)], which indicates the long-range transport of anthropogenic inputs, mostly from northern countries (Glavas 1988;
Camerero and Catalán 1993; LeBolloch and Guerzoni 1995). Although acid
rain does reach southern Europe, soil acidification is less likely to occur than
it does in central and northern Europe because of episodic dust events from
the Sahara. The dust produces red-rains with a high ion concentration that
has the capacity to neutralize acid. In eastern Spain, red rains in rural sites
may account for as much as 50% of the mean annual precipitation, with
enough alkalinity strength to neutralize the input of free acidity of acid precipitation (Rodà et al. 1993).

Overall, the effects of both nitrogen deposition and soil acidification are more of a concern in the northern Mediterranean countries than they are in the southern Mediterranean.

Nitrogen deposition is probably less severe in areas of low population density in other mediterannean-climate regions because air masses move onshore from large fetches of ocean. In closed basins with vehicular traffic, however, there can be sufficient nitrogen deposition to impact vegetation and its diversity. Stewart Weiss (personal communication) describes a loss of biodiversity in grasslands along a nitrogen deposition gradient in the San Francisco Bay area. A rather large shift in composition has been noted in a relatively short time, indicating the great impact that increases in nitrogen can have in the Californian nitrogen-limited ecosystems. In general, in California, as elsewhere, species diversity decreases under increasing N deposition.

Tropospheric Ozone

Levels of ozone two- to threefold higher than those considered damaging for vegetation are measured in the western Mediterranean (Millán et al. 1996). These high levels of ozone of the highly populated Mediterranean coast are transported to 60–100 km inland with the sea breezes almost every day from spring to fall (Millán et al. 1996; see also Klasinc and Cvitas 1996).

One of the most common effects of ozone is a change in species composition because of the high interspecific sensitivity to damage by photochemical pollutants (e.g., ozone). In Israel, high levels of tropospheric ozone have been associated with the general decline of *Pinus halepensis* and the disappearance of the lichen *Xanthoria parietina* from maquis (Naveh et al. 1980). In California, Miller et al. (1982) note that in the mountains near Los Angeles, pines are similarly not competing as well as fir and oaks because of the former's greater sensitivity to ozone injury, particularly in wet years.

Ozone effects are especially damaging for Mediterranean vegetation because drought has been shown to increase species sensitivity to photoinhibition due to ozone exposure, for instance in *P. halepensis* (Wellburn et al. 1996). In fact, the general decline of *P. halepensis* forests in large areas of the Mediterranean has been attributed to the combined detrimental effects of ozone and drought (Klasinc and Cvitas 1996; Wellburn et al. 1996).

Thus, changes in species composition and probably species impoverishment make tropospheric ozone an important factor in altering and diminishing biodiversity. Because it decreases stomatal conductance, elevated CO_2 may ameliorate damage to ozone.

Both coastal California and Chile have ideal conditions for the buildup of tropospheric ozone. They are regions of high sunshine and they have steep coastal mountains that can trap the westward-moving air masses that travel over the ocean to the coastal cities. Tropospheric ozone concentrations have been increasing in the metropolitan areas in Chile, whereas they have been reduced somewhat in California due to pollution control measures. At the same time there has been an increase in tropospheric ozone in interior valley regions as population centers move further from the coast. Air quality in some Californian Central Valley towns is now worse than that of New York, Philadelphia, and Chicago (Lewis et al. 1991). With future warming, and an increase in populations and transportation networks, we would expect to see increasing tropospheric ozone in both regions unless the temperature effect is counterbalanced by technological solutions.

UV-B Radiation

Stratospheric ozone has been declining over the past decades with the consequent increase UV-B radiation reaching the Earth's surface. Near 40 degrees north latitude, stratospheric ozone declined at a rate of 0.8% during the period 1978–1990 (Chandra and Varotsos 1995). Moreover, ozone reduction at temperate latitudes during 1992 was much higher than ever previously recorded (Gleason et al. 1993), which suggests that UV-B doses in regions other than high latitudes may potentially have an impact on ecosystems. There are predictions for substantial increases in UV-B in California during the CO_2 doubling period (Knox, 1991).

Little is known about the effects of increasing UV-B on either biodiversity or ecosystem function, and even less information is available for the Mediterranean ecosystems. Work by Petropoulou et al. (1995) shows the potential beneficial effects of increased UV-B on *Pinus halepensis* and *P. pinea* because of a partial alleviation of the adverse effects of water stress during summer. Because many plants in mediterranean systems have short lifespans, UV-B effects on plant reproductive processes are of potential concern. Musil (1995) has shown delayed flowering, decreased flower production, poorer pollen gemination, and reduced seed set for several Cape dicots exposed to the elevated UV-B expected for a 20% ozone depletion. Geophytic monocots showed negligible effects. No detrimental effects of elevated UV-B on whole plant biomass were found in any species after exposure for one generation; however, longer exposure has been shown to have cumulative effects. Musil (1996) exposed a Cape annual, *Dimorphotheca sinuata*, to ambient and elevated UV-B for three generations. Accumulated UV-B had a greater effect on plant performance than immediate UV-B, causing earlier reproduction and substantial (up to 35%) reductions in dry mass, decreased inflorescence production, pollen tube growth, and germination of seeds. These results suggest that enhanced UV-B may have ecologically significant effects on components of the Cape flora by disrupting reproductive systems.

Fire Climate

The very nature of mediterranean climates, with their summer droughts, makes the vegetation of these regions fire prone. A large literature describes the evolutionary adaptations to such fire frequent habitats (Moreno and Oechel 1994). It is likely that fire frequency and extent will become even more severe in the future due to changes in both climate and land use (Davis and Michaelesen 1995). The most important dimensions for global change and fire climate are how the water balance of the vegetation will change, which is dependent on future precipitation regimes, in part, and on how summer and fall wind patterns will change. With predictions of increased summer wind, fire hazard would certainly be exacerbated. Mediterranean shrublands include species with diverse regeneration responses to frequency, season, and intensity of burns. Patterns of postburn recovery (e.g., Keeley 1998) and analyses of regeneration requirements in particular floras (e.g., Bond 1997) point to significant changes in community composition as a result of climate-driven changes in fire regime. Further, land-use change, bringing people in closer contact with flammable vegetation and the development of plantations (as is occurring in Chile), can increase fire threat and hence vegetation change.

Summary of Other Global Change Impacts

Increasing CO_2, UV-B, tropospheric ozone, and N deposition will all affect the biota of mediterranean-climate regions, probably with somewhat unique effects because of the dissimilar seaonality of these factors among themselves and with biotic growth regimes. All affect biota directly and in a species-specific manner. We would expect large biotic responses to these factors alone without considering climate change. They will, however, interact directly with climate (e.g., temperature increase on ozone production) as well as on processes that are driven by climate change. Changing fire climate, bringing large and more destructive fires, will have a profound effect on the dynamics of mediterranean-climate ecosystems and differentially so because the past history of biotic–fire interactions differs among regions.

Overall Summary

The mediterranean-climate regions of the world are extraordinarily rich in species and in ecosystem types. Further, they have a high degree of endemism, particularly for plants. The nature and diversity of these systems, however, is under considerable threat. These areas have already been subjected to considerable global change in terms of landscape modification, including watershed rearrangements, habitat loss and fragmentation, and the invasion of alien species. In addition, the composition of the atmosphere has changed with increased CO_2 concentrations, and in certain areas, high levels of oxidants. These factors are differentially affecting the composition of the biota. Even without global warming the prognosis for the future for natural systems in many of these regions is poor because of rapid growth of population and their burgeoning economies. The demands on the natural resources, even for recreation alone, is very high with large impacts. Superimposing the threat of climate change on top of these other stressors indicate even greater changes. The least predictable climatic change (i.e., the change in precipitation pattern) is the one that presumably would have the greatest impact on ecosystem distribution and function. Temperature change alone, however, will have a large impact on those mediterranean systems at high elevations where entire ecosystems may be at risk on the lower peaks. A changed fire frequency will also play a large role in shaping the biotic landscape of the future. It is probably safe to say that of all of the world's ecosystems those of mediterranean regions may see the greatest impact on biodiversity from global change because the diversity and endemism are so high in these regions, as is potential population growth and land-use change. The kinds of global changes anticipated are many and are interactive.

References

Armesto JJ, Vidiella P, Jiménez HE (1995) Evaluating causes and mechanisms of succession in the mediterranean regions of Chile and California. In: Arroyo MTK, Fox M, Zedler P (eds) *Ecology and Biogeography of Mediterranean Ecosystems in Chile, California and Australia*, pp. 43–88. Springer-Verlag, New York.

Armesto JJ, Rozzi R, Smith-Ramírez C, Arroyo MTK (1988) Conservation targets in South American temperate forests. Science 282:1271–1272.

Arratia G (1981) Generos de peces de aguas continentales de Chile. Publicacion Ocasional No 34. Museo Nacional de Historia Natural, Santiago.

Artigas JN (1970) Los Asilidos de Chile (Diptera—Asilidae). Gayana, Zoología 17:1–472.

Arroyo MTK, Cavieres L (1997) The mediterranean type-climate flora of central Chile—what do we know and how can we assure its protection? Noticiero de Biología 5:48–56.

Arroyo MTK, Donoso C, Murúa R, Pisano E, Schlatter R, Serey I (1996) *Toward an Ecologically Sustainable Forestry Project. Concepts, Analysis and Recommendations. Protecting Biodiversity and Ecosystem Processes in the Río Cóndor Project, Tierra del Fuego*. Departamento de Investigación y Desarrollo, Universidad de Chile, Santiago.

Arroyo MTK, Simonetti J, Rozzi R, Salaberry M, Marquet P (1998) Central Chile. In: Gil P, Goettsch R, Mittermeier C, Mittermeier R (eds) *Hotspots: Earth's Biologically Wealthiest and Most Threatened Ecosystems*. CEMEX, México.

Bacilieri R, Bouchet MA, Bran D, Granjanny M, Maistre M, Perret P, et al. (1993) Germination and regeneration mechanisms in Mediterranean degenerate forests. Journal of Vegetation Science 4:241–246.

Bairlein F (1991) Body mass of garden warblers (*Sylvia borin*) on migration: a review of field data. Vogelwarte 36:48–61.

Bakun A (1990) Global climate change and intensification of coastal ocean upwelling. Science 247:198–201.

Balletto E, Casale A (1991) Mediterranean insect conservation. In: Collins NM, Thomas JA (eds) *The Conservation of Insects and Their Habitats*, pp. 121–142. Academic Press, London.

Barbéro M (1989) Menaces pesant sur la flore méditerranéenne française. In: *Plantes sauvages menacées de France, bilan et protection*, pp. 11–22. Actes du colloque de Brest, BRG.

Bartholomew B (1970) Bare zone between California shrub and grassland communities: the role of animals. Science 27:505–517.

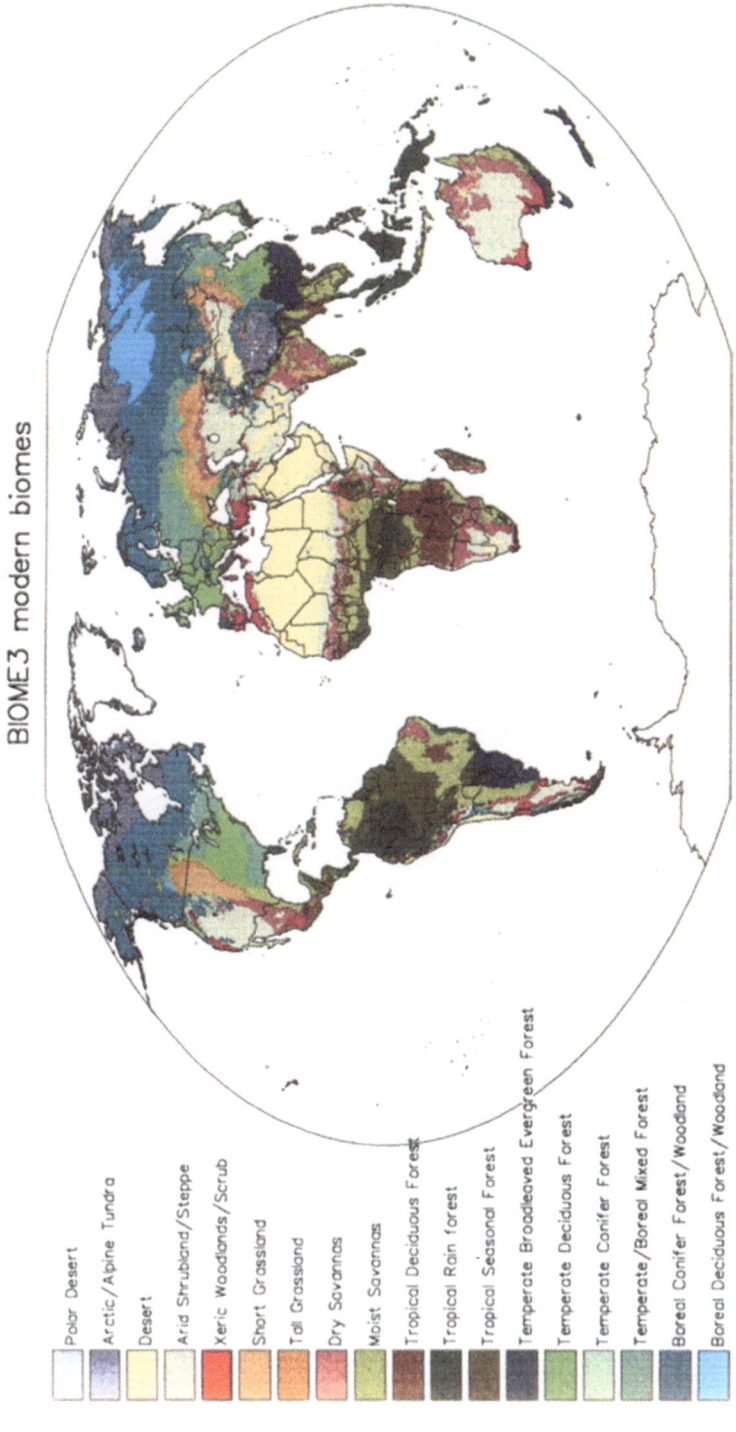

BIOME3 modern biomes

Polar Desert
Arctic/Alpine Tundra
Desert
Arid Shrubland/Steppe
Xeric Woodlands/Scrub
Short Grassland
Tall Grassland
Dry Savannas
Moist Savannas
Tropical Deciduous Forest
Tropical Rain forest
Tropical Seasonal Forest
Temperate Broadleaved Evergreen Forest
Temperate Deciduous Forest
Temperate Conifer Forest
Temperate/Boreal Mixed Forest
Boreal Conifer Forest/Woodland
Boreal Deciduous Forest/Woodland

Figure 2.3. Global biome distribution under present day climate as simulated by BIOME3.

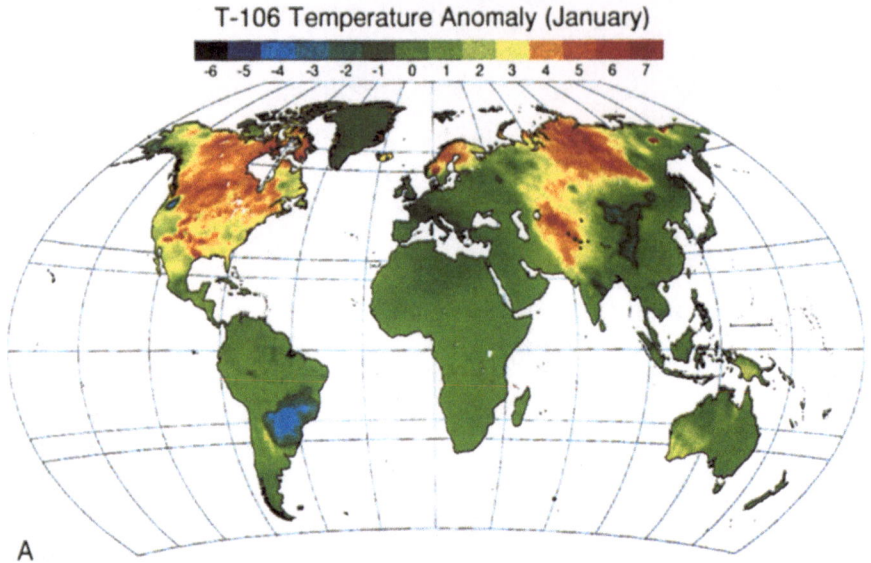

T-106 Temperature Anomaly (January)

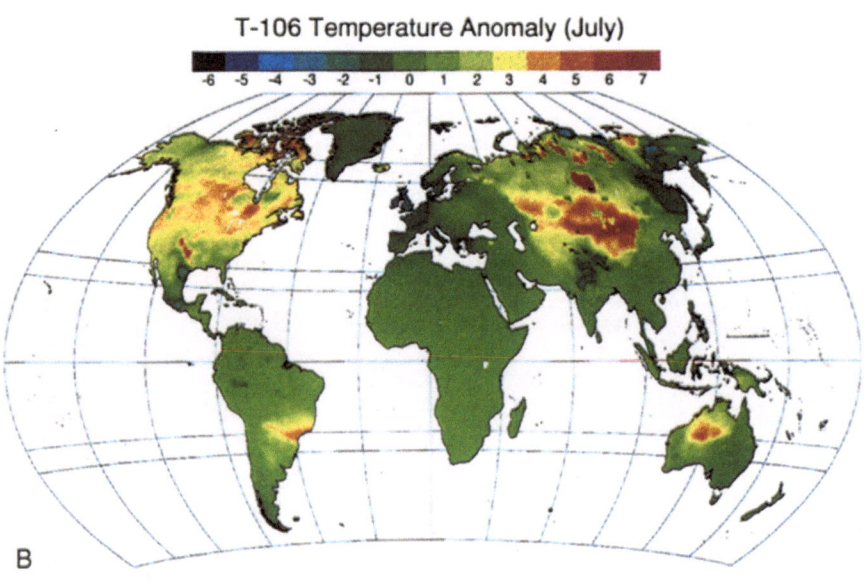

T-106 Temperature Anomaly (July)

T-106 Precipitation Anomaly (cm/yr)

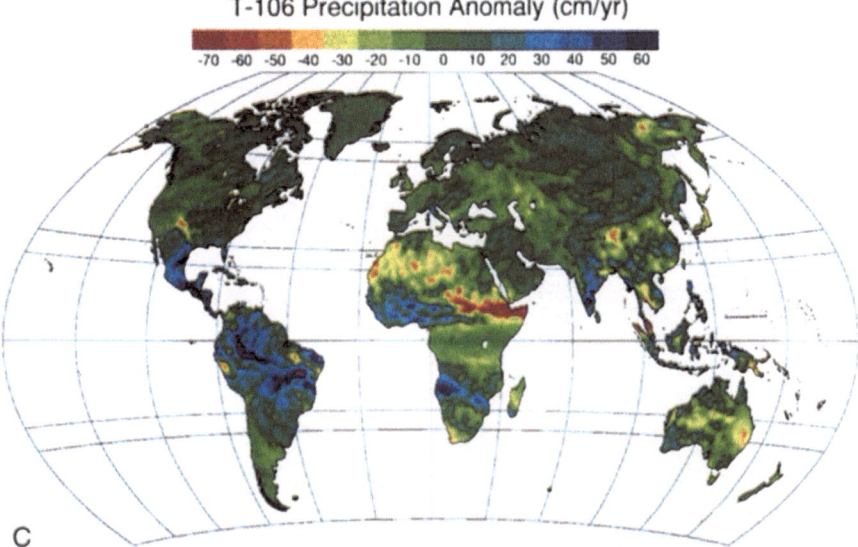

C

Figure 2.4. Global anomalies (differences between simulated current and 2XCO$_2$ climates) using the Hamburg ECHAM3 atmospheric general circulation model: (A) Temperature anomaly (January), (B) Temperature anomaly (July), (C) Precipitation anomaly (cm/yr).

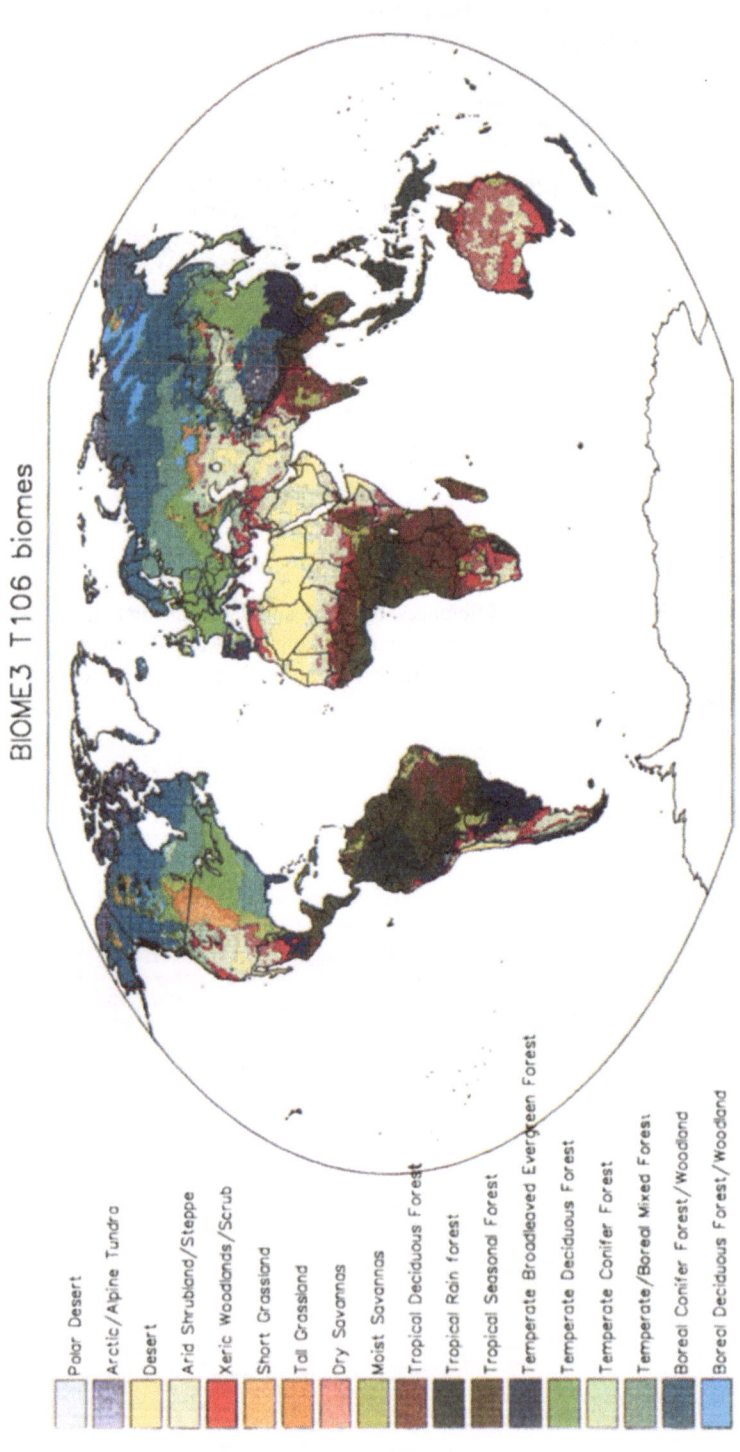

Figure 2.5. Global biome distribution under a 2xCO₂ climate (Hamburg ECHAM3 climate model) as simulated by BIOME3 including direct effects of CO₂.

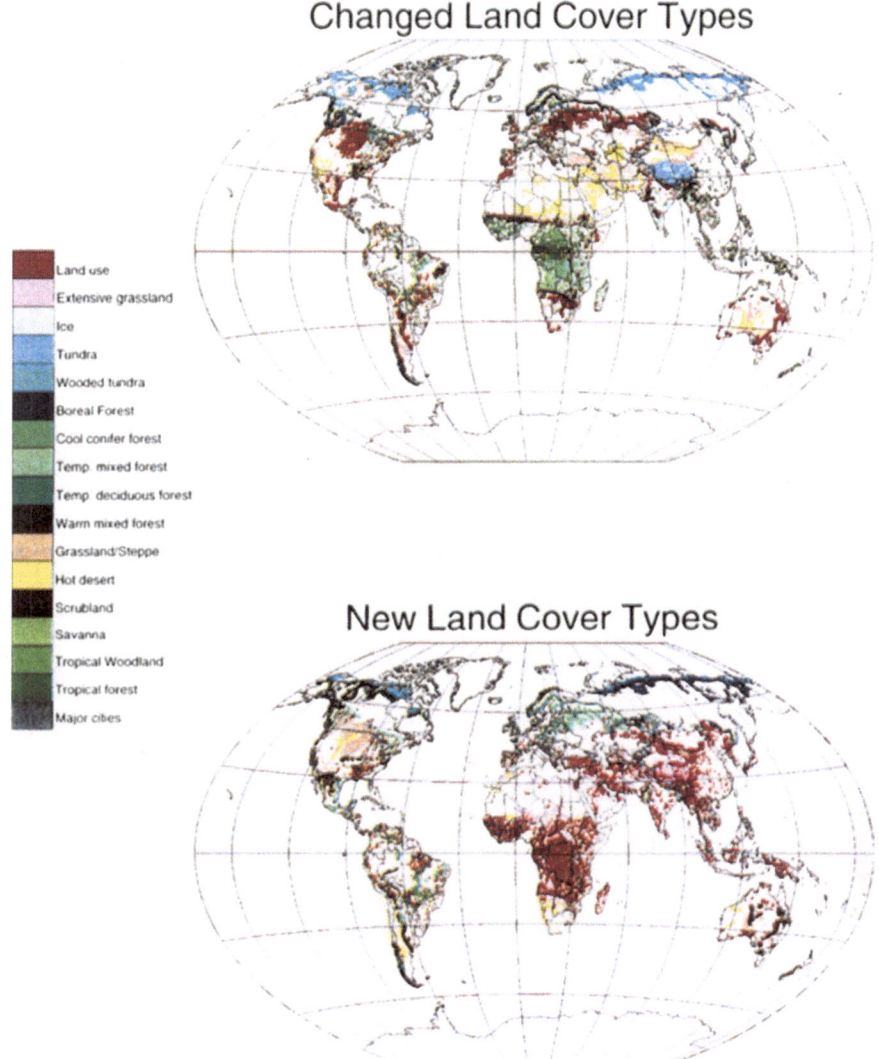

Figure 3.3. Change in actual land cover in 2100 for scenario A. The top panel presents the changes of the original vegetation. Red denotes agricultural land that is abandoned. The lower panel presents the future land cover. Red denotes the expansion of agricultural land, while orange depicts regrowth after clear-cut or abandonment.

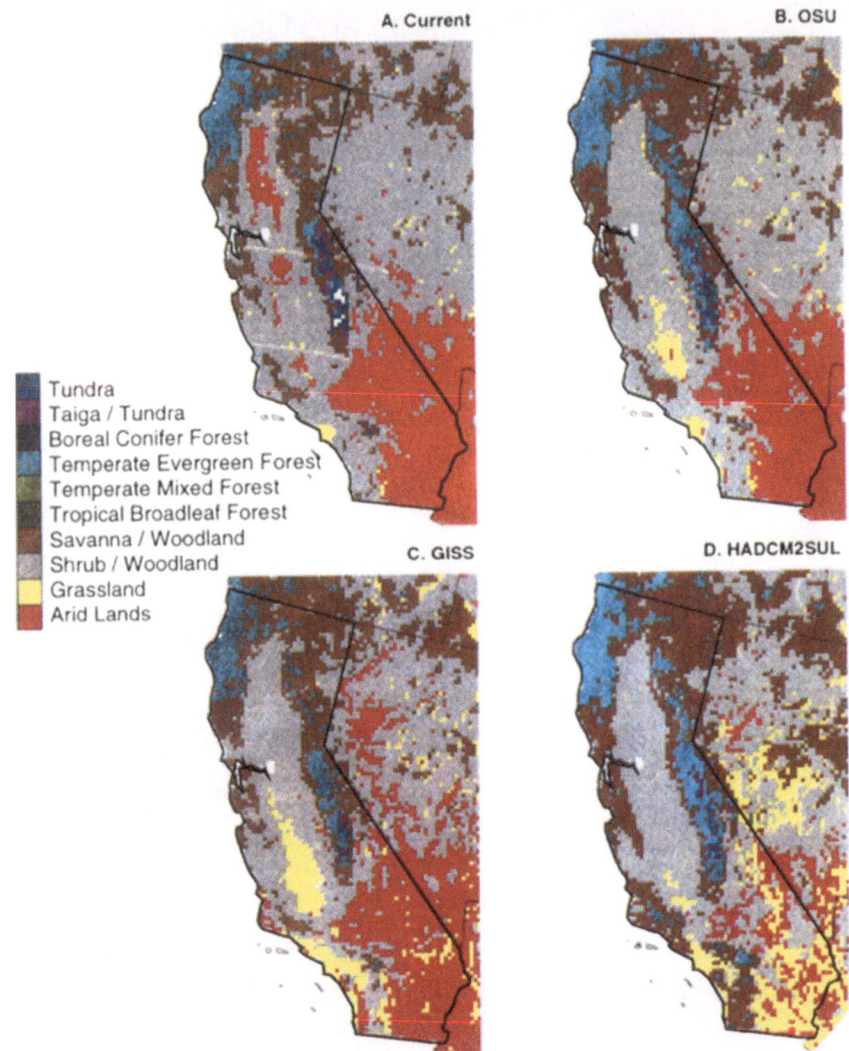

Figure 9.1. Modelled responses of California vegetation to different climate model projections for a doubled CO_2 world. (A) Present vegetation, (B) response to the OSU climate model, (C) to GISS model, and (D) to the Hadley Center model (HADCM2SUL).

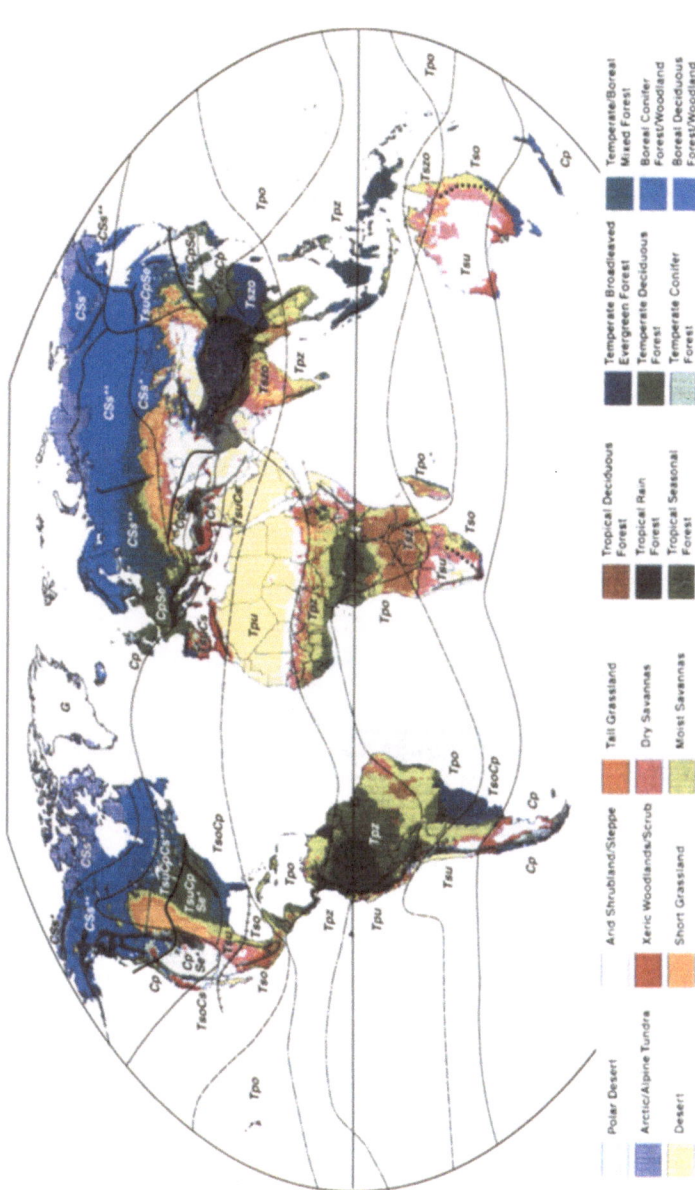

Figure 14.1. Hydroclimatology of floods (after Hayden, 1988, Figure 9, p. 23) combined with BIOME 3 map. After Hayden, the symbols for the flood zones include: T—barotropic; C—barocline; p—perennial; s—seasonal; z—Inter-Tropical Convergence Zone (ITCZ); o—organized convection; u—unorganized convection; S—snow cover; G—glacial; s—seasonal snow cover; e—ephemeral snow cover; *—snow cover 10 to 50 days; and **—snow cover 50 cm or more. The solid and dashed lines are the poleward limit of barotropic conditions in summer and winter, respectively. The average position of the ITCZ is shown by the dash-dot line for January and July. The cross-hatched line indicates for North America the equatorward limit of frontal cyclones. The double-dot-dash line indicates regions with the highest snow fall or duration of snow cover. The thick, solid line indicates the equatorward limit of snow cover for 50 days or more. Mountain regions are shown in solid gray shading. (Cartography kindly provided by D. Lawson.)

Biodiversity Scenarios

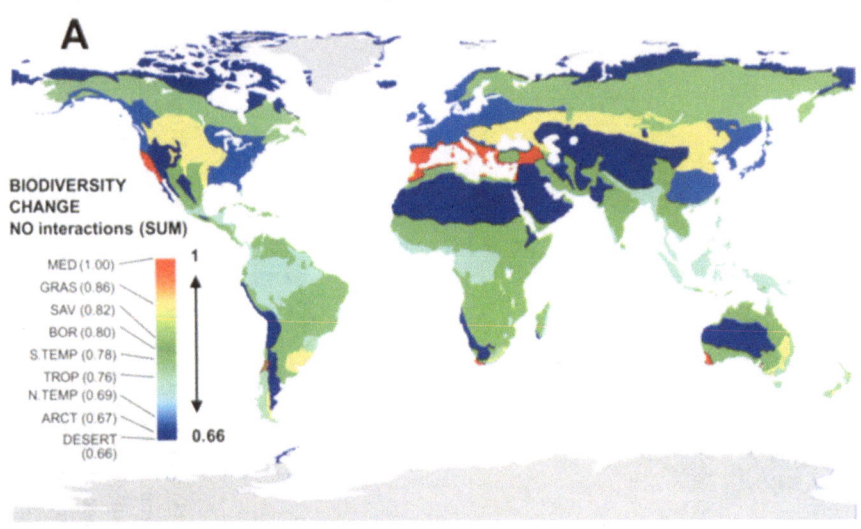

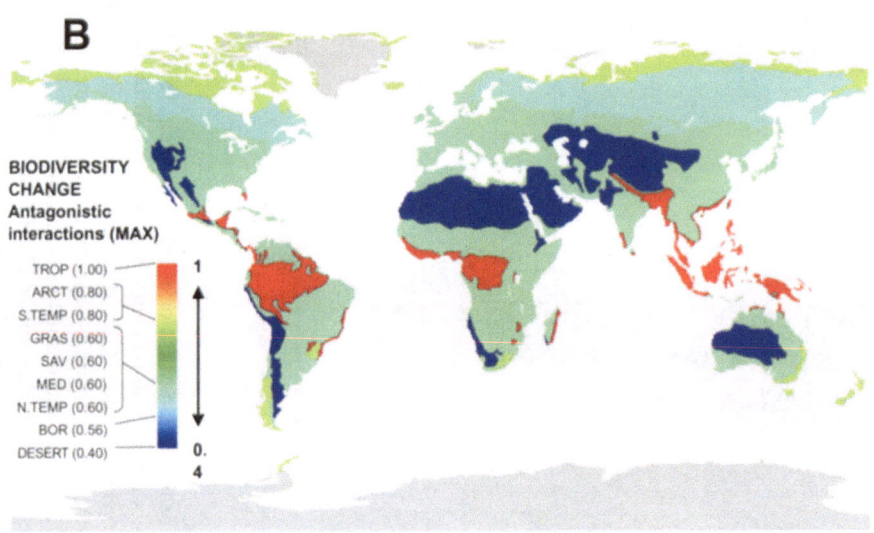

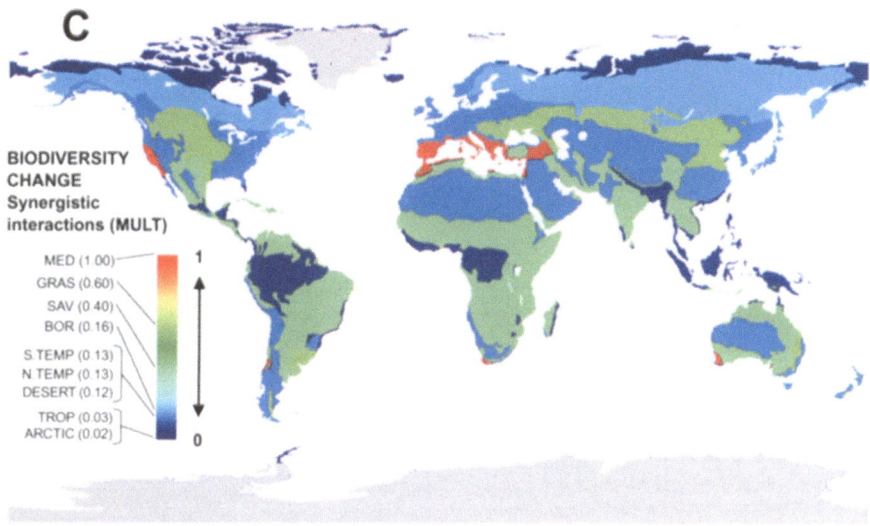

Figure 15.3. Maps of three scenarios of the expected change in biodiversity for the year 2100. Scenario A assumes that there are no interactions among drivers of biodiversity change, and consequently total change is calculated as the sum of the effects of each driver which in turn result from multiplying the expected change in the driver for a particular biome (Table 15.1a) times the effect of the driver which is also a biome specific characteristic (Table 15.1b). Scenario B assumes that total biodiversity change equals the change resulting from the driver that is expected to have the largest effect and is calculated as the maximum of the effects of all the drivers. Scenario C assumes synergistic interactions among the drivers, and consequently the total change is calculated as the product of the changes resulting from the action of each driver. The different colors represent the expected change in biodiversity from moderate to maximum for the different biomes of the world ranked according to the total expected change. The numbers in parentheses represent the total change in biodiversity relative to the maximum value projected for each scenario. The biomes are MED Mediterranean ecosystems, GRAS Grasslands, SAV Savannas, BOR Boreal forest, S. TEMP Southern temperate forest, TROP Tropical forest, N. TEMP Northern temperate forest, ARCT Arctic ecosystems, DESERT Desert. Values for alpine, stream, and lake ecosystems are not shown. (Redrawn with permission from Sala et al. 2000. Copyright 2000 American Association for the Advancement of Science.)

Beard JS (1965) A descriptive catalogue of West Australian plants. Society for Growing Australian Plants, Sydney.

Beard JS (1990) *Plant life in Western Australia*. Kangaroo Press, Kenthurst, New South Wales.

Beard JS, Sprenger BS (1984a) Geographical data from the vegetation survey of Western Australia. Vegetation Survey of Western Australia Occassional Paper No 2. Vegmap Publications, Applecross, Western Australia.

Beard JS, Sprenger BS (1984b) Geographical data from the vegetation survey of Western Australia. Part 1—Area calculations. Vegmap Publications, Applecross, Western Australia.

Beerling DJ, Chaloner WG (1993) Stomatal density responses of Egyptian Olea europea L. leaves to carbon dioxide change since 1327 BC. Annals of Botany 71:431–435.

Birks HJ (1990) Changes in vegetation and climate during the Holocene of Europe. In: Boer MM, de Groot RS (eds) *Landscape Ecological Impact of Climatic Change*, pp. 133–158. IOS Press, Amsterdam.

Blondel J, Aronson J (1995) Biodiversity and ecosystem function in the Mediterranean basin: human and non-human determinants. In: Davis GW, Richardson DM (eds) *Mediterranean-Type Ecosystems: Functions of Biodiversity*, pp. 43–119. Springer-Verlag, Heidelberg.

Bond P, Goldblatt T (1984) Plants of the Cape Flora. A descriptive catalogue. Journal of South African Botany (Suppl.) 13:1–455.

Bond WJ (1994) Do mutualisms matter? Assessing the impact of pollinator/disperser disruption on plant extinctions. Philosophical Transactions of the Royal Society 344:83–90.

Bond WJ (1995) Effects of global change on plant-animal synchrony: implications for pollination and seed dispersal in Mediterranean habitats. In: Moreno J, Oechel W (eds) Global change and mediterranean type ecosystems, pp. 181–202. Springer-Verlag, New York.

Bond WJ (1997) Functional types for predicting changes in biodiversity: a case study in Cape Fynbos. In: Smith TM, Shugart HH, Woodward FI (eds) *Plant Functional Types*, pp. 174–194. Cambridge University Press, Cambridge.

Botkin DB, Nisbet RA, Bicknell S, Woodhouse C, Bentley B, Ferren W (1991) Global climate change and California's natural ecosystems. In: Knox JB, Foley A (eds) *Global Climate Change and California*, pp. 123–149. University of California Press, Berkeley, California.

Bowler JM (1982) Aridity in the late Tertiary and Quaternary of Australia. In: Barker WR, Greenslade PJM (eds) *Evolution of the Flora and Fauna of Arid Australia*, pp. 35–45. Peacock Publications, Adelaide.

Burbidge AA, McKenzie NL (1989) Patterns in the modern decline of Western Australia's vertebrate fauna: causes and conservation implications. Biological Conservation 50:143–198.

Burrows ND, Ward B, Robinson AD (1995) Jarrah forest fire history from stem analysis and anthropological evidence. Australian Forestry 58:7–16.

Camarero L, Catalan J (1993) Chemistry of bulk precipitation in the central and eastern Pyrenees, northeast Spain. Atmospheric Environment 27:83–94.

Chandra S, Varotsos CA (1995) Recent trends of the total column ozone: implications for the Mediterranean region. International Journal of Remote Sensing 16:1765–1769.

Churchill D (1968) The distribution and prehistory of *Eucalyptus diversicolor* F. Muell., *E. marginata* Donn. ex Sm., and *E. calophylla* R. Br. in relation to rainfall. Australian Journal of Botany 16:125–151.

Cody ML (1986) Diversity, rarity and conservation in mediterranean-climate regions. In: Soulé ME (ed) *Conservation Biology: The Science of Scarcity and Diversity*, pp. 122–152. Sinauer, Sunderland, MA.

Cody ML, Fuentes ER, Glanz W, Hunt JR, Moldenke AR (1977) Convergent evolution in the consumer organisms of mediterrean Chile and California. In: Mooney HA (ed) *Convergent Evolution in Chile and California*, pp. 144–192. Dowden Hutchinson and Ross, Stroudsburg, PA.

Cogger HG (1986) *Reptiles and Amphibians of Australia*. Reed, Sydney.

Cohen AN, Carlton JT (1998) Accelerating invasion rate in a highly invaded estuary. Science 279:555–558.

CONAF (1997) Catastro y Evaluación. Recursos Vegetacionales Nativos de Chile. Resultados Finales Síntesis. CONAF, Chile.

Cowling RM, Esler KJ, Midgley GF, Honig MA (1994) Plant functional diversity, species-diversity and climate in arid and semiarid Southern Africa. Journal of Arid Environments 27:141–158.

Cowling RM, Rundel PW, Lamont BB, Arroyo MK, Arianoutsou M (1996) Plant diversity in mediterranean-climate regions. Trends in Ecology and Evolution 11:362–366.

Crivelli AJ (1995) Are fish introductions a threat to endemic freshwater fishes in the Northern Mediterranean region? Biological Conservation 72:311–319.

Crivelli AJ, Maitland PS (1995) Endemic freshwater fishes of the Northern mediterranean Region: Introduction. Biological Conservation 72:121–122.

D'Antonio CM, Vitousek PM (1992) Biological invasions of exotic grasses, the grass/fire cycle and global change. Annual Review of Ecology and Systematics 23:63–87.

Dargavel J (1995) *Fashioning Australia's Forests*. Oxford University Press, Melbourne.

Davis FW, Michaelsen J (1995) Sensitivity of fire regime in chaparral ecosystems in climate change. In: Moreno JM, Oechel, WC (eds) *Global Change and Mediterranean-Type Ecosystems*, pp. 435–456. Springer-Verlag, New York.

Davis SD, Heywood VS, Herrera O, MacBryde J, Villa-Lobo S, Hamilton AC (1997) Centers of plant diversity. A guide and strategy for their conservation. IUCN, Gland.

Dell B, Malajczuk N (1989) Jarrah dieback—a disease caused by *Phytopthora cinammomi*. In: Dell B, Havel JJ, Malajczuk N (eds) *The Jarrah Forest: A Complex Mediterranean Ecosystem*, pp. 67–87. Kluwer Academic Publishers, Dordrecht.

Diaz M, Telleria JL (1994) Predicting the effects of agricultural changes in central Spanish croplands on seed-eating overwintering birds. Agriculture, Ecosystems and Environment 49:289–298.

di Castri F (1981) Mediterranean-type shrublands of the world. In: di Castri F, Goodall DW, Specht RL (eds) *Mediterranean-Type Shrublands*, pp. 1–52. Elsevier, Amsterdam.

di Castri F (1991) The biogeography of mediterranean animal invasions. In: Groves RH, di Castri F (eds) *Biogeography of Mediterranean Invasions*, pp. 439–452. Cambridge University Press, Cambridge.

Dixon B, Keys K, Hopper S, Wycherley P (1995) A fifty year record of fire management in Kings Park bushland. Burning our bushland. Urban Bushland Council (WA), Perth.

Dodd J, Moore JH (1993) Introduction and status of Kochia scoparia in Western Australia. In: Proceedings of the Tenth Australian Weeds Conference and Fourteenth Asian Pacific Weed Society Conference, Volume 1, pp. 496–500. Weed Society of Queensland, Brisbane.

Dukes JS, Mooney HA (1999) Global change may increase the success of biological invaders. Trends in Ecology and Evolution 14:135–139

Ellenberg H Jr (1986) Veranderungen der Flora Mitteleuropas unter dem Einfluá von Düngung und Immissionen. Schweizerische Zeitschrift Für Forstwesen 136:19–36.

Ellery WN, Scholes RJ, Mentis MT (1991) An initial approach to predicting the sensitivity of the South African-grassland biome to climate change. South African Journal of Science 87:499–503.

Euston-Brown D (1995) Environmental and Dynamic Determinants of Vegetation Distribution in the Kouga and Baviaanskloof Mountains, Eastern Cape. MSc Thesis, University of Cape Town, Cape Town.

Farina A (1989) Bird community patterns in Mediterranean farmlands: a comment. Agriculture, Ecosystems and Environment 27:177–181.

Field CB, Chapin FS III, Chiariello NR, Holland EA, Mooney HA (1996) The Jasper ridge CO_2 experiment: design and motivation. In: Koch GW, Mooney HA (eds) *Carbon Dioxide and Terrestrial Ecosystems*, pp. 212–145. Academic Press, San Diego.

Friend JA (1990) The numbat *Myrmecobius fasciatus* (Myrmecobiidae): History of decline and potential for recovery. Proceedings of the Ecological Society of Australia 16:369–377.

Froend RH, McComb AJ (1991) An account of the decline of Lake Towerrinning, a wheatbelt wetland. Journal of the Royal Society of Western Australia 73:123–128.

Froend RH, Heddle EM, Bell DT, McComb AJ (1987) Effects of salinity and waterlogging on the vegetation of Lake Toolibin, Western Australia. Australian Journal of Ecology 12:281–298.

García de Cotaza V, Nobel PS (1990) Wordwide environmental productivity indices and yield predictions for CAM plant, Opuntia ficus-indica, including effects of doubled carbon dioxide levels. Agriculture and Forest Meteorology 49:261–280.

García P, Escribano M, Rodríguez A (1994) The Spanish pastureland ecosystem: current status and incidence of common agricultural policy reform. Avances en Alimentación y Mejora Animal 34:123–133.

Gentilli J (1989) Climate of the jarrah forest. In: Dell B, Havel JJ, Malajczuk N (eds) *The Jarrah Forest. A Complex Mediterranean Ecosystem*, pp. 23–40. Kluwer, Dordrecht.

George RJ, McFarlane DJ, Speed RJ (1995) The consequences of a changing hydrologic environment for native vegetation in south Western Australia. In: Saunders DA, Craig J, Mattiske L (eds) *Nature Conservation 4: The Role of Networks*, pp. 9–22. Surrey Beatty and Sons, Chipping Norton, NSW.

Glavas S (1988) A wet-only precipitation study in a Mediterranean site, Patras, Greece. Atmospheric Environment 22:1505–1508.

Gleason F, Bhartia PK, Herman JR, McPeters R, Newman P, Stolarsky RS, et al. (1993) Record low global ozone in 1992. Science 260:523–526.

Greuter W (1995) Origin and pecularities of Mediterranean island floras. Ecologia Mediterranea 21:1–10.

Groombridge B (ed) (1992) *Global biodiversity. Status of the Earth's Living Resources.* Chapman and Hall, London.

Guillerm JL, Maillet J (1982) Western Mediterranean countries of Europe. In: Holzner W, Numata M (eds) *Biology and Ecology of Weeds*, pp. 227–243. Junk Publishers, The Hague.

Hallam SJ (1975) *Fire and Hearth: A Study of Aboriginal Usage and European Usurpation in South-Western Australia.* Institite of Aboriginal Studies, Canberra.

Hättenschwiler S, Miglietta F, Raschi A, Körner C (1997) 30 years of in situ tree growth under elevated CO_2: a model for future forest responses? Global Change Biology 3:463–471.

Hajek E, Gross P, Espinosa G (1990) Problemas Ambientales de Chile. AID-Pontificia-Universidad Católica de Chile, Santiago.

Hewitson BC, Crane RG (1996) Climate downscaling: techniques and application, Climate Research 7:85–95.

Hingston FJ, Galaitis V (1976) The geographic variation of salt precipitated over Western Australia. Australian Journal of Soil Research 14:319–335.

Hobbs RJ (1992) Function of biodiversity in Mediterreanean ecosystems in Australia: definitions and background. In: Hobbs RJ (ed) *Biodiversity in Mediterranean ecosystems in Australia*, pp. 1–25. Surrey Beatty and Sons, Chipping Norton, NSW.

Hobbs RJ (1993a) Dynamics of weed invasion: implications for control. In: Proceedings of the Tenth Australian Weeds Conference and Fourteenth Asian Pacific Weed Society Conference. Volume 1, pp. 461–465. Weed Society of Queensland, Brisbane.

Hobbs RJ (1993b) Effects of landscape fragmentation on ecosystem processes in the Western Australian wheatbelt. Biological Conservation 64:193–201.

Hobbs RJ, Atkins L (1988) Effect of disturbance and nutrient addition on native and introduced annuals in the Western Australian wheatbelt. Australian Journal of Ecology 13:171–179.

Hobbs RJ, Mooney HA (1997) Broadening the extinction debate—population deletions and additions in California and Western Australia. Conservation Biology 12:271–283.

Hobbs RJ, Saunders DA (1993) Effects of landscape fragmentation in agricultural areas. In: Moritz C (ed) *Conservation Biology in Oceanea*. Surrey Beatty and Sons, Chipping Norton, NSW.

Hobbs RJ, Richardson DM, Davis GW (1995) Mediterranean-type ecosystems: opportunites and constraints for studying the function of biodiversity. In: Davis GW, Richardson DM (eds) *Mediterranean-Type Ecosystems. The Function of Biodiversity*, pp. 1–41. Springer-Verlag, Berlin.

Hobbs RJ, Saunders DA, Lobry de Bruyn LA, Main AR (1993) Changes in biota. In: Hobbs RJ, Saunders DA (eds) *Reintegrating Fragmented Landscapes. Towards Sustainable Production and Nature Conservation*, pp. 65–106. Springer-Verlag, New York.

Holm L (ed) (1997) *World Weeds: Natural Histories and Distributions*. John Wiley and Sons, Chichester.

Holzapfel C, Schmidt W, Shmida A (1992) Effects of human-caused disturbances on the flora along a Mediterranean-desert gradient. Flora 186:261–270.

Hopper SD (1979) Biogeographical aspects of speciation in the southwest Australian flora. Annual Review of Ecology and Systematics 10:399–422.

Hopper SD (1992) Patterns of plant diversity at the population and species level in south-west Australian mediterranean ecosystems. In: Hobbs RJ (ed) *Biodiversity of Mediterranean Ecosystems in Australia*, pp. 27–46. Surrey Beatty and Sons, Chipping Norton, NSW.

Hopper SD, Harvey MS, Chappill JA, Main AR, Main BY (1996) The Western Australian biota as Gondwanan heritage—a review. In: Hopper SD, Chappill DA, Harvey MS, George AS (eds) *Gondwanand Heritage: Past, Present and Future of the Western Australian Biota*, pp. 1–46. Surrey Beatty and Sons, Chipping Norton, NSW.

Hsieh WW, GJ Boer (1992) Global climate change and ocean upwelling. Fisheries and Oceanography 1:333–338.

Hudson DA (1997) Southern African Climate Change Simulated by the GENESIS GCM. South African Journal of Science 93:389–403.

Humphries SE (1993) Environmental impact of weeds. In: Proceedings of the Tenth Australian Weeds Conference and Fourteenth Asian Pacific Weed Society Conference. Volume 2, pp. 1–11. Weed Society of Queensland, Brisbane.

Humphries SE, Groves RH, Mitchell DS (1991) Plant invasions of Australian ecosystems. A status review and management directions. In: Kowari 2. *Plant Invasions. The Incidence of Environmental Weeds in Australia*, pp. 1–127. Australian National Parks and Wildlife Service, Canberra.

Hutchison MJ, Armstrong PH (1993) The invasion of a south-western Australian river system by *Perca fluviatillis*: History and probable causes. Global Ecology Biogeography Letters 3:77–89.

Ibanez JJ, Garcia A, Monturial F (1989) Edaphic heterogeneity induced by dehesa management of mediterranean forest. Anales de Edafologia y Agrobiology

48:433–444. In: Muñoz SM, Núñez H, Yáñez J (eds) Libro Rojo de los Sitios Prioritarios para La Conservación de la Diversidad Biológica, pp. 149–160. Corporación Nacional Forestal, Santiago.

INFOR (1997) Estadísticas Forestales 1996. Boletín Estadístico No 50. INFOR-CORFO, 117 pp.

Ingraham NL, Matthews RA (1995) The importance of fog-drip water to vegetation: Point-Reyes peninsula, California. Journal of Hydrology 164:269–285.

Joffre R, Vacher J, de los Llanos C, Long G (1988) The dehesa: an agrosilvapastoral system of the Mediterranean region with special reference to the Sierra Morena area of Spain. Agroforestry Systems 6:71–96.

Johnson SD (1992) Plant-animal relationships. In: Cowling RM (ed) *The Ecology of Fynbos. Nutrients, Fire and Diversity*, pp. 175–205. Oxford University Press, Cape Town.

Johnson SD, Bond WJ (1993) Red flowers and butterfly pollination in the fynbos of South Africa. In: Arianoutsou M, Groves RH (eds) *Plant Animal Interactions in Mediterranean-Type Ecosystems*, pp. 137–148. Kluwer Academic Publishers, Dordrecht.

Keighery GJ (1991) Environmental weeds of Western Australia. In: Kowari 2. *Plant Invasions. The Incidence of Environmental Weeds in Australia*, pp. 180–188. Australian National Parks and Wildlife Service, Canberra.

Keighery GJ (1995) How many weeds? In: Burke G (ed) *Invasive Weeds and Regenerating Ecosystems in Western Australia*. Institute for Science and Technology Policy, Murdoch University, Perth.

Keeley JE (1998) Postfire ecosystem recovery and management: the October 1993 large fire episode in California. In: Moreno JM (ed) Large forest fires, pp. 69–90. Backhuys, Leiden.

Keeley SC, Mooney (1993) Vegetation in western North America past and future. In: Mooney HA, Fuentes ER, Kronberg BI (eds) *Earth System Responses to Global Change: Contrasts Between North and South America*. Academic Press, San Diego.

Kennedy M (ed) (1990) *Australia's Endangered Species*. Simon and Schuster, Brookville, NSW.

Kinnear JE, Onus ML, Bromilow RN (1988) Fox control and rock wallaby population dynamics. Australian Wildlife Research 15:435–450.

Klasinc L, Cvitas T (1996) The photosmog problem in the Mediterranean region. Marine Chemistry 53:111–119.

Knox JB (1991) Global climage change: impacts on California. An introduction and overview. In: Knox JB, Foley A (eds) *Global Climate Change and California*, pp. 1–25. University of California Press, Berkeley.

Koch JM, Bell DT (1980) Post-fire succession in the Northern Jarrah Forest of Western Australia. Australian Journal of Ecology 5:9–14.

Lamont BB, Hopkins AJM, Hnatiuk RJ (1984) The flora—composition, diversity and origins. In: Pate JS, Beard JS (eds) *Kwongan: Plant Life of the Sandplain*, pp. 27–50. University of Western Australia Press, Nedlands, WA.

Le Bolloch O, Guerzoni S (1995) Acid and alkaline deposition in precipitation on the western coast of Sardinia, central Mediterranean. Water Air and Soil Pollution 85:2155–2160.

Le Houérou HN (1992) Vegetation and land use in the Mediterranean Basin by the year 2050: a prospective study. In: Jeftic L, Milliman JD, Sestini G (eds) *Climatic Change and the Mediterranean. Environmental and Societal Impacts of Climatic Change and Sea Level Rise in the Mediterranean Region*, pp. 175–232. Edward Arnold, London.

LeMaitre DC, Van Wilgen BW, Chapman RA, McKelly DH (1996) Invasive plants and water-resources in the Western Cape Province, South Africa: Modeling the consequences of a lack of management. Journal of Applied Ecology 33:161–172.

Léon C, Lucas G, Singhe H (1985) The value of information in saving threatened mediterranean plants? In: Gomez-Campo C (ed) *Plant Conservation in the Mediterranean Area*, pp. 177–196. Junk Publishers, Dordrecht.

Lewis L, Rains W, Kennedy L (1991) Global climate change and California agriculture. In: Knox JB, Sheuring A (eds) *Global Climate Change and California*, pp. 97–122. University of California Press, Berkeley.

Linder HP, Meadows ME, Cowling RM (1992) History of the Cape flora. In: Cowling R (ed) *The Ecology of Fynbos. Nutrients, Fire and Diversity*, pp. 113–134. Oxford University Press, Cape Town.

Luo Y, Nobel PS (1993) Growth characteristics of newly initiated cladones of *Opuntia ficus-indica* affected by shading, drought and elevated CO_2. Physiologia Plantarum 87:467–474.

Lyons TJ, Schwerdtfeger P, Hacker JM, Foster IJ, Smith RCG, Xinmei H (1993) Land-atmosphere interaction in a semiarid region: the Bunny Fence Experiment. Bulletin of the American Meteorological Society 74:1327–1334.

Macdonald IAW (1989) Man's role in changing the face of southern Africa. In: Huntley BJ (ed) *Biotic Diversty In Southern Africa*, pp. 51–57. Oxford University Press, Cape Town.

Maheras P (1988) Changes in precipitation conditions in the western Mediterranean over the last century. Journal of Climatology 8:179–189.

Main AR (1982) Rare species: precious or dross? In: Groves RH, Ride WDL (eds) *Species at Risk: Research in Australia*, pp. 163–174. Australian Academy of Science, Canberra.

Malanson GP, O'Leary JF (1995) The coastal sage scrub-chaparral boundary and response to global climatic change. In: Moreno JM, Oechel WC (eds) *Global Change and Mediterranean-Type Ecosystems*, pp. 203–224. Springer-Verlag, New York.

Maldonado S, Arroyo MTK, Marticorena C, Muñoz M, León P (1995) Utilidad de las bases de datos para estudios en biodiversidad: evaluación preliminar de algunos parámetros en las Asteraceaes de Chile central (30–40°S). In: Linares E, Dávila P, Chiang F, Bye R, Elias S (eds) Conservación de Plantas en Peligro de Extinción: Diferentes Enfoques, pp. 25–32. Univ Nac Auton Mexico, México DF.

Manders PT, Richardson DM, Masson PH (1992) Is fynbos a stage in succession to forest? Analysis of the perceived ecological distinction between two communities. In: van Wilgen BW, Richardson DM, Kruger FJ, van Hensbergen HJ (eds) *Fire in South African Mountain Fynbos*, pp. 81–107. Springer-Verlag, Heidelberg.

Marticorena C (1990) Contribución a la estadística de la flora vascular de Chile. Gayana Botany 47:85–113.

Matthei O (1995) Manual de las Malezas que Crecen en Chile. Alfabeta Impresores, Santiago.

McBean GA (1996) Factors controlling the climate of the West Coast of North America. In: Lawford RG, Alaback PB, Fuentes E (eds) *High-Latitude Rainforests and Associated Ecosystems of the West Coast of the Americas*, pp. 27–41. Springer-Verlag, New York.

McFarlane DJ, George RJ, Farrington P (1993) Changes in the hydrologic cycle. In: Hobbs RJ, Saunders DA (eds) *Reintergrating Fragmented Landscapes. Towards Sustainable Production and Nature Conservation*, pp. 146–186. Springer-Verlag, New York.

McFarlane DJ, Howell MR, Ryder AT, Orr GJ (1992) The effect of agricultural development on the physical and hydraulic properties of four Western Australian soils. Australian Journal of Soil Research 30:517–532.

McGrath M (1985) Fire planning and management: an overview. In: Ford JW (ed) Fire ecology and management in Western Australian ecosystems, pp. 219–221. WAIT Environmental Studies Group, Bentley, Western Australia.

Merino O, Villar R, Martin A, Garcia D, Merino J (1995) Vegetation response to climatic changes in a dune ecosystem in southern Spain. In: Moreno JM, Oechel WC (eds) *Global Change and Mediterranean-Type Ecosystems*, pp. 225–238. Springer-Verlag, New York.

Millan M, Salvador R, Mantilla E, Artinano B (1996) Meteorology and photochemical air pollution in the Southern Europe: experimental results from EC research projects. Atmospheric Environment 30:1909–1924.

Miller PR, Taylor OC, Wilhour RG (1982) Oxidant air pollution effects on a western coniferous forest ecosystem, pp. 82–276. Environmental Research laboratory; EPA report no EPA-600/D. Corvallis: US Environmental Protection Agency.

Milton SR, Yeaton RJ, Dean WRJ, Vlok JHJ (1997) Succulent karoo. In: Cowling RM, Richardson DM, Pierce SM (eds) *Vegetation of Southern Africa*, pp. 131–166. Cambridge University Press, Cambridge.

Mittermeier RA, Myers N, Thomsen JB, da Fonseca GA, Olivieri S (1998) Biodiversity hotspots and major tropical wilderness areas: approaches to setting conservation priorities. Conservation Biology 12:516–520.

Montenegro G, Ginocchio R (1995) Ecomorphological characters as a resource for illustrating growth-form convergence in matorral, chaparral, and mallee. In: Arroyo MTK, Zedler PH, Fox M (eds) Ecology and biogeography of mediterranean ecosystems in Chile, California and Australia, pp. 160–176. Springer-Verlag, New York.

Mooney HA (ed) (1977) Convergent evolution of Chile and California-mediterranean climate ecosystems. Dowden, Hutchinson and Ross, Stroudsburg, PA.

Mooney HA (1988) Lessons from mediterranan-climate regions. In: Wilson EO (ed) *Biodiversity*, pp. 157–165. National Academy Press, Washington, D.C.

Mooney H, Canadell J, Chapin FS, Ehleringer J, Körner Ch, McMurtrie R, et al. (1998) Ecosystem physiology responses to global change. In: Walker BH, Steffen WL, Canadell J, Ingram JSI (eds) *The Terrestrial Biosphere and Global Change: Implications for Natural and Managed Ecosystems: A Synthesis of GCTE and Related Research*, pp. 141–189. Cambridge University Press, Cambridge.

Moreno JM, Oechel WC (eds) (1994) *The Role of Fire in Mediterranean-Type Ecosystems*. Springer-Verlag, New York.

Moreno JM, Vazquez A, Velez R (1998) Recent history of forest fires in Spain. In: Moreno JM (ed) *Large Forest Fires*, pp. 159–186. Backhuys, Leiden.

Mus M (1995) Conservation of flora in the Balearic Islands. Ecologia Mediterranea 21:185–194.

Musil CF (1995) Differential effects of elevated ultraviolet-B radiation on the photochemical and reproductive performances of dicotyledonous and monocotyledonous arid-environment ephemerals. Plant, Cell and Environment 18:844–854.

Musil CF (1996) Accumulated effect of elevated ultraviolet-B radiation over multiple generations of the arid-environment annual Dimorphotheca sinuata DC. (Asteraceae). Plant, Cell and Environment 19:1017–1027.

Navas M-L, Guillerm J-L, Fabreguettes J, Roy J (1995) The influence of elevated CO_2 on community structure, biomass and carbon balance of mediterranean old-field microcosms. Global Change Biology 1:325–335.

Naveh Z, Whittaker RH (1979) Structural and floristic diversity of shrublands and woodlands in northern Israel and other Mediterranean areas. Vegetatio 41:171–190.

Naveh Z, Liebermann A (1994) Dynamic conservation management of Mediterranean landscapes. In: Naveh Z, Liebermann A (eds) *Landscape Ecology: Theory and Application*, pp. 256–338. Springer-Verlag, Berlin.

Naveh Z, Steinberger EH, Chaim S, Rotmann A (1980) Photochemical air-pollutants-a threat to Mediterranean coniferous forests and upland ecosystems. Environmental Conservation 7:301–309.

Neilson RP (1995) A model for predicting continental-scale vegetation distribution and water balance. Ecological Applications 5:362–385.

Neilson RP, Prentice IC, Smith B, Kittel T, Viner D (1998) Simulated changes in vegetation distribution under global warming. In: Watson RT, Zinyowera MC, Moss RH (eds) *The Regional Impacts of Global Change: An Assessment of Vulnerability*, pp. 439–456. Cambridge University Press, Cambridge.

Nulsen RA (1993) Changes in soil properties. In: Hobbs RJ, Saunders DA (eds) *Reintegrating Fragmented Landscapes: Towards Sustainable Production and Nature Conservation*, pp. 107–145. Springer-Verlag, New York.

Nulsen RA, Bligh KJ, Baxter IN, Solin EJ, Imrie DH (1986) The fate of rainfall in a mallee and heath vegetated catchment in southen Western Australia. Australian Journal of Ecology 11:361–371.

Oosterbroek P (1994) Biodiversity of the Mediterranean region. In: Forey PL, Humphries CJ, Vane-Wright RI (eds) *Systematics and Conservation Evaluation*, pp. 289–307. Clarendon Press, Oxford, UK.

Oosterbroek P, Arntzen JW (1992) Area-cladograms of circum-Mediterranean taxa inrelation to Mediterranean palaegeography. Journal of Biogeography 19:3–20.

Paerl HW (1993) Emerging role of atmospheric nitrogen deposition in coastal eutrophication: Biogeochemical and trophic perspectives. Canadian Journal of Fisheries and Aquatic Sciences 50:2254–2269.

Paerl HW (1995) Coastal eutrophication in relation to atmospheric nitrogen deposition: current perspectives. Ophelia 41:237–259.

Palutikof JP, Wigley TML (1996) Developing climate change scenarios for the Mediterranean region. In: Jeftic L, Keckes S, Pernetta JC (eds) *Climate Change and the Mediterranean*, pp. 27–56. Arnold, London.

Parmesanc C (1996) Climate and species' range. Nature 382:765–766.

Parodi LR (1935) Relaciones de la Agricultura prehispánica con la agricultura argeninta actual. Anal Acad Nac Agr y Vet Buenos Aires 1:115–167.

Parsons WT, Cuthbertson EG (1992) *Noxious Weeds of Australia*. Inkata Press, Melbourne.

Pate JS, Hopper SD (1993) Rare and common plants in ecosystems, with special reference to the South-west Australian flora. In: Schulze E-D, Mooney HA (eds) *Ecosystem Function of Biodiversity*, pp. 293–325. Springer-Verlag, Heidelberg.

Peña LE, Ugargte P (1997) Las mariposas de Chile. Editorial Universitaria, Santiago.

Peñuelas J, Matamala R (1990) Changes in nitrogen and sulfur leaf content, stomatal density and specific leaf area of 14 plant species during the last three centuries of carbon dioxide increase. Journal of Experimental Botany 41:1119–1124.

Petropoulou T, Kyparissis A, Nikolopoulos D, Manetas Y (1995) Enhanced UV-B radiation alleviates the adverse effects of summer drought in two Mediterranean pines under field conditions. Physiologia Plantarum 94:37–44.

Pigott JP, Loneragan WA (1995) Fire and human disturbance regimes and impacts on plant communities at the Star Swamp Bushland Reserve. In: Harris J (ed) Burning our bushland. Proceedings of a conference about fire and urban bushland, pp. 19–24. Urban Bushland Council (WA), West Perth.

Piñol J, Terradas J, Lloret F (1998) Climate warming, wildfire hazard, and wildfire occurrence in coastal eastern Spain. Climate Change 38:345–357.

Piñol J, Terradas J, Avila A, Roda F (1995) Using catchments of contrasting hydrological conditions to explore climate change effects on water and nutrient flows in Mediterranean forests. In: Moreno JM, Oechel WC (eds) *Global Change and Mediterranean-Type Ecosystems*, pp. 371–385. Springer-Verlag, New York.

Pittock AB (1988) Actual and anticipated changes in Australia's climate. In: Pearman GI (ed) *Greenhouse: Planning for Climatic Change*, pp. 35–51. CSIRO, Melbourne.

Quézel P (1985) Definition of the Mediterranean region and the origin of its flora. In: Gomez-Campo C (ed) *Plant conservation in the Mediterranean area*, pp. 9–24. Junk Publishers, Dordrecht.

Quézel P (1995) La flore bu bassin méditerranéen: origine, mise en place, endémisme. Ecologia Mediterranea 21:19–39.

Quézel P, Médail F (1995) La région circum-méditerranéenne, centre mondial majeur de biodiversité végétale. Actes des 6èmes rencontres de l'Agence Régionale pour l'Environnement Provence-Alpes-Côte d'Azur, pp. 152–160. Colloque sceintifique international Bio'Mes, Gap.

Ramade F (1990) The conservation of a specific diversity: its ecological significance and implications for the implementation of nature and resource protection. Courrier de la Nature 130:16–33.

Rebelo AG (1997) Conservation. In: Cowling RM, Richardson DM, Pierce SM (eds) *Vegetation of Southern Africa*, pp. 571–590. Cambridge University Press, Cambridge, UK.

Rejmánek M, Randall JM (1994) Invasive alien plants in California: 1993 summary and comparison with other areas in North America. Madroño 41:161–177.

Richardson DM, Williams PA, Hobbs RJ (1994) Pine invasions in the Southern Hemisphere: determinants of spread and invasibility. Journal of Biogeography 21:511–527.

Richardson DM, Van Wilgen BW, Higgins SI, Trindersmith TH, Cowling RM, McKell DH (1996) Current and future threats to plant biodiversity on the Cape Peninsula, South Africa. Biodiversity Conservation 15:607–647.

Rodà F, Bellot J, Avila A, Escarré A, Piñol J, Terradas J (1993) Saharan dust and the atmospheric inputs of elements and alkalinity to Mediterranean ecosystems. Water Air and Soil Pollution 66:277–288.

Rodríguez-Jiménez AJ (1988) Fenología de una comunidad de anfibios asociada a cursos fluviales temporales. Doñana Acta Vertebrata 15:29–43.

Roumet C, Bel MP, Sonie L, Jardon F, Roy J (1996) Growth response of grasses to elevated CO_2: a physiological plurispecific analyses. New Phytologist 133:595–603.

Rundel PW, Montenegro G, Jaksic FM (eds) (1998) Landscape disturbance and biodiversity in mediterranen-type ecosystems. Springer-Verlag, Berlin.

Salama RB, Bartle GA, Farrington P (1994) Water use of plantation *Eucalyptus camaldulensis* estimated by groundwater hydrograph separation techniques and heat pulse method. Journal of Hydrology 156:163–180.

Sanchez-Arcilla A, Jimenez JA, Stive MJR, Ibanez C, Pratt N, Day JW Jr, et al. (1996) Impacts of sea-level rise on the Ebro Delta: a first approach. Ocean Coastal Management 30:197–216.

Saunders DA, Hobbs RJ (1992) Impact on biodiversity of changes in land-use and climate. In: Hobbs RJ (ed) *Biodiversity in Mediterranean Ecosystems in Australia*, pp. 61–75. Surrey Beatty and Sons, Chipping Norton, NSW.

Saunders DA, Ingram J (1995) *Birds of Southwestern Australia. An Atlas of Changes in the Distribution and Abundance of the Wheatbelt Fauna.* Surrey Beatty and Sons, Chipping Norton, NSW.

Schodde R (1981) Bird communities of the Australian mallee: composition, derivation, distribution, structure and seasonal cycles. In: di Castri F, Goodall DW, Specht RL (eds) *Mediterranean-Type Shrublands*, pp. 387–416. Elsevier, Amsterdam.

Siegfried WR (1989) Preservation of species in southern African nature reserves. In: Huntley BJ (ed) *Biotic Diversity in Southern Africa.* Oxford University Press, Cape Town.

Smith RCG, Huang X, Lyons TJ, Hacker JH, Hick P (1992) Change in land surface albedo and temperature in southwestern Australia following the repacement of native perennial vegetation: satellite observations. Paper No IAF-92-0117 Forty-third Congress of the International Astronautical Federation, Washington, DC.

Snelling RR, Hunt JH (1975) The Ants of Chile (Hymenoptera: Formicidae). Revista Chileña de Entomología 9:63–129.

Stock WD, Allsopp N (1992) Functional perspective of ecosystems. In: Cowling R (ed) *The Ecology of Fynbos. Nutrients, Fire and Diversity*, pp. 241–259. Oxford University Press, Cape Town.

Strahan R (ed) (1983) *The Complete Book of Australian Mammals*. Australian Museum/Angus and Robertson, Sydney, NSW.

Tellería JL, Santos T, Sánchez A, Galarza A (1992) Habitat structure predicts bird diversity distribution in Iberian forests better than climate. Bird Study 39:63–68.

Tingay A (1985) Contemporary views of the voluntary conservation movement on the use of fuel reduction burns as a land management technique. In: Ford JW (ed) *Fire Ecology and Management in Western Australian Ecosystems*, pp. 215–218. WAIT Environmental Studies Group, Bentley, WA.

Toro H (1986) Lista preliminar de los Apidos Chilenos (Hymenoptera: Apoidea). Acta Entomólogica Chileña 13:121–132.

Trenberth KE (1993) North-south comparisons: climate controls. In: Mooney HA, Fuentes ER, Kronberg BI (eds) *Earth System Responses to Global Change*, pp. 35–59. Academic Press, San Diego.

Trollope WSW (1973) Fire as a method of controlling Macchia (Fynbos) vegetation on the Amatole mountains of the eastern Cape. Proceedings of the Grasslands Society of South Africa 8:35–41.

Twigg LE, King DR (1991) The impact of fluoroacetate-bearing vegetation on native Australian fauna: a review. Oikos 61:412–430.

Tzanoudakis D, Panitsa M (1995) The flora of the Greek islands. Ecología Mediterranea 21:195–212.

Underwood RJ, Sneeuwjagt RJ, Styles HG (1985) The contribution of prescribed burning to forest fire control in Western Australia. In: Ford JW (ed) *Fire Ecology and Management in Western Australian Ecosystems*, pp. 153–170. WAIT Environmental Studies Group, Bentley, WA.

Van der Hane T, Van den Berk V (1994) The Mediterranean migration route: network of wetlands for waterbirds. Limosa 67:159–162.

Van Wilgen BW, Cowling RM, Burgers CJ (1996) Valuation of ecosystem services: a case study from South African fynbos. BioScience 46:184–189.

Villagrán C (1994) Quaternary history of the mediterranean vegetation of Chile. In: Arroyo MT, Zedler P, Fox M (eds) *Ecology and Biogeography of Mediterranean Ecosystem in Chile, California and Australia*, pp. 3–20. Springer-Verlag, New York.

Villagrán C, Hinojosa LF (1997) Historia de los bosques del sur de Sudamérica, II: Análisis fitogeográfico. Revista Chileña de Historia Natural 70:241–267.

Wedin DA, Tilman D (1996) Influence of nitrogen loading and species composition on the carbon balance of grasslands. Nature 274:1720–1723.

Wellburn FAM, Lau K-K, Milling PMK, Wellburn AR (1996) Drought and air pollution affect nitrogen cycling and free radical scavenging in *Pinus halepensis* (Mill.). Journal of Experimental Botany 47:1361–1367.

Westman WE (1988) Species richness. In: Specht RL (ed) *Mediterranean Type Ecosystems: A Data Source Book*, pp. 88–89. Kluwer Academic Publishers, Dordrecht.

Westman WE, Malanson GP (1992) Effects of climate change on mediterranean-type ecosystems in California and Baja California. In: Peters RL, Lovejoy TE (eds) *Global Warming and Biological Diversity*. Yale University Press, New Haven.

Whetton PH, Hennessy KJ, Allan RJ (1994) Regional climate change. In: Mitchell CD, Hennessy KJ, Pittock AB (eds) *The Greenhouse Effect: Regional Implications for Western Australia. Final Report 1992–93*, pp. 28–39. CSIRO Division of Atmospheric Research, Melbourne.

Williams JE, Gill AM (1995) *The Impact of Fire Regimes on Native Forests in Eastern New South Wales*. New South Wales National Parks and Wildlife Service, Hurtsville, NSW.

Williams KJ, Hobbs RJ, Hamburg S (1987) Invasion of annual grassland in northern California by *Baccharis pilularis* ssp. consaguinea. Oecologia 72:62–66.

Wills RT (1993) The ecological impact of *Phytopthora cinnamomi* in the Stirling Range National Park. Australian Journal of Ecology 18:145–159.

Wills RT, Keighery GJ (1994) Ecological impact of plant disease on plant communities. Journal of the Royal Society of Western Australia 77:127–131.

Yates CJ, Hobbs RJ (1997) Woodland restoration in the Western Australian wheatbelt: a conceptual framework using a state and transition model. Restoration Ecology 5:28–35.

10. Deserts

Laura Foster Huenneke

Arid and semi-arid lands cover about one third of the earth's terrestrial surface. They are commonly perceived as hostile and barren places; in fact, they are heavily used by human populations. Direct and indirect effects of human activities in arid ecosystems will become increasingly important over the next few decades, and these effects will act on existing patterns of biodiversity in dryland regions. Extreme and severe environments have led to evolution of considerable biodiversity and unique biotic adaptations, some of which are of significant potential value in human terms (Huenneke and Noble 1996). Global changes in climate, land use, population density, and consumption threaten to cause two types of changes in arid and semi-arid systems: changes within existing deserts and conversions to or from other cover types. In this chapter I will consider first the patterns and extent of diversity in arid ecosystems, ranging from true hot deserts to temperate semi-deserts and shrub-steppe. I will then discuss the major mechanisms by which global change should affect drylands. I will outline the scenarios for change and the predictions for conversion among types and discuss how these changes might affect diversity of desert organisms at local and regional levels, with a focus on patterns of species richness. Finally, I will briefly discuss how these changes in biodiversity might be expected to affect ecosystem function.

Patterns of Diversity: The Basics of Desert Biology

Critical Characteristics of Desert Environments and Desert Organisms

Aridity is the defining characteristic of deserts. Precipitation inputs are determined by location within a continent relative to atmospheric sources of moisture; distance from oceans and rainshadow effects constitute major causes of aridity (Cooke et al. 1993). Altitude and local topography, however, modify regional influences to create small areas of greater or lesser moisture availability (e.g., see the discussion by MacMahon and Wagner [1985] for North America). Hence, local patterns are highly variable and not predictable from large-scale geographic trends.

Precipitation in some deserts is strongly seasonal despite the unpredictability of amount (e.g., the winter-dominant Mojave Desert of western North America), but precipitation in other deserts is not seasonal at all (e.g., the Sahara: Grenot 1974). The strong distinction between mediterranean vegetation and that of true deserts highlights the importance of predictable versus nonpredictable precipitation. An extreme example of dependable (but low quality) resources is fog in coastal deserts. Fog inputs, though small in amount, have an entirely different impact than the same amount of water received in unpredictable rain events. Extreme unpredictability or rare weather events (e.g., unusually heavy freezes) can shape the entire nature of a desert community (e.g., in the Sonoran: Hastings and Turner 1965) by eliminating growth forms or physiological types.

Environmental variability is a major contributor to the harshness of desert regions (Noy-Meir 1973; Louw and Seely 1982). Interannual variability of both precipitation and temperature, as measured by coefficient of variation and other indexes of variability, is high (Shmida et al. 1985). Noy-Meir (1973) conceptualized desert ecosystems as following a pulse-reserve pattern, where episodic inputs (e.g., pulses of precipitation) lead to strong responses in desert organisms, leading in turn to increases in the "reserves" of an individual or of a population that allow its persistence through subsequent harsh periods.

Shmida et al. (1985) suggested that there is a tradeoff in arid lands between resource quality and resource availability or predictability. High-quality resources might result from an unusually heavy local rain event, for example, and the resultant pulse of plant productivity. More typical and more widespread would be years/areas with low rainfall, when moisture and/or food resources would be scarce (but predictably so). As referred to earlier, fog inputs might be predictable, but limited. Shmida et al. (1985) speculated that most desert taxa are specialized either for poor quality, predictable resources (e.g., termites, tenebrionids, CAM xerophytes, some ants, and ungulates), or for higher-quality but episodic resources. Examples of organisms specialized on the latter would include: annuals (Inouye 1991) and geophytes; estivating rodents, amphibians, and arthropods; and crustaceans and other inhabitants of ephemeral water bodies. For example, poikilotherms such as lizards can become inactive during harsh conditions, essentially specializing on favorable periods (Pianka 1979). Pianka (1979) predicted that some species may persist by occurring only in better habitats in tough years, but expanding to use a broader array of habitats in favorable years. Whitford (1976) documented such immigration and local extinctions of cricetid rodents as the suitability of habitat patches varied over time. Nomadism or migratory behaviors, which are well represented in desert ungulates, birds, and locusts, is another adaptation to the occurrence of rich resources in small pockets of a larger desert. All of these behaviors—opportunistic population flushes, estivation or dormancy, nomadic movements—exacerbate the difficulty of detecting and sampling biological diversity in arid environments.

Biotic relationships in drylands may enhance local species richness considerably. Large and/or long-lived plants ameliorate the harsh physical environment through modification of both soil and microclimate. These effects may lead to facilitation as well as competition among plant neighbors (Pugnaire et al. 1996). When large plants create conditions suitable for the establishment of seedlings, this facilitative role has been called the *nurse plant phenomenon* (Franco and Nobel 1989; Suzan et al. 1996). Long-lived woody plants certainly modify the environment in multiple ways that shape the distribution and behavior of other organisms (Schlesinger et al. 1996; Shachak and Lovett 1998). Animals may also create favorable microenvironments for other species (e.g., when rodents construct burrow systems used by other animals [Reichman and Smith 1990]), or when animal-created soil disturbance regulates vegetation patterns (Shachak et al. 1991; Miller et al. 1994; Weltzin et al. 1997). Whitford (1993) speculated that animal activities in deserts tend to reinforce the patchiness of soil resources that characterizes desertified systems.

Diversity: Environmental Gradients

Contrary to popular stereotype, deserts are not necessarily species-poor, relative to other temperate environments (Polis 1991). For some taxonomic groups, diversity increases with moisture at this lower end of the moisture gradient (e.g., on a regional or continental scale plant species diversity usually increases as arid lands grade into grasslands, woodlands, or thorn-scrub vegetation [Shmida et al. 1985]). Even though some animal groups follow this pattern, others do not. Both Pianka (1979) and Polis (1991) pointed out that the diversity of some animal taxa decreases with aridity, that of other taxa increases with aridity, and the richness of some groups increases with aridity on one continent but decreases on another. Brown (1975) demonstrated that local rodent diversity may increase with precipitation, but rodent species richness decreases with precipitation over a larger (continental) scale. Hence, it is not possible to generalize about the responses of animal diversity to some deterministic, directional change in moisture availability. Over geographic gradients, plant diversity seems more related to floristic boundaries and biogeographic considerations (e.g., local centers of endemism) than to environmental variables (e.g., MacMahon and Wagner 1985; Mourelle and Ezcurra 1996).

Highly local influences may constrain species richness more directly than any geographic factor. For example, soil texture and its effects in modifying moisture availability have frequently been cited as strong influences on plant species diversity (e.g., in South African Karoo: Werger 1986; see also Orians and Solbrig 1977). In many semi-arid and arid landscapes, geomorphology and local topography are the primary correlates of plant species richness because of their strong control of runoff, runon, and infiltration patterns— and, hence, moisture availability (McAuliffe 1994).

Diversity: Species-Area Considerations

It is impossible to generalize about diversity-area relationships across ecosystems varying from barren Sahara plains devoid of vegetation to diverse semi-desert grasslands. Sampling difficulties undoubtedly contribute to the dearth of published examples of species-area curves for desert regions. The impact of temporal variation on sampling can be seen in the long-term data sets gathered by the Jornada Long-Term Ecological Research program in the northern Chihuahuan desert of New Mexico. At the scale of study sites or stands (areas $70\,m \times 70\,m$, sampled with 49 1-m^2 quadrats), grasslands dominated by perennial *Bouteloua* and *Sporobolus* spp. are significantly richer in plant species than are sites dominated by shrubs (*Larrea tridentata*, *Prosopis glandulosa*, or *Flourensia cernua*). These patterns of relative species richness are consistent whether one examines data from a single sampling season or species lists accumulated over several years of sampling (Huenneke, unpublished analysis). At any one sample date, however, or even pooling data from multiple samples during a single year, we rarely encounter more than 30–35% of the plant species known to have been observed at a site over time. After 10-plus years of sampling, we continue to encounter new plant species on each site.

This intrinsic variability and importance of ephemeral species suggests that it would be difficult to construct meaningful species-area curves or measurements of diversity for purposes of comparison among sites. Even data collected carefully during a time of high resource availability (and high diversity) could miss a substantial portion of a site's flora (or fauna), and comparisons between sites or regions could scarcely be made unless one understood the conditions under which the different areas were sampled. It might be preferable to sample repeatedly over time, and to use cumulative species numbers, but I know of virtually no examples where repeated sampling has been combined with spatial sampling to construct species-area curves.

Diversity: Functional Considerations

Because of harsh environmental conditions and presumed tight linkages between physiology and survival in desert environments, arid communities have often been evaluated by classifying species with physiological or functional approaches. For plants, photosynthetic pathway has been assumed to be a critical characteristic. Many studies have reported strong associations between climate (especially seasonality of precipitation) and dominance of particular photosynthetic pathways in a flora (e.g., MacMahon and Wagner 1985; Vogel et al. 1986; Werger 1986). Photosynthetic pathway is often correlated with other aspects of physiology and growth form (e.g., cacti and CAM, warm-season grasses and the C_4 pathway). Because semi-arid ecosystems often contain very different proportions of species with CAM, C_3, and C_4 pathways, it is easy to assume that the relative abundances of functional types in a given site would be altered dramatically if the seasonality and/or amount of precipitation were to change with climatic changes.

Another functional approach used commonly in analyzing semiarid ecosystems is based on plant growth forms or architectural types (e.g., perennial grasses versus shrubs versus succulents). Although there has been tremendous interest in understanding the displacement of grasses by shrubs, and in understanding the ecosystem implications of those changes (Reynolds et al. 1997; Sala et al. 1997; Kieft et al. 1998), it is becoming clear that there are significant differences within as well as among those functional types. For example, individual perennial grass species vary substantially in productivity and response to drought (Herbel and Gibbens 1996), and there is increasing interest in identifying and modeling these species-level differences. The explosion of interest in functional group models of response to disturbance or to environmental factors (Diaz et al. 1999) is particularly relevant to dry rangelands, where both drought and grazing impacts are expected to affect community structure.

Distributional Categories of Desert Species

For this chapter, I distinguish three groups of desert organisms that will be based on their distributional patterns. These three groups are likely to respond differently to the various drivers of global change, and, therefore, deserve different management and forecasting strategies. The three categories are: organisms of desert plateaus and basins, with widespread distribution in continuous habitat; endemics of special substrates and unique habitats; and organisms dependent on wadis, playas, springs, riparian areas, or other sources of water. I will describe each category here briefly before considering their potential sensitivities and responses to global changes.

Basin/Widespread Species

Most deserts have formed on either shield and plateau surfaces or in basin and range geomorphic regions (West 1983a, MacMahon and Wagner 1985, Cooke et al. 1993, for western United States). In either case there are extensive "floors" or continuous surfaces typically covering more than 50% of the region's surface area, with few barriers to dispersal or gene flow. Just one or two plant species often dominate the community, sometimes over very extensive ranges (MacMahon and Wagner 1985, western United States; Sahara descriptions by Grenot 1974). In northern Chihuahuan desert ecosystems, the dominant shrub in either *Larrea*, *Prosopis*, or *Flourensia* shrublands often comprises 80–95% of the site's total live plant biomass (Jornada Long-Term Ecological Research program, unpublished data). These widespread dominant woody plants (frequently legumes: e.g., Braun Wilke 1982) are conspicuous and key species, strongly influencing soil characters and providing dependable resources for herbivores and decomposers. Many desert organisms are geographically widespread (Shmida et al. 1985); however, they do not necessarily possess good dispersal abilities (e.g., herpetofauna, West 1983a; plants, Ellner and Shmida 1981). A comparison of small mammal assemblages (Kelt et al. 1996) determined that deserts on four continents uniformly supported a low diversity of small mammals in any one place, but considerable turnover or geographic variation in mammal composition from site to site. This pattern would suggest that even animal species occurring in widespread habitats are not necessarily wide-ranging themselves.

Habitat Specialists

Most deserts and semideserts include at least some specialized habitats supporting so-called azonal vegetation. These include dune fields, gypsum substrates, saline or hypersaline areas, and lava flows. Unusual environmental conditions have led to the evolution of unique floral assemblages and, sometimes, even vertebrate species (e.g., mobile sand dunes are centers of diversity for highly specialized animals as well as plants [Seely 1991]). These small, disjunct pockets of specialized habitat with their high endemism harbor a significant and unique fraction of the diversity of arid regions (MacMahon and Wagner 1985; Shmida et al. 1985).

Riparian or Water-Dependent

A small percentage of the surface area of desert regions supports permanent or intermittent water resources: springs/oases, permanent streams, arroyos or wadis, playas, or ephemeral lakes. These locations offer predictable (in space, if not in time) resources for organisms that are critical in keeping landscape-scale productivity and diversity higher than equivalent areas without such "hot spots" of water availability (Shmida et al. 1985). In addition, these habitats may support unique species, adding significantly to the diversity of the overall landscape. For example, desert springs in some regions contain high numbers of endemic fish and aquatic invertebrates. Riparian habitats often support a disproportionately high fraction of the flora; for example, of 479 vascular plants listed in the flora of the Jornada Plain in New Mexico's Chihuahuan desert, 80 (17%) are listed as occurring only in washes, playas, and other relatively moist habitats, although these habitats cover far less than 17% of the total area (Allred 1997).

Human Impacts and Current Conservation Status

Huenneke and Noble (1996) reviewed the ways in which human activity has altered diversity in dryland ecosystems. Many of these impacts are related to traditional extensive use of semiarid regions for livestock grazing: introduction of both livestock and feral grazers to some regions lacking an evolutionary history of large-hooved grazers; removal of predators and of burrowing and herbivorous native mammals; introduction of nonnative plants, either deliberately for improved forage, or accidentally. Another effect on extensively used lands is the depletion of trees or large shrubs taken for fuelwood or other purposes. For other than these conventional uses, drylands have traditionally been viewed as barren "wastelands" of no economic or aesthetic value; therefore, in industrialized nations they have become favored sites for "nuisance installations" (e.g., military, waste disposal, or power plant sites) (West 1983b). The advent of irrigation and other water supplies has allowed conversion of arid lands for cultivation and for urban growth of human populations. Thus, it is only relatively recently that direct habitat conversion and fragmentation have become important over large areas of arid and semi-arid regions.

Human impacts do tend to reduce diversity, at least at some scales. Some species recognized as promoting the diversity and function of dryland ecosystems remain the targets of eradication programs due to perceived conflicts with human interests, primarily grazing (e.g., North American prairie dogs: Miller et al. 1994). Livestock grazing in semi-arid grasslands has been observed to reduce plant diversity at the small scale (Fuhlendorf and Smeins 1999), although not necessarily at the landscape scale (Stohlgren et al. 1999).

The record of reserve establishment in desert regions has been spotty because there is little popular appreciation for the diversity and richness of aridland biota. Parks or preserves are as likely to be established to protect physical features or landscapes (e.g., the White Sands National Monument of New Mexico, which encompasses a large gypsum dune system) as to protect biological elements. Drylands are well represented in national parks and biosphere reserves (recognized by the UNESCO Man and Biosphere program) in North and South America and in Australia, for example, but the record for Asian deserts is much less impressive. The establishment of desert preserves in California and Utah was accompanied by bitter wrangling over political and economic issues. The lack of protection of many desert biotas persists despite a growing recognition of the extremely high diversity and relative endangerment of groups such as rodents, bats, and cacti (Ceballos and Brown 1995; Hernandez and Barcenas 1996).

Scenarios of Change for Arid Lands

Satisfying predictions of the magnitude, or even direction, of climatic change in desert regions have been hard to obtain. Because precipitation inputs are so dependent upon local topography and flows of moist air from sometimes distant oceanic regions (Wurtele 1987; Gasse et al. 1990), different models may predict either significant increases or significant decreases in precipitation depending on the specific path of storm tracks or high-altitude "conveyors" of moist air. (The seasonality of precipitation might also be altered, with or without changes in total average amount.) It has generally been assumed that warming will initially outpace the consequent increase in hydrologic cycling and precipitation, and that many regions will experience at least transient periods of intensified aridity (Mitchell and Warrilow 1987; Rind et al. 1990). Increased precipitation is ultimately forecast, however, for many of today's semiarid regions. This general trend is countered by an almost certain continuation of desertification, driven by more intense use of rangelands and by attempts to convert arid and semiarid lands for cultivation. Projected increases in environmental variability, including increased frequency and intensity of drought, could also push some systems toward arid conditions (Mabbutt 1989).

The land-use–land-cover change projections of the IMAGE model (see Chap. 3) forecast a range of futures for the total area of hot deserts and scrubland, with trends varying by continent and by projections of population increase. Contrary to general expectations, few regions show an overall trend of short-term increase in area of dry zones followed by a decrease (with increasing precipitation); instead, some show strong unidirectional increase in arid area (e.g., Latin America), whereas others demonstrate a continuous decrease in arid area (e.g., the Asian regions). In particular, large areas currently in agricultural and pastoral (i.e., extensive grazing) use are predicted to experience conversion to hot desert or scrubland (i.e., desertification), but even larger areas of desert and scrubland are predicted to become grasslands or even agricultural areas as precipitation inputs rise, and growing human populations require expanding cultivation.

Within today's arid lands, the obvious drivers of change are those influencing (directly or indirectly) moisture and temperature stress: precipitation, temperature, CO_2 (through its effect on stomatal conductance and plant water relations), and wind speed. A more subtle example might be light intensity. Most deserts are regions of clear air and cloud-free days, with plants adapted to extremely high PAR (high light compensation and saturation points). If precipitation increases were accompanied by increased cloud cover, photosynthetic performance of some desert plants could conceivably be compromised despite the improved water balance. Land-use changes, driven both by increased human populations and by changes in the ways in which people use arid landscapes (e.g., conversion to irrigation), could have extremely strong effects—but, realistically, this will happen only in those regions where economics and/or energy use enables the human population to obtain water from below ground or surface sources (e.g., in the southwestern United States and arid Mexico, where a number of cities have grown extremely rapidly from 1970 to the present). Meanwhile, it is reasonable to expect continued pressure to use semi-arid and arid rangelands for production of animal protein. Finally, increased transportation and human activity will inevitably lead to increased rates of accidental (and perhaps also deliberate) species introductions. Because a significant fraction of introductions results in invasions of native systems (Lonsdale 1994; Williamson 1996), increased rates of introductions will probably lead to declines in native species diversity.

The second category of change to be concerned with is the formation of "new" arid ecosystems through desertification, whether caused by climatic shifts or by anthropogenic degradation of semi-arid systems. Even though consensus on a definition of desertification has been elusive (e.g., degradation to non-economic use? or simply decrease in net primary production?), the process is recognized as an important and undesirable consequence of poorly managed human activities (Jain 1986; Meyer and Turner 1992). Desertification has already become an intense problem, with about 70% of drylands globally (excluding hyperarid deserts, which are considered to be immune from further degradation) considered to be desertified at this point (Daily 1995). Jain (1986) rated vulnerability to desertification to be high for 20–25% of the surface area of Africa (45% of Australia). A conventional view is that desertification occurs in developing countries, whereas deserts are actually lost by conversion to other uses in industrialized nations (e.g., Safriel 1987). Attention to desertification in "northern" nations, however, demonstrates that this is an oversimplified view. Given the view that desert organisms do not have particularly well-developed dispersal mechanisms (see earlier) and empirical evidence that abandoned agricultural land in desert regions is extremely difficult to restore (Jackson et al. 1991; Roundy and Biedenbender 1995), one cannot assume that recently desertified areas will be quickly colonized by species suited to arid conditions. Shrublands of recent origin in the northern Chihuahuan desert of New Mexico, derived by shrub encroachment into degraded semi-arid grasslands in the past century, support fewer plant species than desert shrublands in the same region occupying landscape positions that have long been dominated by shrubs (Huenneke, unpublished Jornada Long-Term Ecological Research program data). The desertified sites contain very few species other than those found in grasslands; that is, they appear to be derived simply by the impoverishment and the change in relative abundances of the grassland flora, rather than by addition or substitution of species typical of desert scrub. The inference is that desertified areas are slow to accumulate desert plant species. (Bird and small mammal diversity, however, is as high in desertified sites as it is in desert grassland: Whitford 1993, 1997.)

Impacts of Environmental Changes: Possible Approaches to Prediction

There are at least three possible approaches to attempting to predict the effects of environmental change on the biota of desert regions: the biogeographic approach, the deterministic approach, and the life history/probabilistic approach. As will be discussed here, each of these unfortunately has some severe constraints limiting its usefulness for our objectives.

The Biogeographic Approach

Application of biogeographic principles would focus on changes in the areal extent and the location of arid regions, and on the opportunities for migration, colonization, and persistence of desert organisms in existing and newly arid regions. It is appealing to make the generalization that many desert organisms are widespread, and that geographic barriers to migration are few in the continuous, broad landscapes that characterize many desert regions. As reviewed earlier, however, many groups of arid-land species have poor dispersal capabilities, and are not good candidates for migration or colonization. Brown (1993) pointed out that areas containing considerable topographic diversity (e.g., valleys or mountain ranges) should retain diverse microclimates (and thus diverse organisms), whereas broad plains or plateaus might experience more dramatic shifts in environmental conditions without any local refugia (Brown 1993). Migration is also relevant in the consideration of desertification. It is likely that newly desertified areas, formed by unsuccessful attempts at cultivating marginal land, may be located far from potential sources of colonists. For example, predicted occurrences of desert in Europe would be close in air miles to desert organisms of northern Africa, but the Mediterranean Sea is surely a major barrier to the dispersal of dryland plants and animals, other than birds and flying insects.

Another biogeographic approach would be to examine the likely outcomes of habitat fragmentation, as previously continuous arid regions become more intensively settled and converted by human action. Two major areas of uncertainty prevent the useful application of this approach. First, as reviewed earlier, detection and sampling of many important groups of desert organisms are so problematic that the necessary species–area relationships are not understood. Second, human intrusion into and conversion of drylands have been limited enough that we have no empirical information on how desert organisms respond to fragmentation. This is clearly an area for future research. An additional problem is related to the observation by Brown et al. (1995) that most species are distributed very patchily: Within a species' range, abundance is high at only a few locations ("hot spots"), and individuals are sparse or absent across much of the rest of the range. Projections of the impact of loss or shift of range, then, depend on knowing whether hot spots or less critical portions of the range are affected. This information is rarely available, even for common and well-studied taxa, much less for sparsely distributed and poorly known desert organisms. At present, then, it is not possible to use species–area relationships to make any quantitative estimates of species' losses.

The Deterministic Approach

Another attractive idea is to reason from current correlations of diversity with environmental gradients, to predict how diversity will change with changes in environmental variables. For example, because plant diversity generally increases with increasing precipitation along geographic gradients, one would predict that projected increases in precipitation should lead to increased plant species diversity. One problem with this approach is the overwhelming importance of interactions among environmental factors. One could potentially make predictions about the likely impact of increased precipitation, or of increased temperature, or of increased CO_2 concentrations; but it is far more difficult, perhaps currently impossible, to put together a convincing projection of how even a single plant species will respond to all three simultaneously. This, however, is the most likely scenario for many desert regions. Some claim that rising CO_2 is already altering the competitive balance between C_3 and C_4 plants (paleohistoric evidence, Cole and Monger 1994; experimental results, Johnson et al. 1993; Polley et al. 1994). It is not clear, however, how hydrological changes (either greater water availability or more frequent, intense droughts) will affect this balance (Long and Hutchin 1991; Chaves and Pereira 1992; Field et al. 1992).

Another problem with this approach is the recognition that in the future, as in the past, species are likely to respond individualistically to climate change rather than as integrated assemblages (FAUNMAP Working Group 1996). Thus, one cannot simply make a prediction about the future environmental conditions at a site, seek an area in today's world with similar conditions, and then assume that a similar biotic assemblage will appear in the future at that site. The attempt seems particularly dangerous because, as reviewed earlier, different desert taxa respond differently to the same environmental gradient.

A more sophisticated and interesting approach was used by Rochefort and Woodward (1992), who used known tolerances of vascular plant families for low temperatures and low precipitation to model plant family distributions in a future world of doubled CO_2, increased precipitation, and moderate warming. They predicted that today's dry regions should experience a substantial increase in family diversity, largely as a result of increased precipitation and reduction of evapotranspiration due to higher CO_2 levels. Their analysis, however, ignored the impact of high temperatures; temperatures constrained family diversity only through tolerances of minimum temperatures or through a minimum required heat sum for growth or reproduction. In hot deserts, though, tolerance of high leaf and stem temperatures is already a significant constraint to plant diversity, and I would expect a decline in the diversity of plants able to tolerate such conditions in a scenario of future, even higher, temperatures even if precipitation were to increase. (In fact, the projected decline in evapotranspiration attributed to elevated CO_2 could exacerbate temperature effects, as reduced stomatal conductance would reduce evaporative cooling and increase the potential for reaching lethal leaf temperatures.)

Yet another problem in predicting responses to global change agents is the fact that populations of long-lived organisms may appear to be resisting the impact of change long after stresses have actually reached a critical level; individual longevities, or the "reserves" characteristic of some desert species, might mask the fact that a local environment had crossed some threshold

(Milchunas and Lauenroth 1995). The time lag in detectable response to environmental change, then, makes it difficult to build predictive models of such responses—or even to design monitoring programs to attempt to detect such changes before the system crosses the threshold (Herrick and Whitford 1995).

The Life History or Probabilistic Approach

Individual-based, spatially explicit models of forest tree populations have been suggested as suitable starting points for making predictions of global change impacts (e.g., Shugart et al. 1992). There has been limited application of population modeling to desert dynamics (e.g., McAuliffe 1988). The environmental tolerances of annuals, as expressed in germination/dormancy and reproductive success, have been used to model species' appearances and performances (e.g., Freas and Kemp 1983; Bachelet et al. 1988; Pake and Venable 1995). It may be possible to use a similar approach to model the performance, recruitment, and long-term community dynamics of perennial plants, including scenarios with altered environmental conditions, but the data requirements of such a project would be enormous. A parallel for animal populations (e.g., for estivating amphibians) would be to assess physiological performance, survivorship, and reproductive success in years with different environmental conditions, and then use stochastic models to explore the population impacts of various sequences of years under future climate regimes.

Even though these approaches seem reasonable, it is important to remember that their product would be a probabilistic description of species' appearances and performances over time. Even with such predictions, we would face a daunting statistical problem in comparing our observed sequences of species' appearances with the projections. If a given plant species is abundant only once every 7 years, on average, how are we to detect a climatic response that results in its appearance once every 12 years on average instead?

In summary, biogeographic approaches are probably not applicable to most desert regions; deterministic approaches cannot succeed because of strong and unpredictable interactions of environmental factors; and probabilistic or stochastic approaches are far from development and would be difficult to apply to such variable systems. Researchers have had difficulty understanding historical vegetation changes in semi-arid regions and distinguishing climatic signals from human impacts (e.g., Neilson 1986; Dregne and Tucker 1988; Schlesinger et al. 1990; Tucker et al. 1991; Conley et al. 1992; Bahre and Shelton 1993), demonstrating the immense difficulty of predicting response to future interactions of climate and land use.

Impacts of Environmental Changes: Projected Changes in Diversity

As the preceding comments suggest, I am reluctant to make quantitative predictions about gains or losses in aridland biodiversity, both because the localized expression of precipitation and land-use change is so uncertain and because the underlying nature of organismal response to environment is not well understood. In Table 10.1, I summarize my qualitative judgments about the probable responses of desert organisms to single drivers of environmental change. (I reiterate that interactions among these drivers will be critically important in arid systems, limiting the usefulness of the single-factor approach of Table 10.1.) In constructing the table, I focused on species of widespread desert habitats (i.e., not narrow habitat specialists or species restricted to aquatic or riparian locales). I also assumed that human activities will continue to be primarily extensive in nature (e.g., livestock grazing, although the intensity of that use will assuredly increase as human population increases). That is, I assumed that water availability will constrain the conversion of deserts to more intensively used cover types (urban or cultivated). Finally, I assumed that precipitation will increase overall, in concert with temperature and CO_2. I will discuss the general nature of the predictions here; I will add comments relevant to habitat specialists and riparian species later, and close the section with comments on the general nature of environmental response in desert communities.

Habitat loss (e.g., through conversion to agriculture) would certainly represent a strong negative influence on species diversity, but uncertainties in the social/political/economic arena about availability of water make it difficult to forecast the extent of such conversion. Similar uncertainties surround attempts to predict fragmentation of desert habitats; there is also little basis for predicting the response of desert species or functional groups to fragmentation. In contrast, we can be certain that grazing, recreational use, and harvest of wood or of economically valuable species will intensify in most arid regions as human populations continue to grow. There is some understanding of the impacts of grazing and tree removals on species and functional group diversity, and on ecosystem functions and services (e.g., productivity) (Huenneke and Noble 1996).

Uncertainty is high regarding the response of desert diversity to increasing nitrogen availability and to other forms of pollution. Although nitrogen can be limiting to plant growth in arid lands, water availability interacts strongly with nitrogen (e.g., Gutierrez and Whitford 1987; Fisher et al. 1988), and it is difficult to predict responses to nitrogen when precipitation is variable. At least one long-term fertilization experiment in the Chihuahuan Desert produced reduced plant species richness with nitrogen additions (Wondzell et al. 1987); however, there is growing evidence that exotic species may respond to increased nitrogen and water availability in semi-arid systems (e.g., Milchunas and Lauenroth 1995). I have assumed that increasing precipitation will make desert ecosystems more sensitive to N inputs, but it is doubtful whether many desert regions will see localized increases in urbanization and cultivation leading to increased nitrogen deposition.

As discussed earlier, the uncertain interactions of increased CO_2 concentrations with moisture, temperature, and seasonality of precipitation compromise any prediction of the direction of the response of species or of plant

Table 10.1. Widespread species of desert habitats

Δ In diversity	Change in land use							Change in climate		
	Hab. loss	Fragm.	Intensif.	N rise	Pollut.	CO_2 rise	Intr.spp	Δ Precip.	Δ Temp.	Δ In seas.
Local loss of spp.	5C	3B	3C	2B	2A	3A	4C	4C	3B	5C
Spp. extinctions	4B	2B	3C	2B	1A	2A	3B	3C	2B	4C
Spp. additions	1B	2B	1B	3B	1C	3A	5C	4C	1C	3A
Δ Spp. rel. abund.	5C	3B	4B	4C	4C	4A	5C	4C	3B	5C
Loss funct. grps.	5C	2C	4C	2A	1A	2A	4B	3C	3B	4C
Gains funct. grps.	3C	1B	1B	2A	1A	2A	3B	4C	1C	3B
Δ F.G. rel. abund.	5C	2B	5C	4B	3A	4A	4B	5C	4C	5C
Δ Landscape div.	4C	4C	1B	1B	2A	2A	3B	4C	3A	3B
Direct Δ function	4C	2B	4C	4B	3B	3A	5C	5C	4C	5C
Indirect Δ function	4B	2A	3B	2A	2A	3A	4C	4B	3B	4B
Δ Ecosys. services	5C	2B	4B	2A	2A	3A	4C	4B	3B	3B
Expected Δ drivers	3A	2B	5C	2A	2B	5C	3B	5A	4C	3A
Lag in diversity Δ	5C	2B	4C	3B	3A	3A	4C	3B	3B	2B

Drivers of global change abbreviated as: Hab. loss, habitat loss or conversion; Fragm., fragmentation; Intensif., intensification of current human land use patterns (grazing and wood harvest); N rise, increase in nitrogen deposition and availability; Pollut., increase in anthropogenic pollution; CO_2 rise, doubling of atmospheric CO_2 concentrations; Intr.spp, introductions of invasive nonnative plants and animals; Δ Precip., change in precipitation (assumed to increase); Δ Temp., change in ambient temperatures (assumed to increase); and Δ In seas., change in seasonality.

Changes in diversity abbreviated as follows: Local loss of spp., elimination of current desert species on local scale; Spp. extinctions, global extinctions of desert species; Spp. additions, local additions to flora or fauna through species migration or survival in more benign conditions; Δ Spp. rel. abund., change in species' relative abundances; Loss funct. grps., local disappearance of functional groups, plant growth forms, or physiological categories; Gains funct. grps., local additions of functional groups; Δ F.G. rel. abund., change in relative abundances of functional groups; Δ Landscape div., change in landscape diversity; Direct Δ function, direct changes in ecosystem function caused by environmental change; Indirect Δ function, indirect changes in ecosystem function due to alterations in diversity; Δ Ecosys. services, change in ecosystem services due to alterations in diversity; Expected Δ drivers, relative magnitude of change expected in environmental condition; Lag in diversity Δ, time required for diversity to change in response to change in environmental condition.

Key: Numeric values represent intensity or magnitude of response, from 1 = low to 5 = high. (For Lag in diversity Δ, 1 = low to 5 = high. 3 = 10–50 years, 4 = <10 years, 5 = instantaneous.) Letters represent certainty of understanding, from A = low to C = high.

functional groups. One can say no more than that there is the potential for substantial shifts in the relative abundance of plants with different photosynthetic and water-use patterns.

Nonnative plant species are noted for more aggressive invasion into semiarid temperate regions than into hot deserts (e.g., Mack 1986); however, where invasive plants do spread, they can have significant impacts on the diversity and functioning of native systems. One example is where planted exotics have promoted fire in previously fuel-limited drylands (Anable et al. 1992; D'Antonio and Vitousek 1992). Kerley and Erasmus (1992) noted that increased fire in deserts can eliminate mammals that are dependent on aboveground nests or shelters of wood and other plant debris by destroying the shelters and the materials for their construction. Arid lands also have a record of successful invasions by vertebrates (especially feral livestock; e.g., Freeland 1990). Increased transport and human activity in deserts, including deliberate forage plant introductions and other grazing-related disturbances, will increase opportunities for species introductions and the potential for negative impacts on native diversity. If low moisture availability has limited the invasion of some plants into semi-arid systems, then increases in precipitation could increase the impacts of nonnatives.

As discussed earlier, it is difficult indeed to predict how desert biotas will respond to climatic changes. Indeed, although I am confident that changes in precipitation (both amount and seasonality) could have large effects on desert species, I am not at all confident of the sign of the effect, positive or negative, for the biota in aggregate!

Discussion and Conclusions

What, then, can we conclude about how desert ecosystems are likely to respond to global change? Although I have been pessimistic about formal attempts to predict diversity changes, I do feel there is value in examining the various distributional categories of species described early in this chapter and considering how each is likely to respond.

First, the pulse-reserve organization of desert systems lends stability to component populations. Species of widespread desert habitats are often dependent on some sort of reserve (or, alternatively, dormant state) to survive dry periods. Such reserves include the grain stores of rodents and harvester ants, and the eggs or resting stages of poikilotherms. Any environmental change that threatens those reserves could destabilize populations and lead, potentially, to local extinctions. Brown et al. (1997) point out that unusually wet winters can reduce granivores, presumably when infiltrating water ruins seed caches, and Muth (1980) suggested that unusually wet or unusually dry soils can be fatal to incubating lizard eggs. Hence, changes in precipitation regimes might lead to abrupt reductions in many desert species. The novelty of fire in arid systems, and its elimination of woody litter (shelter for animals, organic inputs to the soil), might be an equivalent dramatic perturbation.

Second, the fact that desert regions are sometimes dominated by one or a few widespread species suggests that the system could be vulnerable to disruption by anything changing the behavior of those dominants; that is, the system may demonstrate nonlinear or abrupt, discontinuous responses to change (Myers 1995; Brown et al. 1997). If the invasion of shrub species can alter ecosystem processes so dramatically (Schlesinger et al. 1990, 1996), then the loss of a single dominant woody species could lead to equally dramatic shifts in ecosystem function. The importance of "nurse plant" effects or other types of facilitation in these harsh physical environments would exacerbate the destabilization of a system where the dominant species responded negatively to some environmental shift.

Third, species endemic to local, specialized habitats will be much more vulnerable to environmental changes than would the species discussed earlier. Both direct habitat loss (e.g., through conversion) and climatic change would be devastating to substrate specialists that are unable to locate and migrate to suitable habitat patches in a favorable climate (Pimm et al. 1995). Extensive mining of gypsum in western North America, where gypsum soils and dunes support many endemic plants and animals, is an example of habitat loss driven by economic pressures.

Finally, riparian and aquatic habitats are without doubt the most threatened components of arid regions. Human and livestock impacts are often most intense in these areas (Grenot 1974; Kauffman and Krueger 1984; Hanan et al. 1991). With growing human populations one may reasonably expect increased impacts of grazing animals, water withdrawal and diversion, and even recreational visitors. Furthermore, because both permanent and ephemeral water sources in arid regions are usually dependent on contemporary precipitation and runoff, these habitats will be extremely vulnerable to climatic change. The linkage between aquatic resources and climate is an area where considerable progress has been made in modeling approaches (DeAngelis and Cushman 1990; Chang et al. 1992). For example, Chiew et al. (1995) simulated the effects of climate change on watersheds in Australia; runoff and soil moisture appeared to be far more sensitive to precipitation than to temperature change. It is interesting that this study compared the projections of different global circulation models and found that different models all predicted large effects on arid watersheds, but that different models predicted change in different directions! Another aspect of the vulnerability of riparian habitats is their susceptibility to biological invasions; watercourses and springs in arid regions have experienced high rates of invasion (e.g., by *Tamarix* spp.) relative to other portions of the desert landscape. It is easy to predict that such impacts will intensify as human activity increases.

In summary, then, it is likely that desert organisms will show dramatic responses to climatic and land-use change, although it is difficult to forecast those responses quantitatively. Alterations of dominant plants, plant functional types, and soil organisms are likely to have strong ecosystem consequences [e.g., on potential productivity and rain use efficiency of the vegetation (Le Houérou 1984), on soil properties and erosion (Kovda et al. 1979; Schlesinger et al. 1990), and on organic matter inputs (Grenot 1974; West 1979; Seely 1991)]. Those alterations will also influence ecosystem services to humans (e.g., forage and fuel yields). The future of the world's arid lands is likely to be important on a global scale because of the influence of deserts on the globe's albedo, dust production, and biogeochemical cycling of several minerals (Schlesinger et al. 1990; Graetz 1991). Thus, it is critical that we improve our understanding of arid land ecosystems and our management of their most vulnerable components.

Acknowledgments. I appreciate the invitation by Terry Chapin and Osvaldo Sala to participate in the workshop and to prepare this chapter. Manuscript preparation and results from the Jornada Long-Term Ecological Research program were supported by the Jornada LTER grant (NSF DEB 94-111971) and by the Global Change research program of the USGS Biological Resources Division (formerly the National Biological Service). Helpful comments by the following colleagues improved a previous version of this chapter: Jim Brown, Bob Gibbens, Jordan Golubov, Kris Havstad, Rob Jackson, Graham Kerley, Maria del Carmen Mandujano, Mitch McClaran, Graham Roy, Bill Schlesinger, and Walt Whitford.

References

Allred KW (1997) *A Field Guide to the Flora of the Jornada Plain, Second Edition.* New Mexico State University Range Herbarium, Las Cruces, NM.

Anable ME, McClaran MP, Ruyle GB (1992) Spread of introduced Lehmann lovegrass, *Eragrostis lehmanniana*, in southern Arizona, USA. Biological Conservation 61:181–188.

Bachelet D, Wondzell SM, Reynolds JF (1988) A simulation model using environmental clues to predict phenologies of winter and summer annuals in the northern Chihuahuan Desert. In: Marani A (ed) *Advances in Environmental Modelling*, pp. 235–260. Elsevier, Amsterdam.

Bahre CJ, Shelton ML (1993) Historic vegetation change, mesquite increases, and climate in southeastern Arizona. Journal of Biogeography 20:489–504.

Braun Wilke RH (1982) Net primary productivity and nitrogen and carbon distribution in two xerophytic communities of central-west Argentina. Plant and Soil 67:315–323.

Brown JH (1975) Geographical ecology of desert rodents. In: Cody M, Diamond J (eds) *The Ecology and Evolution of Communities*, pp. 315–341. Belknap Press/Harvard University Press, Cambridge, MA.

Brown JH (1993) Assessing the effects of global change on animals in western North America. In: Mooney HA, Fuentes ER, Kronberg BI (eds) *Earth System Responses to Global Change*, pp. 267–284. Academic Press, San Diego.

Brown JH, Mehlman DW, Stevens GC (1995) Spatial variation in abundance. Ecology 76:2028–2043.

Brown JH, Valone TJ, Curtin CG (1997) Reorganization of an arid ecosystem in response to recent climate change. Proceedings of the National Academy of Sciences 94:9729–9733.

Ceballos G, Brown JH (1995) Global patterns of mammalian diversity, endemism, and endangerment. Conservation Biology 9:559–568.

Chang LH, Hunsaker CT, Draves JD (1992) Recent research on effects of climate change on water resources. Water Resources Bulletin 28:273–286.

Chaves MM, Pereira JS (1992) Water stress, CO_2 and climate change. Journal of Experimental Botany 43:1131–1139.

Chiew FHS, Whetton PH, McMahon TA, Pittock AB (1995) Simulation of the impacts of climate change on runoff and soil moisture in Australian catchments. Journal of Hydrology 167:121–147.

Cole DR, Monger HC (1991) Influence of atmospheric CO_2 on the decline of C4 plants during the last deglaciation. Nature 368:533–536.

Conley W, Conley MR, Karl TR (1992) A computational study of episodic events and historical context in long-term ecological processes: climate and grazing in the northern Chihuahuan Desert. Coenoses 7:1–19.

Cooke R, Warren A, Goudie A (1993) Desert Geomorphology. UCL Press, London.

D'Antonio CM, Vitousek PM (1992) Biological invasions by exotic grasses, the grass/fire cycle, and global change. Annual Review of Ecology and Systematics 23:63–87.

Daily GC (1995) Restoring value to the world's degraded lands. Science 269:350–354.

DeAngelis DL, Cushman RM (1990) Potential applications of models in forecasting the effects of climate change on fisheries. Transactions of the American Fisheries Society 119:224–239.

Dregne HE, Tucker J (1988) Green biomass and rainfall in semi-arid sub-Saharan Africa. Journal of Arid Environments 15:245–252.

Ellner S, Shmida A (1981) Why are adaptations for long-range seed dispersal rare in desert plants? Oecologia 51:133–144.

FAUNMAP Working Group (1996) Spatial response of mammals to Late Quaternary environmental fluctuations. Science 272:1601–1606.

Field CB, Chapin FS III, Matson PA, Mooney HA (1992) Responses of terrestrial ecosystems to the changing atmosphere: a resource-based approach. Annual Review of Ecology and Systematics 23:201–235.

Fisher FM, Zak JC, Cunningham GL, Whitford WG (1988) Water and nitrogen effects on growth and allocation patterns of creosote bush in the northern Chihuahuan desert. Journal of Range Management 41:387–391.

Franco AC, Nobel PS (1989) Effect of nurse plants on the microhabitat and growth of cacti. Journal of Ecology 77:870–886.

Freas KE, Kemp PR (1983) Some relationships between environmental reliability and seed dormancy in desert annual plants. Journal of Ecology 71:211–217.

Freeland WJ (1990) Large herbivorous mammals—exotic species in northern Australia. Journal of Biogeography 17:445–450.

Fuhlendorf SD, Smeins FE (1999) Scaling effects of grazing in a semi-arid grassland. Journal of Vegetation Science 10:731–738.

Gasse F, Tehet R, Durand A, Gibert E, Fontes J-C (1990) The arid-humid transition in the Sahara and the Sahel during the last deglaciation. Nature 346:141–146.

Graetz RD (1991) The nature and significance of the feedback of changes in terrestrial vegetation on global atmospheric and climatic change. Climatic Change 18:147–173.

Grenot CJ (1974) Physical and vegetational aspects of the Sahara Desert. In: Brown GW Jr. (ed) Desert Biology, Volume II, pp. 103–164. Academic Press, NY.

Gutierrez JR, Whitford WG (1987) Chihuahuan Desert annuals: importance of water and nitrogen. Ecology 68:2032–2045.

Hanan NP, Prevost Y, Diouf A, Diallo O (1991) Assessment of desertification around deep wells in the Sahel using satellite imagery. Journal of Applied Ecology 28:173–186.

Hastings JR, Turner RM (1965) *The Changing Mile: An Ecologicakl Study of Vegtation Change with Time in the Lower Mile of an Arid and Semi-Arid Region.* University of Arizona Press, Tucson, AZ.

Herbel CH, Gibbens RP (1996) Post-drought vegetation dynamics on arid rangelands of southern New Mexico. Bulletin 776, New Mexico State University Agricultural Experiment Station, Las Cruces, NM.

Hernandez HM, Barcenas RT (1996) Endangered cacti in the Chihuahuan Desert, II. Biogeography and conservation. Conservation Biology 10:1200–1209.

Herrick JE, Whitford WG (1995) Assessing the quality of rangeland soils: challenges and opportunities. Journal of Soil and Water Conservation 50:237–242.

Huenneke LF, Noble I (1996) Ecosystem function of biodiversity in arid ecosystems. In: Mooney HA, Cushman JH, Medina E, Sala OE, Schulze E-D (eds) *Functional Roles of Biodiversity: A Global Perspective*, pp. 99–128. SCOPE/UNEP. John Wiley and Sons, Chichester.

Inouye RS (1991) Population biology of desert annual plants. In: Polis GA (ed) *The Ecology of Desert Communities*, pp. 27–54. University of Arizona Press, Tucson, AZ.

Jackson LL, McAuliffe JR, Roundy BA (1991) Desert restoration. Restoration and Management Notes 9:71–79.

Jain JK (1986) *Combating Desertification in Developing Countries. UNEP, Scientific Reviews on Arid Zone Research*, Volume 4. Scientific Publishers, Jodhpur, India.

Johnson HB, Polley HW, Mayeux HS (1993) Increasing CO_2 and plant–plant interactions: effects on natural vegetation. Vegetatio 104/105:157–170.

Kauffman JB, Krueger WC (1984) Livestock impacts on riparian ecosystems and streamside management implications: a review. Journal of Range Management 37:430–438.

Kelt DA, Brown JH, Heske EJ, Marquet PA, Morton SR, Reed JRW, et al. (1996) Community structure of desert small mammals: comparisons across four continents. Ecology 77:746–761.

Kerley GIH, Erasmus T (1992) Fire and the range limits of the bush Karoo rat *Otomys unisulcatus*. Global Ecology and Biogeography Letters 2:11–15.

Kieft TL, White CS, Loftin SR, Aguilar R, Craig JA, Skaar DA (1998) Temporal dynamics in soil carbon and nitrogen resources at a grassland–shrubland ecotone. Ecology 79:671–683.

Kovda VA, Samoilova EM, Charley JL, Skujins JJ (1979) Soil processes in arid lands. In: Goodall DW, Perry RA, Howes KMW (eds) *Arid-Land Ecosystems: Structure, Functioning, and Management*, Volume 1, pp. 439–470. International Biological Programme, Cambridge University Press, Cambridge, UK.

Le Houérou HN (1984) Rain use efficiency: a unifying concept in arid-land ecology. Journal of Arid Environments 7:213–247.

Long SP, Hutchin PR (1991) Primary production in grasslands and forests with climate change: an overview. Ecological Applications 1:139–156.

Lonsdale WM (1994) Inviting trouble: introduced pasture species in northern Australia. Australian Journal of Ecology 19:345–354.

Louw GN, Seely MK (1982) *Ecology of Desert Organisms.* Longman, London.

Mabbutt JA (1989) Impacts of carbon dioxide warming on climate and man in the semi-arid tropics. Climatic Change 15:191–221.

Mack RN (1986) Alien plant invasion into the intermountain west: a case history. In: Mooney HA, Drake JA (eds) *Ecology of Biological Invasions of North America and Hawaii*, pp. 191–213. Springer-Verlag, New York.

MacMahon JA, Wagner FH (1985) The Mojave, Sonoran and Chihuahuan Deserts of North America. In: Evenari M, Noy-Meir I, Goodall DW (eds) *Hot Deserts and Arid Shrublands, A. Ecosystems of the World*, Volume 12A, pp. 105–202. Elsevier, Amsterdam.

220 L.F. Huenneke

McAuliffe JR (1988) Markovian dynamics of simple and complex desert plant communities. American Naturalist 131:459–490.

McAuliffe JR (1994) Landscape evolution, soil formation, and ecological patterns and processes in Sonoran Desert bajadas. Ecological Monographs 64:111–148.

Meyer WB, Turner BL II (1992) Human population growth and global land-use/cover change. Annual Review of Ecology and Systematics 23:39–61.

Milchunas DG, Lauenroth WK (1995) Inertia in plant community structure: state changes after cessation of nutrient-enrichment stress. Ecological Applications 5:452–458.

Miller B, Ceballos G, Reading R (1994) The prairie dog and biotic diversity. Conservation Biology 8:677–681.

Mitchell JFB, Warrilow DA (1987) Summer dryness in northern mid-latitudes due to increased CO_2. Nature 330:238–240.

Mourelle C, Ezcurra E (1996) Species richness of Argentine cacti: a test of biogeographic hypotheses. Journal of Vegetation Science 7:667–680.

Muth A (1980) Physiological ecology of desert iguana (*Dipsosaurus dorsalis*) eggs: temperature and water relations. Ecology 61:1335–1343.

Myers N (1995) Environmental unknowns. Science 269:358–360.

Neilson RP (1986) High-resolution climatic analysis and southwest biogeography. Science 232:27–34.

Noy-Meir I (1973) Desert ecosystems: environment and producers. Annual Review of Ecology and Systematics 4:25–51.

Orians GH, Solbrig OT (1977) A cost-income model of leaves and roots with special reference to arid and semiarid areas. American Naturalist 111:677–690.

Pake CE, Venable DL (1995) Is coexistence of Sonoran Desert annuals mediated by temporal variability in reproductive success? Ecology 76:246–261.

Pianka ER (1979) Diversity and niche structure in desert communities. In: Goodall DW, Perry RA, Howes KMW (eds) *Arid-Land Ecosystems: Structure, Functioning, and Management*, Volume 1, pp. 321–341. International Biological Programme, Cambridge University Press, Cambridge, UK.

Pimm SL, Russell GJ, Gittleman JL, Brooks TM (1995) The future of biodiversity. Science 269:347–350.

Polis GA (1991) Desert communities: an overview of patterns and processes. In: Polis GA (ed) *The Ecology of Desert Communities*, pp. 1–26. University of Arizona Press, Tucson, AZ.

Polley HW, Johnson HB, Mayeux HS (1994) Increasing CO_2: comparative responses of the C4 grass *Schizachyrium* and grassland invader *Prosopis*. Ecology 75:976–988.

Pugnaire FI, Haase P, Puigdefábregas J (1996) Facilitation between higher plant species in a semiarid environment. Ecology 77:1420–1426.

Reichman OJ, Smith SC (1990) Burrows and burrowing behavior by mammals. In: Genoways HH (ed) *Current Mammalogy*, Volume 2, pp. 197–244. Plenum Press, New York.

Reynolds JF, Virginia RA, Schlesinger WH (1997) Defining functional types for models of desertification. In: Smith TM, Shugart HH, Woodward IA (eds) *Plant Functional Types: Their Relevance to Ecosystem Properties and Global Change*, pp. 195–216. Cambridge University Press, Cambridge, UK.

Rind D, Goldberg R, Hansen J, Rosenzweig C, Ruedy R (1990) Potential evapotranspiration and the likelihood of future drought. Journal of Geophysical Research (Sect. D) 95:9983–10004.

Rochefort L, Woodward FI (1992) Effects of climate change and a doubling of CO_2 on vegetation diversity. Journal of Experimental Botany 43:1169–1180.

Roundy BA, Biedenbender SH (1995) Revegetation in the desert grassland. In: McClaran MP, Van Devender TR (eds) *The Desert Grassland*, pp. 265–303. University of Arizona Press, Tucson, AZ.

Safriel UN (1987) The stability of the Negev Desert ecosystems: why and how to investigate it. In: Berkofsky L, Wurtele MG (eds) *Progress in Desert Research*, pp. 133–144. Rowman and Littlefield Publishers, Totowa, NJ.

Sala OE, Lauenroth WK, Golluscio RA (1997) Plant functional types in temperate arid regions. In: Smith TM, Shugart HH, Woodward IA (eds) *Plant Functional Types: Their Relevance to Ecosystem Properties and Global Change*, pp. 217–233. Cambridge University Press, Cambridge, UK.

Schlesinger WH, Reynolds JF, Cunningham GL, Huenneke LF, Jarrell WM, Virginia RA, et al. (1990) Biological feedbacks in global desertification. Science 247:1043–1048.

Schlesinger WH, Raikes JA, Hartley AE, Cross AF (1996) On the spatial pattern of soil nutrients in desert ecosystems. Ecology 77:364–374.

Seely MK (1991) Sand dune communities. In: Polis GA (ed) *The Ecology of Desert Communities*, pp. 348–382. University of Arizona Press, Tucson, AZ.

Shachak M, Lovett GM (1998) Atmospheric deposition to a desert ecosystem and its implications for management. Ecological Applications 8:455–463.

Shachak M, Brand S, Gutterman Y (1991) Porcupine disturbance and vegetation pattern along a resource gradient in a desert. Oecologia 88:141–147.

Shmida A, Evenari M, Noy-Meir I (1985) Hot desert ecosystems: an integrated view. In: Evenari M, Noy-Meir I, Goodall DW (eds) *Hot Deserts and Shrublands, A. Ecosystems of the World*, Volume 12A, pp. 379–387. Elsevier, Amsterdam.

Shugart HH, Smith TM, Post WM (1992) The potential for application of individual-based simulation models for assessing the effects of global change. Annual Review of Ecology and Systematics 23:15–38.

Stohlgren TJ, Schell LD, Vanden Heuvel B (1999) How grazing and soil quality affect native and exotic plant diversity in Rocky Mountain grasslands. Ecological Applications 9:45–64.

Suzan H, Nabhan GP, Patten DT (1996) The importance of *Olneya tesota* as a nurse plant in the Sonoran Desert. Journal of Vegetation Science 7:635–644.

Tucker CJ, Dregne HE, Newcomb WW (1991) Expansion and contraction of the Sahara Desert from 1980 to 1990. Science 253:299–301.

Vogel JC, Fuls A, Danin A (1986) Geographical and environmental distribution of C3 and C4 grasses in the Sinai, Negev, and Judean deserts. Oecologia 70:258–265.

Weltzin JF, Archer S, Heitschmidt RK (1997) Small-mammal regulation of vegetation structure in a temperate savanna. Ecology 78:751–763.

Werger MJA (1986) The Karoo and Southern Kalahari. In: Evenari M, Noy-Meir I, Goodall DW (eds) *Hot Deserts and Arid Shrublands, B. Ecosystems of the World*, Volume 12B, pp. 283–359. Elsevier, Amsterdam.

West NE (1979) Formation, distribution and function of plant litter in desert ecosystems. In: Goodall DW, Perry RA, Howes KMW (eds) *Arid-Land Ecosystems: Structure, Functioning, and Management*, Volume 1, pp. 647–659. *International Biological Programme*. Cambridge University Press, Cambridge, U.K.

West NE (1983a) Overview of North American temperate deserts and semi-deserts. In: West NE (ed) *Temperate Deserts and Semi-Deserts. Ecosystems of the World*, Volume 5, pp. 321–330. Elsevier, Amsterdam.

West NE (1983b) Comparisons and contrasts between the temperate deserts and semi-deserts of three continents. In: West NE (ed) *Temperate Deserts and Semi-Deserts. Ecosystems of the World*, Volume 5, pp. 461–472. Elsevier, Amsterdam.

Whitford WG (1976) Temporal fluctuations in density and diversity of desert rodent populations. Journal of Mammalogy 57:351–369.

Whitford WG (1993) Animal feedbacks in desertification: an overview. Revista Chilena de Historia Natural 66:243–251.

Whitford WG (1997) Desertification and animal biodiversity in the desert grasslands of North America. Journal of Arid Environments 37:709–720.

Williamson M (1996) *Biological Invasions.* Chapman and Hall, London.

Wondzell SM, Cunningham GL, Bachelet D (1987) A hierarchical classification of landforms: some implications for understanding local and regional vegetation dynamics. Strategies for classification and management of native vegetation for food production in arid zones. In: Gonzales EF, Vicente CE, Moir WE, Aldon H (eds) General Technical Report, U.S. Forest Service, Rocky Mountain Forest and Range Experiment Station.

Wurtele MG (1987) The meteorology of desertification. In: Berkofsky L, Wurtele MG (eds) *Progress in Desert Research*, pp. 245–259. Rowman and Littlefield Publishers, Totowa, NJ.

11. Temperate Forests of North and South America

Juan J. Armesto, R. Rozzi, and J. Caspersen

Temperate regions of the world sustain forests comparable in extension and biomass to that of tropical forests. Although they are less speciose than tropical forests, temperate forests harbor notable numbers of endemics, especially in remote and isolated regions of the southern hemisphere (Armesto et al. 1996a; Arroyo et al. 1996). They also contain some of the oldest living organisms on Earth. Some living conifers in temperate forests are frequently older than 1000 years, and species such as *Fitzroya cupressoides* live for 3000 years or more (Lara and Villalba 1993; Hill and Enright 1995). These long-lived temperate trees provide an invaluable and irreplaceable record of climatic changes during the last millenium. In addition, temperate forests are currently the major source of timber and related wood products worldwide. Thus, their long-term sustainability is of utmost economic importance, both regionally and globally.

Deforestation in temperate regions equals or exceeds that of tropical regions because large territories of Asia, Europe, and North America were converted long ago to pasture, croplands, or intensively managed for timber production. According to Norton (1996) less than 1% of some eucalyptus forest associations remain unlogged in productive areas of New South Wales, Australia, and some of the most productive temperate deciduous forests in south-central Chile have been reduced to small patches (Donoso and Lara 1996). These large-scale changes in temperate forest cover (Table 11.1) entail negative impacts on regional and global biodiversity (Armesto et al. 1992;

Table 11.1. Percentage of forest cover change estimated for regions of the world predominantly covered by temperate and boreal forest between 1700 and 1980

	Area 1700	Area 1980	Net change 1700–1980
North America	1016	942	−7.3%
Europe	230	199	−7.8%
Ex Soviet Union	1138	941	−17.3%
China	135	58	−57.0%
Southern Chile	15	8	−46.7%
Australia	28	10.7	−38.2%

Forest area in million hectares. Estimates for the regions in the northern hemisphere from Richards 1993; for southern Chile from Armesto et al. 1994. Estimates for Australia are for wet-cool productive environments, original and remaining area as of 1995 (Norton 1996). Note that boreal forests, which represent a larger percentage of all forests in the northern hemisphere, are lacking in the southern hemisphere. Because boreal forests have been less cleared than temperate forests, percentages of forest cover lost are higher than that shown for the northern hemisphere.

Berg et al. 1994; Ehrlich 1996) and have complex effects on the C and N cycles in the biosphere (Vitousek 1994; Asner et al. 1997).

In this chapter we will present our views about the current status and future scenarios for biological diversity in the temperate forests of austral South America. These forests are conspicuous hot spots because of their high endemism and endangered biodiversity (Armesto et al. 1996a; Arroyo et al. 1996; Armesto et al. 1998). Because of their limited area, even though southern latitude forests play a minor role on C and N fluxes in the biosphere, their regional and global influence might increase in the future as vast forested regions of the northern hemisphere become increasingly affected by enhanced N deposition and N saturation, impairing their role as carbon sinks (Asner et al. 1997). We will focus on historical and regional factors that influence taxonomic diversity and discuss the relative impact that present drivers of global change will have on southern latitude ecosystems (see Arroyo et al. 1993). In order to place South American temperate forests in a broader context, however, we will first discuss global biogeographic patterns in the phenology, physiognomy, and floristic diversity of temperate forests worldwide. In this comparative context, we will discuss the causes and consequences of functional diversity and compare the potential responses of North and South American temperate forests to the major drivers of global change. The purpose of this analysis is to provide an overview of the factors that are more likely to endanger biological diversity and ecosystem function in temperate forests in the first decade of the twenty-first century. This information can be used as a reference for land planners and local forest managers.

Climatic Factors Governing the Distribution, Phenology, and Physiognomy of Temperate Forests Worldwide

Temperate forests presently occur in the northern and southern hemispheres in humid to semi-arid climates with occasional or prolonged freezing temperatures, approximately between 35 and 60 degrees of latitude (Walter 1973). Such conditions are found in the meridional and nemoral climate zones, which span a broad portion of the globe in the middle latitudes (Ovington 1983; Rohrig and Ulrich 1991). Periodic frost distinguishes the meridional climate zone from the tropics as the region in which temperatures occasionally fall as low as −10°C. Further poleward the nemoral climate zone is characterized by a pronounced cold season in which the average temperature remains below 10°C for several months of the year. In the northern hemisphere the nemoral climate zone is bounded in the north by the boreal climate zone, where the average temperature remains below 10°C for more than 4 months of the year (Walter 1973). Climatic conditions are generally milder and temperatures more constant in South America than are those at comparable latitudes in the northern hemisphere due to the moderating influence of expansive southern oceans (Arroyo et al. 1996). This results in longer growing seasons at corresponding latitudes of the southern hemisphere, which in turn influence phenology and production. In addition, because of the limited amount of continental land present above 50 degrees latitude in the southern hemisphere, there is an absence of boreal climates and true boreal forest or tundra ecosystems.

Within this broad climatic range, temperate forests exhibit striking diversity in structure, function, and floristic composition. This diversity reflects the contrasting climatic, geologic, and evolutionary histories of temperate biotas in the two hemispheres (Axelrod et al. 1991). Nevertheless, broad patterns in phenology and physiognomy serve to characterize the functional diversity of temperate forests worldwide. Temperate forests of the world fall into four broad categories: (1) broadleaf evergreen rain forests, (2) broadleaf deciduous forests, (3) broadleaf sclerophyll forests and (4) evergreen conifer forests (Ovington 1983; Rohrig and Ulrich 1991).

Broad-leaved evergreen rain forests are most widespread in the southern hemisphere, particularly in Australia, New Zealand, and Chile. They are also distributed locally in eastern Asia and the Himalayas (Ovington 1983). Deciduous forests cover broad areas in the northern hemisphere, including Europe, the Near East, eastern North America, and eastern Asia, as well as less extensive areas in southern Chile (Schulze et al. 1996). Broadleaf sclerophyll forests are distributed in a discontinuous band from the Mediterranean, through the Middle East, to the Himalayas and southeastern Asia, as well as in California, central Chile, Australia, and South Africa (Di Castri et al. 1981). Finally, the largest region of the temperate zone dominated by evergreen conifers is in northwestern North America, but evergreen conifer forests are locally present in montane environments of southern South America, Tasmania, and New Zealand (Hill and Enright 1995).

Global patterns in the physiognomy and phenology of temperate forests reflect latitudinal and regional patterns in the seasonality of temperature and rainfall. The phenology of temperate forests is largely governed by the length of the growing season. Broadleaf evergreen trees are the dominant life form in meridional climate regions with a continuous growing season, whereas broadleaf deciduous trees are the dominant life form in nemoral climate regions with a pronounced cold season (Ovington 1983; Rohrig and Ulrich 1991). Due to the asymmetric distribution of land and sea between the northern and southern hemispheres, broadleaf deciduous forests prevail in the cold continental climates of the northern hemisphere, whereas broadleaf evergreen forests prevail in the mild maritime climates of the southern hemisphere in South America, Australia, and Tasmania. Increases in global temperature will presumably have a significant impact on deciduous forests by extending the length of the growing season, particularly in the northern hemisphere.

The physiognomy and production of temperate forests is largely governed by the seasonal distribution of rainfall. Broad mesophyllic leaves are the dominant leaf forms in humid climates in which rainfall is evenly distributed throughout the year. Forests are dominated by broad-leaved evergreen trees in regions of northwestern North America, southern South America, New Zealand, and Tasmania that receive between 1500 and 6000 mm of rainfall annually with limited summer drought (Alaback 1991). These massive rainforests are also characterized by a richer array of plant life forms and epiphytic cryptogams than other temperate forests (Arroyo et al. 1996; Galloway 1996). In climates where the rain falls primarily during the winter months, summer drought favors xerophytic leaf forms. In warm mediterranean climates, broadleaf sclerophyll forests are dominated by hardwood trees with small, leathery leaves and waxy surfaces (Ovington 1983). In the cool climate of northwestern North America, needleleaf conifers dominate regions with wet winters and dry summers (Waring and Franklin 1979) and become increasingly dominant near the timberline. Southern hemisphere conifer forests are found in limited areas both in dry-summer environments as well as in the rain forest region (Hill and Enright 1995).

The functional diversity of temperate forests is manifest through the effects of phenology and physiognomy on biogeochemical and biophysical processes. Phenology exerts a profound influence on decomposition and production through the effect of leaf lifespan on C and N cycles (Reich et al. 1997). The high lignin contents and C–N ratios of long-lived leaf litter depress the rate of decomposition as compared with deciduous leaf litter (Pastor and Post 1986; Schulze et al. 1996). As a consequence production can vary tremendously among deciduous and evergreen stands. Leaf form exerts a profound influence on the exchange of energy and matter between the canopy and atmosphere through its effect on boundary layer conductance and surface roughness (Woodwell and Mackenzie 1995).

Given the remarkable functional diversity of temperate trees, we face a considerable challenge in anticipating how global climate change will alter the floristic composition and distribution of temperate forests. Global warming is expected to be greatest in the midlatitudes of the northern hemisphere due to the positive feedback between melting snow and decreased surface albedo (Groisman et al. 1994). A 3–4°C increase in winter and spring temperatures has increased the length of the growing season in boreal regions of North America by a week in the 1990s alone (Keeling et al. 1996; Myneni et al. 1997). These changes may lead to the expansion of temperate deciduous tree species into the boreal zone in the northern hemisphere. In the southern hemisphere, where temperatures will rise more gradually, other imminent threats to biodiversity may be more dire than the consequences of climate change.

Today, South American temperate forests are highly endangered because of their reduced and isolated geographical range (Armesto et al. 1996a, 1998). Rapid deforestation and widespread introduction of exotic timber species are the major immediate threats to biological diversity in these forests, acting at a much faster rate than other drivers of global change (e.g., climate warming or industrial pollutants) (see later). These processes are reducing biological diversity and jeopardize the potential for native species to respond to future climate change.

Functional Contrasts Between North and South American Temperate Forests

Plant–Animal Interactions

A major functional difference between northern and southern temperate forests of the Americas is the important role played by plant–animal interactions in the reproductive biology of plants in southern latitude forests (Armesto and Rozzi 1989; Willson 1991; Armesto et al. 1996b). Thus, a broader array of pollination and seed dispersal interactions link animal and plant populations in southern temperate forests, making these ecosystems more susceptible to changes in land cover, climate, fragmentation, and loss of species than their North American counterparts.

A larger proportion of the plants in southern temperate forests is dependent on animal species for their pollination and seed dispersal than in North American temperate forests (Willson 1991; Arroyo et al. 1993). For instance, nearly 20% of the woody species in temperate forests of Chiloé are pollinated by hummingbirds, compared with less than 5% in Pacific Northwest conifer forests (Smith-Ramírez 1993; Armesto et al. 1996b). Nearly 70% of the woody species in Chiloé have fleshy fruits compared with less than 30% in temperate deciduous forests of eastern North America (Armesto and Rozzi 1989; Willson 1991). Of 51 species of temperate forest plants studied by

Riveros et al. (1996) in southern Chile, 30 species (59%) were totally or highly dependent on animal pollinators for successful fruit set. At the same time, seeds, fleshy fruits, and flower nectar constitute important food resources for many insects, birds, mammals, and some reptiles inhabiting austral forests (Riveros 1991; Smith-Ramìrez 1993; Armesto et al. 1996b; Murà 1996; Willson et al. 1996a,b; Smith-Ramírez and Armesto 1998). The frequency of woody plants dependent to some degree on animals for seed dispersal and pollination is greater in forests between 35° and 42°S (64–72% of endozoochory, and 70–90% of biotic pollination), and decreases, but still remains high, in southern Patagonian forests at 50–55°S (47–52% of endozoochory, and 47–60% biotic pollination).

Although a detailed understanding of these plant–animal interactions is still emerging, some plant and animal species are suspected to play critical roles in the maintenance of species diversity in southern latitude forests (Armesto et al. 1996b; Willson et al. 1996c; Smith-Ramìrez and Armesto 1998). In the face of global change, species assemblages will be highly sensitive to the loss of species that have high levels of connectivity. In addition, these interactions would influence the patterns of forest expansion or recovery following a disturbance because guilds of mutualistic species would be required to colonize together. For instance, in the matorral of central Chile, where forest clearing has greatly reduced the wild populations of animal seed dispersers, seed banks of fleshy-fruited trees are extremely poor, compared with climatically analogous North American forests. Hence, rates of colonization during secondary succession are limited by low seed inputs to open areas (Jimènez and Armesto 1992; Armesto et al. 1995).

Plant Breeding Systems

The fact that many woody species in southern Chilean forests are obligate outcrossers (Riveros et al. 1996), either dioecious or genetically self-incompatible, implies that their potential for population expansion into new environments is limited by their need for biotic pollen vectors and cross pollination. This also suggests that tree species may show low genetic variation, although they may have high levels of heterozygocity (Arroyo et al. 1993). This may hamper the potential for local population differentiation in response to rapid climate change. Genetic bottlenecks associated with reduction and fragmentation of species ranges during the glacial period (see later) might have contributed to further reduce genetic variability of trees in South American temperate forests (Alnutt et al., unpublished data). In some cases, however, extreme longevity may be an important factor in maintaining genetic variability through these bottlenecks, as has been shown for the fragmented populations of the dioecious conifer *Fitzroya cupressoides* in southern Chile and Argentina (Alnutt et al., unpublished data).

We lack comparable information for North American trees. The higher representation of conifers that do not have self-incompatibility mechanisms suggests that obligate outcrossers may be less represented among woody species in North America than in Chilean forests. Deciduous trees, on the other hand, which flower early in the spring, are likely to present some levels of self-incompatibility. Genetic self-incompatibility is frequent among North American temperate forest herbs (Barret and Helernum 1987), in contrast with findings for herbs in temperate rainforests of southern Chile (Riveros 1991). Higher self-incompatibility among herbaceous species in deciduous North American forests may be related to the opening of the canopy in early spring when herb species flower, which is a condition that is absent in South American forests that have a predominantly evergreen canopy. In alpine open-canopy woodlands in central Chile, Arroyo and Uslar (1993) have also shown that many herb species are self-incompatible.

Ecosystem Function

One of the most challenging questions regarding global change concerns the impact that changing climate, loss of species, and modification of landscapes will have on ecosystem function. For southern temperate forests no studies have addressed this question, and the current knowledge of ecosystem processes is limited. Information derived from northern hemisphere forests (e.g., Schulze et al. 1996) may be misleading because of the large difference in the sources and chemical quality of nutrient inputs. In contrast to most North American forests, temperate rain forests of southern South America exhibit atmospheric inputs of nutrients that are among the lowest in the world (Hedin et al. 1995; Galloway et al. 1996). Nitrate inputs from rainfall in southern Chilean forests have been estimated in 0.74 kg/ha/year (Hedin et al., unpublished), compared with average inputs of 4–5 kg/ha/year for North American forests, and up to 20 kg/ha/year in regions subjected to high industrial air pollution (Schlesinger 1991). Although Andean and lowland forests occupy volcanic or glacial soils of young geological age, coastal range forests in Chile occur on old, nutrient-poor soils developed over highly weathered and impermeable bedrock (Pèrez et al. 1991). Despite the limited nutrient supply of nitrate and ammonium from atmospheric sources, old-growth forests of the coastal range of Chile accumulate much nitrogen in their above-ground biomass and in soil organic matter. These pools are comparable to North American old-growth forests that occur on richer soils (Johnson et al., unpublished). This suggests that plants, soils, and microorganisms in southern temperate forests are strong sinks for inorganic nitrogen within the ecosystem (see Hedin et al. 1995; Pèrez et al. 1998). In addition, this indicates that these forests may depend largely on recycled nutrients and biological nitrogen fixation to fulfill their annual growth requirements (Pèrez et al. 1998).

With regard to nutrient capture, the abundance of vascular and non-vascular epiphytes in South American temperate rain forests provides an additional pathway for nutrient input and flux that is rare in some North

American forests due the more limited presence of epiphytic components (Arroyo et al. 1996). Epiphytes are able to scavenge nutrients from stemflow (Serrano et al., unpublished) and transfer them to their biomass. The abundance of epiphytic lichens that contain blue-green algae can be a significant source of nitrogen to the forest ecosystem (Guzman et al. 1990; Galloway 1996). Laboratory assays indicate that free-living bacteria associated with epiphytes and coarse woody debris may represent a meaningful source of newly fixed N (Carmona et al., unpublished). It is likely that the functional role of these free-living N fixers is less critical for North American forests today, especially in areas of high N deposition.

Nitrogen mineralization assays in forest soils in southern Chile (Pèrez et al. 1998) and experiments adding ^{15}N-labeled inorganic N (Perakis, unpublished) suggest that soil microbial populations are actively involved in the process of retention of inorganic nitrogen in soils of southern latitude forests; however, we do not know the sensitivity of these microorganisms to alterations of climate and landscapes associated with global change. A very important observation in this regard is that N acquisition and storage mechanisms, associated with the presence of large amounts of epiphytes and microbial activity, appear to be more developed in structurally complex, old-growth forests that occur on largely organic soil substrates with a high C–N ratio (Pèrez et al. 1998). In old-growth (450–500 years old) *Fitzroya cupressoides* forests, one tree may have 12–15 kg of epiphyte dry mass, mainly vines, mosses, and lichens; however, the epiphytic cover is absent from young, growing stands. It can be predicted that, as these components of biological diversity are lost in simplified or degraded forests due to harvesting, poor management, or disturbance, the biological mechanisms of nutrient retention and recycling can be greatly impaired in South American forests. In contrast, a major threat to the function of northern hemisphere forests seems to be related to the high levels of N deposition (Aber et al. 1989), which have been associated with rapid losses of soil cations, alteration of C–N rations, and nutrient imbalance (see later). In the long term, these effects may lead to significant reductions of forest productivity and, in some cases, to forest decline (Schulze 1989; Aber 1992; Asner et al. 1997).

Concentration of Diversity and Endemism

In contrast to North American temperate forests that maintained intermittent land connections with Eastern Asia and Europe during the entire Pleistocene and most of the Tertiary, South American temperate forests were isolated from other continental masses during most of the Tertiary, and had no connection with other forests within South America during the entire Pleistocene (Axelrod et al. 1991; Villagrán and Hinojosa 1997). Temperate forests in the southern hemisphere are better described as biogeographic relics, representing the discontinuous fragments of the ancient continent of Gondwana. South American temperate forests harbor a flora and fauna rich

Table 11.2. A comparison of floristic diversity (woody species only) in some northern hemisphere temperate forests and South American temperate forests

Forest type and geographic region	No. of species	No. of genera	Species–genus ratio
Deciduous forests, East Asia	876	59	14.8
Deciduous forests, eastern North America	157	40	3.9
Deciduous forest, Europe	106	23	4.6
Deciduous forests, South America	47	24	1.9
Broadleaf rain forests, South America	162	84	1.9

Data from Schulze et al. 1996 and Arroyo et al. 1996.

in endemics, bearing many phylogenetic links with the forest biota of New Zealand and Tasmania, and with the paleoflora and fauna of Antarctica (Axelrod et al. 1991). Temperate rain forests of southern Chile exhibit a higher species richness compared with similar ecosystems in the Pacific coast of North America, New Zealand, and Tasmania (Arroyo et al. 1996). The unique flora and endemism of the southern South American biota, which is a product of its long isolation, is evident in its lower average number of woody species per genus than in temperate forests of North America, Europe, and Asia (Table 11.2).

The history of climatic and geological transformations of southern South America since the mid-Tertiary has resulted in extraordinary levels of endemism in the temperate flora and fauna. Many plant species are unique representatives of their genera or families. About 34% of the genera and 85% of the species in the local woody flora are endemic to southern temperate forest region. Nearly 80% of the endemic genera are monospecific, which probably reflects a richer ancestral biota (Villagrán and Hinojosa 1997). Narrow range endemics are concentrated between 36 and 40°S, which is the area where important forest refugia were located during the last glacial period (see later). This latitudinal range can be defined as the most sensitive area of the Chilean temperate region with regard to biodiversity, but it is at the same time the most densely populated and intensely managed region.

The highest species richness for trees and vines also occurs between 38 and 42°S, in the region of the Valdivian rain forest (Armesto et al. 1992; Arroyo et al. 1996). A similar geographic pattern is observed for vertebrates. From a total of 147 species of vertebrates inhabiting the forest, 66 species (45%) are endemic (Armesto et al. 1996a). For mammals, diversity drops sharply from 19 species at 41°S, to six species in the Andes of mainland Chiloé (Murà 1996). In southern Patagonia, mammalian diversity increases again because a new species invades the forest habitat from the contiguous steppe. Although temperature decreases monotonically poleward, from 35 to 55°S, the latitudinal trend in tree, vine, and fern species richness shows a sharp decline (close to 50% reduction for trees and vines, nearly two-thirds reduction for ferns) at about 43°S (i.e., south of Chiloé island) (Arroyo et al. 1996). This abrupt reduction in species richness coincides with the northernmost point at which the continental ice sheet reached the Pacific Ocean during the last glacial maximum, leaving no land available for the growth of vegetation. The decline in woody species richness with increasing latitude is more gradual in Pacific Northwest forests (Alaback 1991).

Historical Background on South American Temperate Forests

South American forests are concentrated in a narrow longitudinal band, no more than 200 km wide, along the Pacific Ocean coast of southern Chile and the eastern slopes of the Andes, between 35°S in central Chile and 56°S in Tierra del Fuego. Along 20 degrees of latitude, a striking diversity of mixed deciduous and broad-leaved evergreen forests, and discontinuous patches of conifer forests, are associated with pronounced topographic and edaphic gradients, as well as with contrasting disturbance regimes (Donoso 1993; Armesto et al. 1996a). Some of the wettest temperate forests on the world are found in this region, receiving more than 6000 mm of rain per year. Annual rainfall increases with latitude, from about 800 mm at 35°S to more than 4000 mm at around 47°S. It then decreases again as we approach the tip of southern South America. The maritime influence throughout this extensive latitudinal range maintains relatively constant temperatures. The largest difference in mean annual temperatures between forest types along this latitudinal gradient is 6°C (Arroyo et al. 1993). Seasonal differences in temperature are also low, with average temperature ranges of 8°C, except in the seasonally deciduous (Maulino) forest where temperature ranges up to 13°C (Arroyo et al. 1996).

Isolation of Temperate Forests in South America

A global cooling trend began in the mid-Tertiary, coinciding with tectonic processes that gave rise to the Andean mountain range in western South America. As a result austral forests became increasingly isolated from other forests in the continent (Villagrán and Hinojosa 1997). Climatic aridity barriers arose in the north of Chile and in Patagonia as the Andean mountains reached elevations that blocked the trajectory of moist westerly winds, south of 30°S, and of easterly monsoonal fronts, north of 30°S. Aridity conditions and high mountain barriers also interrupted the biotic flow between tropical latitudes in the eastern Andes and western South America for at least the last 3–5 million years. This separation is reflected in the large number of temperate forest taxa (i.e., plants and animals) that have disjunct distributions at the genus level extending to tropical latitudes. Some important plant genera found in southern forests have congeners in the cool highlands of southern

Brazil, eastern Bolivia, northwestern Argentina, or in the Northern Andes, occasionally reaching to Central America (Landrum 1981; Arroyo et al. 1996; Villagrán and Hinojosa 1997). Aridity conditions were intense during glacial times, further constricting the area of southern temperate forests, pushed equatorward by the expansion of glacial ice in the south (Villagrán 1990; Villagrán and Armesto 1993), thus increasing their isolation.

Loss of Biodiversity Through Glacial Times

During a series of alternate cooling and warming periods, mountain glaciers advanced and retreated several times during the Pleistocene, causing the confined area of temperate forests in southern South America to successively shrink and expand. The inflow of species from subtropical latitudes was no longer possible due to arid conditions to the north and mountain barriers to the east, established in the mid- to late Tertiary (Villagrán and Hinojosa 1997). Immigration during the warmer interglacials, therefore, could not compensate for species extinction during the cold periods when forests declined. Increasing aridity and isolation could have led to the extinction of congeners resulting in the large number of monotipic genera in the austral forest flora (Arroyo et al. 1996). The lack of immigration resulted in a net loss of species, particularly of taxa with tropical ancestors. Estimates indicate that about 60% of the paleoflora with tropical affinity became extinct (Villagrán and Hinojosa 1997) due to environmental changes since the end of the Tertiary. Avian diversity associated with southern latitude forests appears to have been greatly reduced during the Pleistocene as well (Vuilleumier 1985).

Forest Refugia During the Last Glaciation

Because the piedmont glaciers covered a large portion of southern South America, forest biodiversity survived in small glacial refugia within and outside the glaciated territories. For most forest species, derived from tropical or warm temperate ancestors, major refugia were located in coastal areas, close to the northern limit of temperate forests today, where precipitation was presumably higher during glacial times, and temperatures were mild due to the oceanic influence. Some areas of the coastal range between 38 and 40°S remained free of ice and the effects of periglacial processes that affected vegetation persistence further south. It can be estimated from reconstruction of the extent of ice fields (Villagrán et al. 1996) that the area covered by temperate forests could have been greatly reduced at the Last Glacial Maximum (LGM), at 21,000 years ago, to about one third of its present range. Small patches of some conifers (*Fitzroya*, *Pilgerodendron*) and *Nothofagus* species probably persisted in the vicinity of glaciers (Villagrán 1991) because of their ability to withstand nutrient-poor substrates and waterlogged soils. The location of forest refugia is confirmed by the peak in taxonomic richness of trees observed between 35 and 40°S, and by the results of palynological studies in the Lake District and Chiloé island, which document a rapid range expansion of tree species from sources located in coastal areas beginning about 12,000 years ago (Villagrán 1991; Villagrán et al. 1996).

Forest Expansion in Holocene Times

Examining the forest recolonization process after the last glaciation can suggest how present-day forests could expand their ranges in response to rapid climatic warming (Moreno, unpublished data). The colonization of Valdivian rainforest taxa in the Lake District and Chiloé was rapid and began 12,000 years ago, immediately following the rise in temperature (Villagrán 1991; Villagrán et al. 1996). Not all species, however, colonized at the same rate because some tree species are recorded in the Andes only in the past 3000–5000 years. Colonization of the southern Patagonian region by *Nothofagus* forests occurred progressively during the last 10,000 years (Villagrán and Armesto 1993; Villagrán et al. 1996).

Present Threats to Biological Diversity in Southern Temperate Forests

Forest Fragmentation and Loss of Forest Cover

Studies of floristic and avian diversity in small, undisturbed islands south of Chiloé that were connected to the mainland during the LGM, and studies of remnant forest fragments in rural landscapes of northern Chiloé island, provide approaches to estimate the expected loss of species that may result from decreasing forest cover (Villagrán et al., unpublished; Rozzi et al. 1996). Two conclusions are derived from this analysis: (1) Wind-dispersed taxa appear to be more sensitive to forest fragmentation and isolation than avian-dispersed, fleshy-fruited species, and (2) other things being equal, reductions in the extent of forest habitat can cause significant losses of biodiversity. Floristic richness falls 6–8% and avian species richness 5–7% for a reduction of 50% in forest area (Villagrán, Rozzi, and Armesto, unpublished data).

Forest fragmentation and logging has led to important decreases in bird species abundance and richness (Willson et al. 1994). Birds that use large emergent trees and coarse woody debris as habitat for feeding, perching, or nesting are particularly sensitive to reductions in forest cover. Snags and woody residues are typical of old-growth forest habitat, but become sparse as landscapes are converted to agriculture or fast-growing commercial tree plantations, leading to a reduction in the amount of habitat suitable to support populations of such forest birds as the Rhynocriptids, *Pteroptochos tarnii*, the large woodpecker, and the owl *Stryx rufipes*.

Present trends of deforestation and replacement of native forests in Chile are estimated to be nearly 1% annually (JJA, unpublished data). This rate underestimates the potential impact of deforestation on biodiversity, however, because, as discussed earlier, biological diversity is concentrated in a narrow latitudinal range, from 35 to 42°S. For this latitudinal range, GIS-based estimates of deforestation rates (Banco Central de Chile 1995) suggest that forest cover has decreased 2–7% per year since 1984. From these rates, we predict that 40–50% of the native forest outside protected areas will be lost in the next 25 years. According to regressions between species and area for forest islands in Chiloé (Villagrán, unpublished data, Rozzi et al. 1996) between 18 and 24 species of woody plants and one to two species of birds could become extinct over the next 25 years due to loss of forest habitat. *Extinction debt* associated with past and current habitat destruction and fragmentation may result in a larger number of extinctions in the coming decades because endangered populations already suffering from inbreeding depression and habitat loss may take time to become extinct.

Pulliam (1988) suggested that species losses from small forest patches could be reduced by "source-sink" dynamics (i.e., by immigration from large patches of undisturbed forest serving as sources). As a consequence, the slope of the regression between area and number of species in fragmented mainland habitats may be lower than it would be in the absence of source habitats. This, in turn, indicates that we may underestimate the species loss that could actually occur after large forest fragments disappear. Endemic understory dwellers are particularly sensitive to reductions of forest cover (Willson et al. 1994).

Endangered Ecosystems

Pronounced topographic and edaphic gradients, from lowland coastal areas to the summits of Andean volcanoes, determine the occurrence of a variety of forest types along the latitudinal range of temperate forests. Distributions of forest types vary from localized (e.g., coastal swamp forests, *Maulino* forest, *Araucaria* forest) to latitudinally extensive and continuous distributions (e.g., *Nothofagus pumilio* forests). Mixed deciduous-evergreen forest types predominantly occupy the Central Valley (e.g., dominated by deciduous *Nothofagus* species). Some of the more localized and discontinuous forest types are presently reduced to small fragments due to land conversion to agriculture and pasture following European settlement. The extirpation of lowland *Fitzoya cupressoides* forests from the Lake District occurred during the last century due to intensive burning and logging (Elizalde 1970). Some individuals of this long-lived conifer became established in these areas more than 3000 years ago (Lara and Villaba 1993), making these nearly extinct ecosystems some of the oldest in the world. Genotypes that endured centuries of climatic change are now lost.

Although indigenous people probably burned forests to open land for crops alongside river valleys and coastal areas and used a variety of forest products, their impact on overall forest cover was probably limited in the temperate forest region (Armesto et al. 1994). Beginning in the sixteenth century, with the settlement of Europeans, increasingly larger areas of temperate forests were logged and burned to provide land for agriculture or pastures and harvested for timber and firewood. The exploitation of native forests, with few exceptions, has always been done with little regard for the capacity of the forest to recover its original composition and structure (Donoso and Lara 1996).

Pine Plantations

The timber industry in Chile developed rapidly in the past 50 years associated with the expansion of *Pinus radiata* plantations (Lara and Veblen 1993). Monocultures of this exotic conifer increased exponentially since the 1940s (Lara et al. 1996), first on abandoned agricultural land, and later at the expense logged or burned native forests. Intensive forestry, based on plantations, has concentrated mainly between 35 and 40°S, coinciding with the region where the temperate forest persisted during the LGM (Villagrán and Hinojosa 1997). In this area, we find the largest number of threatened and endangered woody and animal species (CONAF 1989). The expansion of fast-growing commercial plantations in rural areas has forced farmers to migrate to the nearest cities, with negative consequences for their quality of life (Lara et al. 1996).

Logging for Wood Chips

A factor further contributing to deforestation and loss of old-growth forest habitat that has increased exponentially in the 1990s is the harvesting of timber to produce wood chips, which are exported mainly to Japan (Lara et al. 1996). As the demand for unprocessed wood products increased in the 1990s, small landowners are logging the remaining patches of primary or secondary forest, which are often characterized by low-quality timber, for a marginal profit.

Comparative Analysis of Drivers of Global Change

From the preceding analysis, it is evident that the future scenarios for biodiversity will differ markedly between the geographically continuous and extensive region of north temperate forests and the isolated and reduced area of southern temperate forests. The predictions about biodiversity scenarios for South America should also apply to small areas of temperate forests in Tasmania and New Zealand. Predicting the impact of global change on these isolated southern hemisphere regions is a high-priority task, because they harbor largely endemic species assemblages, with representatives of ancient families and genera having one or a few living species (see Table 11.2). Entire ecosystems in these southern regions are biologically unique and highly susceptible to the effects of global change.

As a useful way to convey the information, which is in most cases still incomplete, about drivers of global change and their expected effects on biological diversity and ecosystem function in North and South American temperate forests, we present a comparative analysis by driver (Table 11.3). This analysis is based on the available information about forest ecosystems for each region and the present knowledge of environmental, biological, and socioeconomic variables associated with recognized drivers of global change. New information, as it becomes available, and changing socioeconomic scenarios will likely alter our current predictions. Ranks are assigned on a relative context (i.e., higher ranks indicate that specific processes or impacts will be comparatively more important in one region than in the other, but they are not directly comparable with other biomes). Response functions were for selected variables relevant for ecosystem managers and regional planners.

Response Functions

We focused on the impact of global change on two components of biological diversity (Table 11.3), species richness, and functional group diversity (Mooney and Schulze 1993). *Functional group diversity* is operationally defined here as the number of different groups of species that are responsible for critical ecosystem processes (e.g., N mineralization, or primary production). Net changes in species richness are the result of several concurrent processes. We will examine here the process of invasion of species from neighboring biomes (e.g., species from Mediterranean or boreal ecosystems), the loss of species due to local or regional extinction, and the invasion by exotic species, either deliberately or unintentionally introduced by people from other parts of the world. We will also consider separately the impact of global change on the loss of endemic species, restricted to small areas within temperate forest ecosystems. A prediction concerning the time lag for expected changes of biological diversity for North and South America, based on the relative impact of drivers of global change in each region, will be provided as a reference (Table 11.3).

Other measurable elements of regional biodiversity examined in Table 11.3 are *landscape heterogeneity*, which is defined as the number and types of different of forests that compose the regional landscape; *ecosystem functioning*, which is defined as the maintenance of productivity and its supporting processes, including nutrient and hydrologic cycles; and *ecosystem services*, which is defined as the maintenance of ecosystem attributes valued by people, (e.g., supply of commodities, clean water, esthetic values, and recreation). The nature of the relationship between biological diversity and the maintenance of ecosystem functions and services has been analyzed in depth in several textbooks and journal articles (Schulze and Mooney 1993; Mooney et al. 1996; Tilman et al. 1996; Chapin et al. 1998). Finally, we provide a qualitative estimate of how much variation can be expected for each driver of global change in the coming decades. These predictions (Table 11.3) are qualitative

Table 11.3. Comparative analysis of major drivers of global change (columns) and response functions (rows) in North (N) and South (S) American temperate forests

Effect or response function	Land-use change		Atmospheric change			Climate change	
	Conversion	Fragmentation	N-saturation	Pollution	CO$_2$ incr.	Δ Temp.	Δ Seasonality
Species diversity							
Gain of spp from other biomes	N2/S1	N4/S2	N1/S1	N1/S1	N3/S2	N3/S2	N3/S4
Loss of species	N1/S4	N4/S4	N3/S1	N5/S1	N3/S1	N2/S1	N4/S3
Loss of endemics	N1/S5	N2/S5	N1/S3	N2/S2	N3/S1	N2/S1	N2/S4
Exotic spp invasions	N2/S4	N4/S3	N4/S1	N4/S1	N3/S2	N2/S3	N3/S4
Δ Relative abundance	N1/S5	N5/S4	N5/S1	N4/S1	N3/S1	N3/S3	N4/S4
Functional group diversity							
Gain of species	N1/S1	N3/S3	N3/S1	N1/S1	N1/S1	N3/S2	N1/S1
Loss of species	N2/S5	N3/S5	N5/S1	N5/S2	N1/S1	N1/S1	N3/S4
Δ Relative abundance	N4/S5	N5/S5	N4/S1	N3/S1	N4/S1	N4/S3	N5/S3
*Time lag in Δ diversity***	N2/S5	N4/S4	N5/S3	N4/S2	N3/S2	N3/2	N3/S2
Δ Landscape heterogeneity	N2/S5	N5/S5	N5/S1	N4/S1	N2/S1	N3/S3	N3/S3
Δ Functioning	N1/S5	N4/S4	N5/S2	N4/S1	N4/S2	N4/S3	N5/S4
Δ Ecosystem services	N2/S5	N4/S3	S5/S2	N4/S1	N2/S2	N5/S2	N4/S3
Expected Δ in drivers	N3/S4	N5/S5	N3/S5	N2/S5	N4/S4	N5/S2	N5/S3

Values for expected effect of drivers on response function range from low (1) to high (5). ** Time lag of change in diversity: 1 = >100 years, 2 = 50–100 years, 3 = 10–50 years, 4 = <10 years, 5 = present.

and are intended to provide guidance to managers and land planners about the foreseeable consequences of global change for biodiversity in temperate forests. References and evidence supporting these predictions are discussed later.

Land-Use Change

Contrasting histories of land use in temperate regions illustrate how rates of global change and expected impacts depend strongly on socioeconomic scenarios. Europeans settled the east coast of North America early in the seventeenth century, leading to a rapid conversion of forests into farmland that continued until the eighteenth century (Foster et al. 1998). In 100 years many areas of less-productive forests and farmland were converted to pastures that continued to be used for grazing well into the nineteenth century. In the late 1800s and early twentieth century, decreasing productivity, development of intensive farming in other regions of the country, and migration of farmers to cities as a result of the industrial revolution led to abandonment of farmland and a reversion of the forest-clearing trend. Much of today's forests in eastern North America are derived from this period, and their land cover is expanding (Table 11.4). The history of land use in this region, however, reflects strongly on the present floristic structure of temperate forests, revealing a reduction of spatial heterogeneity with respect to presettlement forests (Foster et al. 1998; Fuller et al. 1998).

In contrast, forests of southern South America were settled two centuries later than eastern North America. Forest clearing for agriculture is still ongoing, and the forest area is rapidly shrinking (Table 11.4, Lara et al. 1996). Cleared forests that are not suitable for agriculture may be planted with exotic pines for timber (Lara and Veblen 1993), thus losing their potential for recovery or restoration (Armesto 1992). Likewise, Pacific Northwest forests in North America have suffered considerable impact from logging in the recent decades (Ehrlich 1996; Wallin et al. 1996), and their history differs considerably from that of eastern North America. Foresters in the Pacific Northwest, however, promote the recovery of the original tree species in logged areas, which is in contrast with the situation of southern Chile, where substitution of native woods by exotic pines is pursued.

Table 11.4. Changes in land cover from 1830 to 1985 in eastern North America (Massachusetts) and in lowland areas of northern Chiloé Island, southern Chile

Land cover	Eastern North America		Chiloé	
	1830	*1985*	*1834*	*1990*
% Forest	24	53	90	<30
% Agriculture and pasture	65	27	<5	>60

Data for Massachusetts from Foster et al. 1998; for Chiloé, based on Díaz and Armesto, unpublished data.

As a consequence records for the impacts of land-use change on biological diversity and ecosystem function differ greatly between these two regions (Table 11.3). Major and impending impacts of global change on temperate South American forests will result from fragmentation and conversion to monocultures of pine, which are predicted to severely reduce biological diversity in the early twenty-first century (Armesto et al. 1992). Homogenization of landscapes due expanding pine plantations will continue to threaten biological diversity and to impair the recovery of native forests. Conversion to other uses (i.e., cropland, exotic plantations) over most of North America is not a major driver of global change, although logging and fragmentation are a serious threat to biological diversity in Pacific Northwest forests (Wallin et al. 1996), with major impacts likely on species diversity and ecosystem services (Table 11.3). In contrast, eastern North America exhibits a trend toward regional homogenization of forests (Foster et al. 1998), associated with past land use, with predictable losses of functional guilds and impaired ecosystem functions and services, both accentuated by the effects of pollution and nitrogen deposition (Aber 1992; Likens et al. 1996). One of the most important drivers leading to landscape homogenization, changes in species abundances, and local species extirpation in both eastern and western North America is the human alteration and control of historical rates and magnitudes of natural disturbance (e.g., fire regimes) (Wallin et al. 1996; Fuller et al. 1998).

Atmospheric Chemistry

The main drivers of global change in temperate forests are enhanced rates of N deposition, known as *nitrogen saturation* effects (Aber 1992; Vitousek et al. 1997), that eventually alter forest growth and biogeochemical processes, as well as acid rain associated with industrial air pollution (Schlesinger 1991; Likens et al. 1996). These effects are concurrent with the impact of increased global CO_2 levels (Schlesinger 1991). All these effects are more prevalent today in northern hemisphere temperate forests, but their impacts are expected to increase in southern temperate forests in the coming decades (e.g., Galloway et al. 1994).

Nitrogen saturation effects on element cycling, tree growth, and ecosystem processes in northern hemisphere forests are not fully understood, but present evidence suggests that processes such as decomposition and nutrient storage are negatively affected, thereby limiting productivity and the range of ecosystem services that forests provide (Vitousek et al. 1997). For example, losses of soil cations and reduced growth have been associated with enhanced nitrogen deposition (Aber 1992; Likens et al. 1996; Asner et al. 1997). Water quality for human use may also be reduced due to nitrate leaching to ground and surface waters (Vitousek et al. 1997). Alteration of soil chemistry (i.e., lowered pH, low C–N ratios) may have important consequences for biological diversity, leading to disruption of functional guilds of decomposer organisms (e.g., rhizosphere fungi and microbes), as well as to changes in the relative abundances of tree species. Nutrient enrichment by atmospheric deposition of N and other pollutants can also lead to reduced species richness due to the predominance of fast-growing species (Tilman 1982).

In this context southern temperate forests are still relatively free of industrial pollution because of the lower density of contaminating sources and because of the lower land–ocean ratio at high latitudes, so that much of the precipitation derives from moist westerly winds coming from the central Pacific (South America) or Subantarctic Oceans (New Zealand, Tasmania). The chemistry of rainfall in these temperate areas resembles dilute seawater and differs markedly from that of many areas of North America and Europe subjected to higher N deposition (Galloway et al. 1994; Galloway et al. 1996). Although the impact of air pollution is predicted to increase rapidly in the coming decades (Table 11.3), we believe that the potential effects of air pollution on biological diversity in southern forests will show a longer time lag when compared with the rate and effects of current land use patterns (see earlier).

Climatic Change

Major changes in climate predicted by Global Circulation Models (GCMs) indicate that increases in temperature and seasonality will be more pronounced in the northern hemisphere because of the larger land–ocean ratio at high latitudes, in contrast to the distribution of land in southern hemispheric temperate regions. Keeling et al. (1996) and Miynani et al. (1997) have documented using satellite measurements that the length of the growing has increased by about 1 week since the 1960s, with the greatest incidence in regions between 45 and 70°N. Marked warming has occurred in the springtime due to early disappearance of snow. This agrees with GCMs that predict more pronounced warming in continental regions of the northern hemisphere than in the southern hemisphere (Stouffer et al. 1989). The projected rise in global temperatures may be as high as 2 degrees by the year 2030 (Trenberth 1993). The consequences of doubling CO_2 levels for hydrologic cycles and precipitation patterns in northern temperate forests are less predictable (Kellog and Zhao 1988; Rind 1988), but they generally suggest a speeding of the hydrologic cycle that would result in wetter conditions along the eastern coast of North America and drier conditions in the interior of the continent and on the west coast. Overall, warming trends are predicted to have a greater effect on shifting species abundances in forest communities (Table 11.3): Trees presently confined to warmer regions may increase in abundance and expand their range into higher latitudes, whereas cold-tolerant species may decrease in abundance. Invasion of exotic species may be facilitated by a longer growing season.

In southern temperate forests warming trends are expected to occur at a lower rate than they do in the northern hemisphere (Mooney et al. 1993). Analysis of surface temperature records for the twentieth century based on instrumental data (Aceituno et al. 1993) and dendrochronology (Lara and Villalba 1993) reveal a stepwise increase, south of 45°S, and unexpected cooling from 34 to 45°S. They do not show enhanced tree growth as detected in the northern hemisphere. Our prediction regarding the impact of warming in southern temperate forests is that changes will lag behind overriding impacts of land use (see earlier). Seasonality is much lower in southern temperate forests than in temperate North America (Alaback 1991; Arroyo et al. 1996); hence, spring warming will likely be of less consequence for evergreen trees that are adapted to a longer growing season. Based on the fact that native Chilean trees are derived from subtropical and tropical ancestors, Waring and Winner (1996) predict that they are more likely to show improved growth if climatic warming occurs than would the Pacific Northwest forests of North America, which are dominated by conifers derived from a boreal stock. Available evidence, however, does not support this prediction.

The potential exists, however, for increased seasonality of precipitation, which is controlled by the position of the subtropical anticyclone of the southeastern Pacific (Aceituno et al. 1993). A reduction in summer precipitation characterizes the southern temperate forest region, from 36 to 45°S (Arroyo et al. 1996), with summer months receiving 10–25% of the annual precipitation (Alaback 1991). Summer drought is. a consequence of a displacement of the normal westerly flow of moist winds by an increase in the intensity and a southward shift in the position of the subtropical anticyclone. Although current GCMs have not addressed this problem, a shift in the position of the subtropical Pacific anticyclone further to the south might increase the length and intensity of the dry season, with critical consequences for the biological diversity and ecological interactions in temperate forests (Arroyo et al. 1993). Loss of species less tolerant to drought and shifts in abundance of tree species in forests can be predicted as a consequence of increased seasonality of precipitation in southern temperate forests (Table 11.3). Waring and Winner (1996) speculate that the consequences of more pronounced summer droughts are likely to be greater for Pacific Northwest forests because they receive only 5–10% of their annual precipitation during that interval.

Potential for Range Expansion

Keeping other ecological factors constant, climatic warming should provide opportunities for temperate forests to expand to higher (boreal and austral) latitudes. However, because of the distribution of land in the southern hemisphere, this range expansion will be less important than the predicted expansion of North American temperate forests into the vast territory of boreal forests. This expansion will be facilitated by the fact that a majority of trees in northern hemisphere temperate forests are pollinated by wind and their propagules are wind dispersed. In southern South America, at latitudes above 43°S, land is mostly an archipelago of mountains and fjords, where colonization will be difficult. The expansion of Valdivian rain forest taxa over

ocean barriers is complicated by the fact that many plant species have mutualistic partners, insects, or vertebrates that act as pollinators or seed dispersers
(Arroyo et al. 1993); therefore, the entire assemblage must migrate together.
Moreover, because many woody species are obligate outcrossers (Riveros et
al. 1996), colonization events require the establishment of more than one
individual tree or shrub.

Another major problem for the expansion of southern temperate forests is
that forests north of 43°S, which should provide the main sources of propagules and genetic variability for range expansion, have been greatly fragmented
and reduced in area as a result of deforestation and conversion to plantations and agriculture. Population sizes of many species have been reduced,
which decreases genetic variability and limits the potential for range expansion. Palynological data about range expansion in Holocene times (Villagrán
1991; Moreno, unpublished data), however, show rapid colonization of
Valdivian forest species, most likely from fragmented coastal populations that
survived the last glacial period. The paradox today is that we cannot rely on
these sources for future recolonization because they have been decimated by
land conversion. Restoration programs aimed at increasing the area of forest
and the population sizes of sensitive species will consequently be essential as
an insurance against the uncertainties associated with future climatic change.

Research and Policy Needs

Our examination of the scenarios for biological diversity in temperate forests
of South America and their comparison with North America indicates that
current GCMs can provide general predictions about future scenarios, but an
in-depth analysis for each area will reveal considerable differences in the rates
and magnitudes of major impacts and in the response functions of species
and ecosystems. These differences are the result of integrating socioeconomic
trends and evolutionary histories of each region in the analysis, as well as the
historical record of recent disturbance regimes. To our knowledge, this integration has not been fully achieved for any area in the temperate forest region.
Because major differences exist in socioeconomic trends, evolutionary histories, and disturbance regimes between northern and southern hemisphere
temperate forests, and between North and South America in particular, we
expect that the trajectories of change in land use, atmospheric chemistry, and
climate will differ considerably among regions (Table 11.3). Even within
North America, disturbance histories, drivers of global change, and forest
composition are quite different between the eastern and western coasts,
leading to contrasting scenarios for future biodiversity. As a consequence,
broad-range generalizations and commonplace recommendations intended
to help managers and land planners anticipate and prevent future losses of
biological diversity are likely to be inaccurate or laden with uncertainty. Interdisciplinary research programs that seek to integrate knowledge of ecology,

climatic models, socioeconomic trajectories, and evolutionary histories for each geographical area may lead to more accurate regional models that can project environmental changes into the future. Our analysis suggests that even though North American temperate forests are likely to expand both locally and into the boreal region, at least in the northeast, South American temperate forests will continue to shrink because of the different cultural and economic trends in each region, despite predictions that global warming may enhance the growth rates of southern temperate trees.

Another major area of uncertainty exists concerning the long-term consequences of increased N deposition and air pollution for North American temperate forests, both for biological diversity and the terrestrial C and N cycles (Asner et al. 1997; Vitousek et al. 1997). The analysis of the causes and effects of N loading in temperate ecosystems is an urgent task, given the exponential rate of increase of anthropogenic N inputs to the terrestrial and aquatic environments (Vitousek et al. 1997). Because of anticipated industrial growth (Galloway et al. 1994), acid rain and N deposition are likely to add to the current burden on southern temperate forests, leading to further reductions in biodiversity as well as to reduced ecosystem potential to recover from natural or human-caused catastrophes. Losses of biodiversity that affect the soil microflora and fauna may seriously alter biogeochemical pathways and ecosystem resilience (e.g., Vitousek and Matson 1984).

The predicted scenarios for biological diversity (Table 11.3) warn us that South American temperate forests may loose a considerable fraction of their biological diversity in the coming decades and that these losses will constitute an important drain of unique and little-known endemic species, despite all current conservation efforts (Armesto et al. 1998). This is because patterns of land development in southern South America are predicted to follow the same path followed by European settlers of eastern North American forests more than one century ago, despite the evidence for losses of species and soil erosion that have prevented the recovery of North American forests (Foster et al. 1998). Market forces promoting intensive forestry based on the substitution of native woods by single-species plantations of fast-growing exotic trees worsen the scenario of biological diversity in temperate forests of Chile. This type of forestry has proven to be successful in generating economic profits for few, but it has important negative social and environmental impacts at the regional level that have largely been ignored (Lara and Veblen 1993; Lara et al. 1996). Proposed alternatives for managing native forests with a greater variety of species, forest products and services, and fewer losses of wildlife habitat (Donoso and Lara 1996; Franklin and Armesto 1996) have not yet been implemented and are disregarded in favor of fast profits.

What is the main message from this analysis for scientists and ecologists that are actively seeking to predict global change and its consequences for the future of temperate forests? A major conclusion is that our immediate tasks cannot be limited any longer to research and modeling efforts that seek to improve our understanding of the causes and consequences of environmental change. We must also commit time to transmit new and existing knowledge to policy and decision makers effectively (Lubchenco 1998). Both Chilean and other international scientists aware of the imminent threats to biodiversity in temperate South America (Armesto et al. 1998) must take the lead in an effective campaign addressed to the Chilean public, government authorities, and forest managers to eradicate the practice of forestry based on the substitution of species-rich indigenous forests by exotic monocultures. This is a small, but important, step toward reducing the number of factors that imperil regional biodiversity and landscape heterogeneity in the immediate future.

Acknowledgments. We appreciate the financial support by Fondecyt (Chile), grants 1950461 and 1980705, The A.W. Mellon Foundation (United States), Darwin Initiative for the Survival of Species (U.K.), SUCRE project (Inco-DC), and a Fulbright Graduate Fellowship to RR. This is a contribution to the program of the Institute of Ecological Research Chiloé, *Senda Darwin* Biological Station, and to the Program of Native Forest Studies of the University of Chile. JJA acknowledges the support of a John S. Guggenheim Fellowship and the hospitality of the Institute of Ecosystem Studies (New York) during the final preparation of the final manuscript. We thank Mary Arroyo, Terry Chapin, John Silander, Cecilia Smith, and Carolina Villagrán for useful advice, references, and comments on the manuscript.

References

Aber JD (1992) Nitrogen cycling and nitrogen saturation in temperate forest ecosystems. Trends in Ecology and Evolution 7:220–223.

Aber JD, Nadelhoffer JK, Steudler PA, Melillo JM (1989) Nitrogen saturation in northern forest ecosystems: hypotheses and implications. BioScience 39:378–386.

Aceituno P, Fuenzalida H, Rosenbluth B (1993) Climate along the extratropical coast of South America. In: Mooney H, Fuentes E, Kronberg B (eds) *Earth System Responses to Global Change: Contrasts between North and South America*, pp. 61–69. Academic Press, New York.

Alaback PB (1991) Comparative ecology of temperate rainforests of the Americas along analogous climatic gradients. Rev Chil Hist Nat 64:399–412.

Armesto JJ (1992) Mitos y realidades del bosque nativo chileno. Rev Chil Hist Nat 65:173–176.

Armesto JJ, Rozzi R (1989) Seed dispersal syndromes in the rain forest of Chiloé: evidence for the importance of biotic dispersal in a temperate rainforest. Journal of Biogeography 16:219–226.

Armesto JJ, Rozzi R, Smith-Ramírez C, Arroyo MT (1998) Effective conservation targets in South American temperate forests. Science (In press).

Armesto JJ, Donoso C, Villagrán C (1994) Desde la era glacial a la industrial: la historia del bosque templado chileno. Amb Des 10(1):66–72.

Armesto JJ, Vidiella PE, Jiménez HE (1995) A comparative analysis of causes and mechanisms of succession in mediterranean-type regions of Chile and California. In: Arroyo MTK, Fox M, Zedler P (eds) *Ecology and Biogeography of Mediterranean Ecosystems in Chile, California and Australia*, pp. 418–434. Springer-Verlag, Berlin.

Armesto JJ, Villagrán C, Arroyo MTK (eds) (1996a) Ecologìa de los Bosques Nativos de Chile. Editorial Universitaria, Santiago.

Armesto JJ, Smith-Ramìrez C, Sabag C (1996b) The importance of plant-bird mutualism in the temperate rainforest of southern South America. In: Lawford RG, Alaback P, Fuentes ER (eds) *High Latitude Rain Forests and Associated Ecosystems of the West Coast of the Americas: Climate, Hydrology, Ecology and Conservation*, pp. 248–265. Springer-Verlag, Berlin.

Armesto JJ, Smith-Ramìrez C, Leûn P, Arroyo MTK (1992) Biodiversidad y conservaciûn del bosque templado en Chile. Amb Des 8(3):19–24.

Arroyo MTK, Uslar P (1993) Breeding systems in a temperate mediterranean-type climate montane sclerophyllous forest in central Chile. Biological Journal of the Linnean Society 111:83–102.

Arroyo MTK, Armesto JJ, Squeo F, Gutièrrez J (1993) Global change: the flora and vegetation of Chile. In: Mooney H, Fuentes E, Kronberg B (eds) *Earth System Response to Global Change: Contrasts between North and South America*, pp. 239–264. Academic Press, New York.

Arroyo MTK, Riveros M, Peòaloza A, Cavieres L, Faggi AM (1996) Phytogeographic relationships and regional richness patterns of the cool temperate rainforest flora of southern South America. In: Lawford RG, Alaback P, Fuentes ER (eds) *High Latitude Rain Forests and Associated Ecosystems of the West Coast of the Americas: Climate, Hydrology, Ecology and Conservation*, pp. 134–172. Springer-Verlag, Berlin.

Asner GP, Seastedt TR, Townsend AR (1997) The decoupling of terrestrial carbon and nitrogen cycles. BioScience 47:226–234.

Axelrod DI, Arroyo MTK, Raven PH (1991) Historical development of temperate vegetation in the Americas. Rev Chil Hist Nat 64:413–446.

Banco Central de Chile (1995) Proyecto de cuentas ambientales y bosque nativo. Gerencia de Estudios, Santiago, Chile.

Berg A, Ehnstrñ B, Gustaffson L, Hallingbock T, Jonsell M, Weslien J (1994) Threatened plant, animal, and fungus species in Swedish forests: distribution and habitat asssociations. Conservation Biology 8:718–731.

Barrett SCH, Helenurm K (1987) The reproductive biology of boreal forest herbs. I. Breeding systems and pollination. Canadian Journal of Botany 65:2036–2046.

Chapin FS III, et al. (1998) Ecosystem consequences of changing biodiversity. Experimental evidence and a research agenda for the future. BioScience 48:45–52.

CONAF (1989) Libro Rojo de la Flora Terrestre de Chile. Benoit I (ed) Corporaciûn Nacional Forestal, Santiago.

Donoso C (1993) Bosques Templados de Chile y Argentina, Variaciûn, Estructura y Din·mica. Editorial Universitaria S.A., Santiago, Chile.

Donoso C, Lara A (1996) Utilizaciûn de los bosques nativos en Chile: pasado, presente y futuro. In: Armesto JJ, Villagrán C, Arroyo MTK (eds) Ecologia de los Bosques Nativos de Chile, pp. 363–387. Editorial Universitaria S.A., Santiago, Chile.

Ehrlich PR (1996) Conservation of temperate forests: what do we need to know and do? Forest Ecology and Management 85:9–19.

Elizalde R (1970) La Sobrevivencia de Chile. Ministerio de Agricultura, Servicio Agrìcola y Ganadero, Santiago.

Foster DR, Motzkin G, Slater B (1998) Land-use history as long-term broad scale disturbance: Regional forest dynamics in central New England. Ecosystems 1:96–119.

Franklin JF, Armesto JJ (1996) Una alternativa de manejo para los bosques nativos chilenos. Amb Des 12(2):69–79.

Fuller JL, Foster DR, McLachlan JS, Drake N (1998) Impact of human activity on regional forest composition and dynamics in central New England. Ecosystems 1:76–95.

Galloway D (1996) Los liquenes del bosque templado de Chile. In: Armesto JJ, Villagrán C, Arroyo MTK (eds) Ecologìa de los Bosques Nativos de Chile, pp. 101–112. Editorial Universitaria, S.A., Santiago.

Galloway JN, Levy H II, Kasibhatla PS (1994) Year 2020: consequences of population growth and development on deposition of oxidized nitrogen. Ambio 23:120–123.

Galloway JN, Keene WC, Likens GE (1996) Processes controlling the composition of precipitation at a remote southern hemisphere location: Torres del Paine National Park, Chile. Journal of Geophysical Research 101:6883–6987.

Groisman PY, Karl TA, Knight TW (1994) Observed impact of snow cover on the heat balance and the rise of continental spring temperatures. Science 263:198–200.

Guzman G, Quilhot W, Galloway DJ (1990) Decomposition of species of Pseudocyphellaria and Sticta in southern a Chilean forest. Lichenologist 22:325–331.

Hedin LO, Armesto JJ, Johnson A (1995) Patterns of nutrient loss from unpolluted old-growth temperate forests: evaluation of biogeochemical theory. Ecology 76:493–509.

Hill RS, Enright NE (eds) (1995) Ecology of Southern Conifers. Melbourne University Press, Melbourne.

Jimènez HE, Armesto JJ (1992) Soil seed bank of disturbed sites in the Chilean matorral: its importance in early secondary succession. Journal of Vegetation Science 3:579–586.

Keeling CD, Chin JFS, Whorf TP (1996) Increased activity of northern vegetation inferred from atmospheric CO_2 measurements. Nature 382:146–149.

Kellog Z (1988) Sensitivity of soil moisture to doubling CO_2 in climate model experiments. Part I: North America. Journal of Climate 1:349–366.

Landrum L (1981) The phylogeny and phytogeography of Myrceugenia (Myrtaceae). Brittonia 33:105–129.

Lara A, Veblen TT (1993) Forest plantations in Chile: a successful model? In: Mather A (ed) Afforestation. Policies, Planning and Progress, pp. 118–139. Belhaven Press, London.

Lara A, Villalba R (1993) A 3620-year temperature record from Fitzroya cupressoides tree rings in southern South America. Science 260:1104–1106.

Lara A, Donoso C, Aravena JC (1996) La conservaciûn del bosque nativo en Chile: problemas y desafios. In: Armesto JJ, Villagrán C, Arroyo MTK (eds) Ecologìa de los Bosques Nativos de Chile, pp. 335–362. Editorial Universitaria, Santiago.

Lbchenco J (1998) Entering the century of the environment: a new social contract for science. Science 279:491–497.

Likens GE, Discroll CT, Buso DC (1996) Long-term effects of acid rain: response and recovery of a forest ecosystem. Science 272:244–246.

Mooney HA, Fuentes ER, Kronberg B (eds) (1993) Earth System Response to Global Change: contrasts between North and South America. Academic Press, New York.

Mooney HA, Cushman JH, Medina E, Sala OE, Schulze E-D (eds) (1996) SCOPE 55. Functional roles of biodiversity. A global perspective. John Wiley and Sons, New York.

Murà R (1996) Comunidades de mamíferos del bosque templado de Chile. In: Armesto JJ, Villagrán C, Arroyo MTK (eds) Ecologìa de los bosques nativos de Chile, pp. 113–134. Editorial Universitaria, S.A., Santiago.

Myneni RB, Keeling CD, Tucker CJ, Asrar G, Nemani RR (1997) Increased plant growth in the northern high latitudes from 1981 to 1991. Nature 386:698–201.

Norton TW (1996) Conserving biological diversity in Australiaís temperate eucalypt forests. Forest Ecology and Management 85:21–33.

Ovington JD (ed) (1983) Temperate Broad-Leaved Forests. Ecosystems of the World. Elsevier, Amsterdam.

Pastor J, Post WM (1986) Influence of climate, soil moisture, and succession on forest carbon and nitrogen cycles. Biogeochemistry 2:3–27.

Pèrez C, Armesto JJ, Ruthsatz B (1991) Litter decomposition and soils of mixed-conifer forests in the Cordillera de Piuchuè (42°S), Chiloé National Park, Chile. Rev Chil Hist Nat 64:479–490.

Pèrez C, Hedin LO, Armesto JJ (1998) Nitrogen mineralization in two unpolluted, old-growth forests of contrasting structure and biodiversity. Ecosystems (In press).

Pulliam HR (1988) Sources, sinks, and population regulation. American Naturalist 132:652–661.

Reich PB, Walters MB, Ellsworth DS (1997) From tropics to tundra: global convergence in plant functioning. Proceedings of the National Academy of Sciences USA 94:13730–13734.

Richards JF (1993) Land transformation. In: Turner BL II, Clark WC, Kates RW, Richards JF, Matthews JT, Meyer WB (eds) *The Earth as Transformed by Human Action*, pp. 164–178. Cambridge University Press, Cambridge, UK.

Rind (1988) The doubled CO_2 climate and the sensitivity of the modeled hydrologic cycle. Journal of Geophysical Research 93:5385–5412.

Riveros M (1991) Aspectos sobre la biología reproductiva en dos comunidades del sur de Chile. Docoral Thesis, Facultad de Ciencias, Universidad de Chile, Santiago.

Riveros M, Humaòa AM, Arroyo MTK (1996) Sistemas de reproducciûn en especies del bosque Valdiviano (40° Latitud Sur). Phyton 58:167–176.

Rohrig E, Ulrich B (eds) (1991) *Temperate Deciduous Forests. Ecosystems of the World*. Elsevier, Amsterdam.

Rozzi R, Martìnez D, Willson MF, Sabag C (1996) Avifauna de los bosques templados de Sudamèrica. In: Armesto JJ, Villagrán C, Arroyo MTK (eds) Ecologìa de los bosques nativos de Chile, pp. 135–152. Editorial Universitaria, S.A., Santiago.

Schlesinger WH (1991) *Biogeochemistry. An Analysis of Global Change*. Academic Press, New York.

Schulze E-D (1989) Air pollution and forest decline in a spruce (Picea abies) forest. Science 244:776–783.

Schulze E-D, Monney HA (eds) (1993) *Biodiversity and Ecosystem Function*. Springer-Verlag, Berlin.

Schulze E-D, Bazzaz FA, Nadelhoffer KJ, Koike T, Tkatsuki T (1996) Biodiversity and ecosystem function of temperate deciduous broad-leaved forests. In: Mooney HA, Cushman JH, Medina E, Sala O, Schulze E-D (eds) *SCOPE 55. Functional Roles of Biodiversity. A Global Perspective*, pp. 71–98. John Wiley and Sons, New York.

Smith-Ramirez C (1993) Los picaflores y su recurso floral en el bosque templado de la isla de Chiloé, Chile. Rev Chil Hist Nat 66:65–73.

Smith-Ramìrez C, Armesto JJ (1998) Nectarivorìa y polinizaciûn por aves en Embothrium coccineum (Proteaceae) en el bosque templado del sur de Chile. Rev Chil Hist Nat 71:53–65.

Stouffer RJ, Manabe S, Bryan K (1989) Interhemispheric asymmetry in climate responses to a gradual increase in atmospheric CO_2. Nature 342:660–662.

Tilman D (1982) *Resource Competition and Community Structure*. Princeton University Press, Princeton, NJ.

Tilman D, Wedin D, Knops J (1996) Productivity and sustainability influenced by biodiversity in grassland ecosystems. Nature 379:718–720.

Trenberth KE (1993) Northern hemisphere climate change: physical processes and observed changes. In: Mooney H, Fuentes E, Kronberg B (eds) *Earth System Responses to Global Change: Contrasts between North and South America*, pp. 35–59. Academic Press, New York.

Villagrán C (1990) Glacial climates and their effect on the history of vegetation of Chile. A synthesis based on palynological evidence from Isla de Chiloé. Review of Paleobotany and Palynology 65:17–24.

Villagrán C (1991) Historia de los bosques templados del sur de Chile durante el Tardiglacial y el Postglacial. Rev Chil Hist Nat 64:447–460.

Villagrán C, Armesto JJ (1993) Full and late-Glacial paleoenvironmental scenarios for the west coast of southern South America. In: Mooney H, Fuentes ER, Kronberg B (eds) *Earth System Responses to Global Change. Contrasts between North and South America*, pp. 195–207. Academic Press, New York.

Villagrán C, Hinojosa L (1997) Historia de los bosques del sur de Sudamerica. II: Análisis fiogeográfico. Rev Chil Hist Nat 70:241–267.

Villagrán C, Moreno P, Villa R (1996) Antecedentes palinolûgicos acerca de la historia cuaternaria de los bosques Chilenos. In: Armesto JJ, Villagrán C, Arroyo MTK (eds) Ecologìa de los Bosques Nativos de Chile, pp. 51–65. Editorial Universitaria, S.A., Santiago.

Vitousek PM (1994) Beyond global warming: ecology and global change. Ecology 75:1861–1876.

Vitousek PM, Matson P (1984) Mechanisms of nutrient retention in forest ecosystems: a field experiment. Science 255:51–52.

Vitousek PM, Aber J, Howarth RH, Likens GE, Matson PA, Schindler DW, et al. (1997) Human alteration of the global nitrogen cycle: causes and consequences. Issues in Ecology 1:1–15.

Vuilleumier F (1985) Forest birds of Patagonia: ecological geography, speciation, endemism, and faunal history. Ornithological Monographs 36:255–304.

Wallin DO, Swanson FJ, Marks B, Cissel JH, Kertis J (1996) Comparison of managed and pre-settlement landscape dynamics in forests of the Pacific Northwest, USA. Forest Ecology and Management 85:291–309.

Walter H (1973) *Vegetation of the Earth in Relation to Climate and the Eco-Physiological Conditions*. English Universities Press, London.

Waring RH, Franklin JF (1979) Evergreen coniferous forest of the Pacific Northwest. Science 204:1380–1386.

Waring RH, Winner WE (1996) Constraints on terrestrial primary productivity in temperate forests along the Pacific coast of North and South America. In: Lawford RG, Alaback P, Fuentes ER (eds) *High Latitude Rain Forests and Associated Ecosystems of the West Coast of the Americas: Climate, Hydrology, Ecology and Conservation*, pp. 89–102. Springer-Verlag, Berlin.

Willson MF (1991) Dispersal of seeds by frugivorous animals in temperate forests. Rev Chil Hist Nat 64:537–554.

Willson MF, DeSanto TL, Sabag C, Armesto JJ (1994) Avian communities of fragmented south-temperate rainforests in Chile. Conservation Biology 8:508–520.

Willson MF, Sabag C, Figueroa J, Armesto JJ, Caviedes M (1996a) Seed dispersal by lizards in Chilean rain forests. Rev Chil Hist Nat 69:339–342.

Willson MF, Sabag C, Figueroa J, Armesto JJ (1996b) Frugivory and seed dispersal of Podocarpus nubigena in Chiloé. Chile. Rev Chil Hist Nat 69:343–349.

Willson MF, Smith-Ramìrez C, Sabag C, Hernandez JF (1996c) Mutualismos entre plantas y animales en bosques templados de Chiloé. In: Armesto JJ, Villagrán C, Arroyo MTK (eds) Ecologìa de los bosques nativos de Chile, pp. 251–264. Editorial Universitaria, S.A., Santiago.

Woodwell GM, Mackenzie FT (eds) (1995) *Biotic Feedbacks in the Global Climatic System*. Oxford University Press, Oxford, UK.

12. Tropical Forests

Rodolfo Dirzo

Tropical forests are one of the most prominent terrestrial biomes on earth, estimated to have originally covered 1.4 billion ha (Reid 1992). Tropical forests generally occur in frost-free regions between the tropics of Cancer and Capricorn, in areas of relatively high precipitation. Holdridge's life zones system (Holdridge 1967), which is a commonly used vegetation classification protocol, distinguishes numerous kinds of tropical forests based on bioclimatic factors. Other systems (e.g., Miranda and Hernández-X 1963) classify tropical forests based on their physiognomy (e.g., height of the trees) and phenological patterns (e.g., proportion of trees that maintain their foliage throughout the year). Other systems classify tropical forests in very simple terms based on the precipitation they receive as wet, moist, and dry. In the most rigorous of these latter systems, humid tropical forests are defined as those that occur in areas where annual rainfall exceeds potential evapotranspiration for the year.

Overall, tropical forest regions are characterized by a relative lack of seasonal changes in temperature, but there is striking variation in total annual rainfall, from ca. 1–8 m of annual precipitation, and in the length and severity of the dry season. Most tropical regions, even the wettest ones, exhibit dry seasons during which one to several months receive less than 100 mm of precipitation. There are extensive areas that are always wet (all months >100 mm rainfall) only in Malaysia and Indonesia, whereas tropical zones with prolonged dry seasons are typically located away from the more equatorial wet forest regions (Walter 1973). In such tropical and subtropical regions with lengthy dry seasons, rainfall is maximal during the summer months, whereas the "winter" months, which are only slightly cooler than the summer months, are relatively dry.

Tropical forests also occur in mountains at elevations greater than 3000 m above sea level (m a.s.l.). Rainfall increases with elevation except on the lee side of tropical mountains, and clouds are formed on most days around 1500 m a.s.l., which leads to tropical cloud forests between 1000 and 2500 m. Precipitation decreases beyond these elevations, but some low-stature tropical forests (i.e., dominated by scleromorphic shrubs) are still present, given that evapotranspiration decreases with the decline in temperature at these high tropical altitudes.

The tropical forest biome can be further differentiated by factors such as local edaphic conditions (e.g., soil fertility) and meso- and microtopographic position. *Tropical forest*, therefore, is a term that encompasses (or conceals) a complex and diverse variety of vegetation types (e.g., life zones, sensu Holdridge 1967) as well as many associations with distinct physiognomy. For example, Hartshorn (1992) indicates that within the latitudinal and altitudinal limits outlined earlier, 27 life zones can be distinguished within what we generically call humid tropical forests. For example, Costa Rica has 12 distinguishable life zones, which is one more than in the eastern United States, even though Costa Rica is only the size of West Virginia.

The previous description highlights two salient aspects relevant to this chapter: (1) an important (though largely neglected) facet of the biodiversity of the tropical forest biome is the diversity of ecosystem types it includes; (2) biological diversity and most of the natural biological processes and patterns in tropical forests are strongly influenced by three primary drivers: precipitation and its seasonality, soil fertility, and elevation. Finally, given the complex and confusing terminology, this chapter will distinguish among three major types of tropical forests: rain (including moist and wet; i.e., evergreen) forest, dry (i.e., seasonally deciduous) forest, and montane (i.e., cloud) forest. The most studied type by far is the tropical rain forest, so most of the following analysis is based on information derived from this tropical forest type; however, this should not diminish the importance of and need to direct studies to the other two.

Tropical rain forests exert a significant monopoly of the planet's biological diversity, which is often cited to be of the order of at least half the species of organisms on earth (Wilson 1992). Considering all types of tropical forests collectively, the proportion of species is even larger. Such a monopoly of the planet's biodiversity and the significant threats tropical forests are currently experiencing justifies an analysis of the available information regarding the future scenarios of biodiversity of this biome. In this chapter I will first describe the other important facets of the biological diversity of tropical forests. I will then analyze the drivers or correlates of the natural variation in tropical forest biodiversity (i.e., rainfall, soil fertility, and elevation), and finally discuss the most significant current threats to this biome, particularly from the perspective of how the ecological processes driven by the components of its biodiversity are likely to be modified in the future.

The Facets of Tropical Forest Biodiversity

In addition to the diversity of communities encompassed by the tropical forest biome, other facets are prominent, particularly from the phytocentric point of view.

Diversity of Plant Life Forms

The combination of plant life forms provides forests with their characteristic physiognomy and structure. The most obvious diversity of life forms is evident in some harsh environments (e.g., arid lands) (see Chap. 10), yet the comparatively "benign" environments of the tropical biome also exhibit a spectacular diversity. There are at least seven distinguishable categories: dicotyledoneous trees, trees with one or few meristems (e.g., the arborescent palms and pandans), treelets, shrubs, giant herbs, vines, and epiphytes (Ewel and Bigelow 1996). Vines can be of two major, conspicuous types in tropical forests: herbaceous (vines properly) and woody (lianas), whereas two tropical groups are prominent among the "epi" plants: the nonwoody epiphytes (e.g., bromeliads, orchids) and the woody hemiepihytes (e.g., the conspicuous strangler figs). Life forms differ in their contribution to species richness and structural diversity. For example, the epiphytes contain an impressive proportion of the species of humid and cloud tropical forests, particularly due to the diversity within the Orchidaceae and the Polypodiaceae, which are frequently two of the most species-rich families contributing, for example, 70% of the total 950 species in the forest of Los Tuxtlas, Mexico (Ibarra-Manriquez et al. 1997). Regarding structure, trees and lianas are the most conspicuous life forms in most tropical forests. Such physiognomic importance is accompanied by what ecologists typically refer to as "dominance," which stems basically from the stature/massivity, longevity, capture of radiation (via their canopy and subcanopy leaves), and nutrient capture by the roots of these life forms. Palms follow closely to trees and vines/lianas in structural importance, although palms are largely restricted to the understory in several forests.

Light and humidity seem to be two important drivers for the diversity of life forms within tropical forests. In the case of light, this is exemplified by the occurrence of the impressive hemiepiphytic trees (e.g., *Ficus* spp. and several Araliaceae and Clusiaceae), which germinate, establish, and grow initially from the top down, or the evolution of the liana life form in plants phylogenetically unexpected to do so (e.g., some palms like *Desmoncus* spp. or *Chamaedorea elatior*). The importance of humidity as a driver of life-form diversity is shown by the presence of epiphytic cacti (e.g., *Epiphyllum* spp. or

Rhypsallis spp.) as well as the profusion of CAM epiphytes (e.g., some Bromeliaceae) in otherwise humid forests with ca. 5 m of annual rainfall.

Even if plant species diversity shows a clear geographical pattern of increase from the poles to the equator, such a pattern is not as clear in the case of life forms. An important analysis by Box (1981) indicates that tropical regions contain 18 life forms, whereas the subtropics and the temperate regions have 20 and no less than 25, respectively. Nevertheless, the elucidation of life-form diversity patterns may be obscured by the lack of consistent, universally adopted classification systems. In their now-classical system of architectural models for trees, Hallé et al. (1978) documents a maximum diversity in tropical forests, and Gentry (1988) claims that the greatest diversity of life forms within the epiphytes also lies in the tropics. Other significant geographical patterns of life-form diversity include the poor representation of arborescent palms in Africa and Asia in comparison to the Neotropics (Moore 1973), together with the abundance of Bromeliad tank epiphytes in the Neotropics and their marked absence in Asian and African tropical forests.

The basic significance of this facet of tropical biodiversity is at least twofold. On the one hand, life forms represent an expression of the morphofunctional adaptation of plants to the environment and are generally the result of evolutionary forces that lead to morphological and even ecological convergence (Bocher 1977); therefore, their conservation merits concerns of similar magnitude to those afforded to species or other evolutionary lineages. On the other hand, life forms represent functional diversity that may influence ecological processes at several scales of space and time, and their loss in the face of current and future environmental change may trigger changes in the structure, dynamics, and functioning of tropical ecosystems, as aptly discussed by Ewel and Bigelow (1996) (see later).

Life History Diversity

An important driver of the dynamics of tropical forests, particularly humid and cloud forests, is the occurrence of pulses of light availability concurrent with the opening of forest gaps due to the fall of trees or limbs from an otherwise closed canopy. Such natural disturbance has selected for a number of morphofunctional responses of plants and animals with contrasting life histories (particularly evident in the case of plants) co-adapted for colonization during forest gap regeneration (Denslow 1987; Clark and Clark 1992). These successional dynamics result in a diverse spectrum of plant-life histories, ranging from the extreme light-demanding plant species that colonize large gaps, to the extreme shade-tolerant species typical of the mature understory. Denslow (1996) recognizes 12 functional groups that respond to rain forest disturbance dynamics. These include three groups among the herbs and shrubs (i.e., colonizing pioneers, and large- and small-leaved understory plants), two among the treelets (i.e., pioneer and understory), three among the canopy trees (i.e., legumes, palms, and emergents), two among the climbers (i.e., vines and lianas), and two among the epiphytes (nonparasitic herbs, and parasitic and hemi-parasitic trees and shrubs). For each of these groups, she identified their mode of colonization after disturbance and/or their effects on forest dynamics. Some of these life history and functional groups coincide with the life forms identified by Ewel and Bigelow (1996) referred to earlier. The relative representation of each of these functional groups differs among tropical forests, with the most noticeable being the scarcity (species richness and abundance) of palms, most epiphyte types, and large-leaved understory herbs in the drier tropical forests (Denslow 1996). Some types (e.g., canopy palms, hemiepiphytic trees) occur almost exclusively in wet tropical forests. In addition, the species richness within each of the life history groups (as a result of the overall greater species richness) is higher in the tropics than it is in other biomes.

Species Richness and Taxonomic Novelties

Species richness is the most outstanding facet of tropical forest biodiversity. Floristic diversity is remarkable, considering both data for species density (i.e., species richness per unit area) or for relatively complete regional floras. Enumerations of species density include records of more than 300 large tree species per hectare in Western Amazonia and Borneo (Gentry 1988; Ashton 1993, respectively) and the world's record of floristic diversity, with 473 tree species, considering only individuals of more than 10-cm diameter at breast height, in a 1-ha plot in Amazonian Ecuador (Valencia et al. 1994). The single estimate of total vascular plant species diversity (i.e., plants of all sizes) in a tropical forest yielded values of 365, 173, and 169 species in three 0.1-ha samples in Ecuador (Gentry and Dodson 1987).

The estimated richness of 4314 species in the regional flora of Lacandonia, Chiapas, Mexico (Martínez et al. 1994) toward the northernmost limit of tropical rain forest in the Americas underscores the remarkable floristic diversity of tropical forests. Likewise, surveys at more restricted scales in tropical research stations (i.e., a few hundred hectares) [e.g., La Selva (Costa Rica) and Barro Colorado Island (Panama) with 450 and 362 tree species, respectively Hartshorn (1992)], is also indicative of the great regional diversity.

Animal diversity is also impressive and unparalleled in tropical forests; for example, Wilson (1992) includes the accumulation of ca. 30% of the world's bird species in the rain forests and riverine and swamp woodlands of the Amazon Basin, and an additional 16% in the same habitats of Indonesia; 1209 butterfly species within the $55\,km^2$ of the Tambopata Reserve in Peru; and 46 ant species in a single leguminous tree in the same Peruvian locality. The world record of animal diversity is an estimate of 1200 species of beetles associated with the canopy of a single tree species, sampled in Panama (Erwin 1992). Extrapolations from this beetle diversity led Erwin (1992) to hypoth-

esize that tropical arthropod diversity may be of the order of 30 million
species.

Other groups of animals, including vertebrates and invertebrates, present
extremely high species richness, conforming to a strong and well-known biogeographic pattern of decreasing diversity with latitude, whereby the greatest concentration of species occurs in the tropics, particularly in the rain
forests.

Fourteen of the 18 global areas with unusually high levels of plant
endemism lie within tropical rain forests. In an area of less than 0.2% of the
earth's land surface, these forests contain at least 37,000 plant species, which
is equivalent to ca. 15% of all plant species. Plant endemism in neotropical
dry forests is also very high (e.g., on the order of 20% in Mexican deciduous
forests) (Rzedowski 1978, Trejo 1998). In addition, numerous species of cloud
forest plants in some South American forests are endemic to isolated areas
smaller than 10 km² (Gentry 1992). Among the birds of South American
tropical forests, 440 species (i.e., 25% of the total) similarly have ranges less
than 50,000 km² (Terborgh and Winter 1980).

In synthesis, in an excellent analysis of the diversity of tropical
forests, using the four biodiversity measures of Pimm (1984) (i.e., species richness, evenness, connectance and interaction strength), Wright (1996) concluded that "by all definitions, tropical forests have exceptionally high plant
diversities."

Regarding taxonomic novelties, the following examples demonstrate that
there is exceptional diversity despite the limited nature of current biological
inventories. Gentry (1992) found that in the region of Iquitos, Peru, 70% of
the extracted timber comes from a tree first described in 1976. In addition,
two new tropical plant families were described to science in the late 1980s:
Lacandoniaceae, from Mexico (Martínez and Ramos 1989), and Ticoden-
draceae, from Costa Rica (Gómez-Laurito 1989). Gaston (1991) similarly
estimated that the proportion of still undescribed tropical insects ranges from
65 to 99%.

Correlates/Drivers of Natural Variation in Biodiversity

Tropical biodiversity is not uniformly high. In addition to the detectable
latitudinal variation in species richness within the rather restricted range of
the tropics (see Gentry 1988), variation in this and other facets of biodiversity is related to variation in physical attributes of the environment.

Species Richness

Through the application of a standardized sampling protocol in numerous
tropical localities, Gentry (1988) detected a significant variation in α-
diversity associated with the variation in total annual precipitation of the
sampled sites. He found that species richness increases up to fivefold as annual

rainfall increases from around 1000 to 4500 mm. This relationship was found to hold for large trees, small trees, and lianas, as well as for all these plant types. Species richness remained relatively constant above 4500 mm. A similar positive relationship with precipitation was found in Ghana, West Africa (Hall and Swaine 1981).

Work in seasonally deciduous tropical dry forests at higher latitudes has indicated that these forests do not conform to such a relationship, and, surprisingly, a sample of 20 sites surveyed in Mexico (using the same protocol of Gentry) revealed a tendency for these forests to be outliers above the expected regression line of Gentry's relationship (Trejo 1998). A similar situation appears to exist in equivalent latitudes in South America (Gentry, personal communication, 1993). In the case of dry forests, seasonality of rainfall, rather than total annual precipitation, appears to be of major significance for variations in species richness (Trejo 1998).

Significant variation in species richness is also detected with variation in soil fertility (although some exceptions are known). Diversity increases with soil fertility for neotropical understory herbs and shrubs (Gentry and Emmons 1987), as well as trees (Gentry 1988). The same is true for trees in the families Rubiaceae, Meliaceae, and Euphorbiaceae in Borneo (Ashton 1988), although diversity in the Dipterocarpaceae in Borneo is greatest at intermediate soil fertilities (Ashton 1977). In contrast, a negative relationship was found between plant species richness and indexes of soil fertility in Costa Rica and Ghana (Huston 1994), although covariation with precipitation may also be involved in these cases (Wright 1996).

Data are more scarce regarding the variation in species richness with elevation, but the limited evidence suggests a negative relationship in the case of plants (Gentry 1988; Orians et al. 1996), birds (Terborgh 1977), and insects (Janzen 1987).

In synthesis, significant variation in species richness is driven at the proximate level by variation along three major gradients (i.e., precipitation, soil fertility, and elevation). In addition, evidence is strong to suggest that species richness increases with area of forest, both locally and due to species turnover in regional or geographic scales (β and γ diversity) in both tropical rain (Gentry 1988, 1992; Wilson 1992) and seasonally dry (Trejo 1998) forests.

Life Forms

A review of the environmental correlates of life-form diversity (Ewel and Bigelow 1996) suggests that if a classification system is used that emphasizes the most important types of the tropics, precipitation may be an important proximal factor determining plant life-form distribution. The most salient aspect of the data compiled by Ewel and Bigelow (1966) is the marked increase in epiphyte diversity with precipitation, from 1–4% of the species in the driest sites, to 22–25% in the wettest sites. Palms, although of poor relative richness in general, increased from the driest site in which they are present to the wettest sites by a factor of ca. 3. Other life forms showed an erratic pattern of variation in species richness with precipitation. From the raw data presented in Table 6.1 of Ewel and Bigelow, I calculated the diversity index of life forms (Shannon's Index) for each site. (In this analysis, categories corresponded to plant life forms, and proportions were derived from the percentages of species in each category.) Overall, the index was low (due to the predominance of large trees), but ranged from 0.454 to 0.758; however, such variation was not significantly correlated with annual precipitation ($p > 0.20$). It would thus appear that only some life forms (e.g., epiphytes, palms) are sensitive to the variation in precipitation, but more detailed and controlled information is needed to elucidate this relationship.

Variation in elevation (and the physical factors that covary with it) causes variation in the relative representation of life forms. Epiphytes display their greatest abundance and diversity in cloud forests at intermediate elevations (Grubb et al. 1964; Brown et al. 1983), whereas palm diversity decreases with elevation, even though their abundance is quite significant in several midelevation forests. Hemiepiphytic plants reach their greatest abundance at midelevations, and this life form is replaced at lower and higher elevations by lianas (Gentry 1988). The most salient aspect of the variation in life forms with elevation is the peaking of epiphyte abundance, species richness, and structural diversity in midelevations (Brown et al. 1983), which results in increased nutrient capture and retention as well as nutrient transfers at the atmosphere–terrestrial interface (Silver et al. 1996).

Data on the role of soil fertility are very scarce, and even though some tendencies have been observed (see Ewel and Bigelow 1996), no conclusive patterns are warranted with the available information.

The correlations between environment and biodiversity like the ones described earlier are important because they provide a framework and working hypotheses to assess how future environmental changes that affect climate, soil characteristics, and area of habitat may determine the future scenarios of biodiversity in tropical forests, and how this in turn may affect tropical ecological processes.

The Threats to Biodiversity

Climatic Changes and CO_2-Enriched Atmospheres

The land cover projection under changed climate plus physiological effects of the IMAGE model (see Chap. 3) forecasts relatively little change in the area of the tropical rain and seasonal forests. If anything, some increase is evident, particularly for the seasonal forest in Africa, Asia, and South America. In addition to these global predictions, broad by necessity, studies at a more regional scale hint at similar conclusions. In a study of the potential response of Mexican forests to climatic change, Villers and Trejo (1998) developed prospective scenarios using models based on a doubled atmospheric CO_2 concentration (leading to increased temperatures of 2.8–3.2°C and decreased precipitation by 7–20%) and an assumption of a homogeneous temperature increase of 2°C and a homogeneous decrease of precipitation by 10%. There was considerable consistency of results among predicted scenarios, and the salient forecasts were that even though tropical cloud forests would experience coverage reductions (of 45–75% depending on the model), the tropical forests of warm climates would be the least affected: Dry forests would remain about the same and rain forests are even predicted to increase slightly in their coverage.

The lack of reduction in coverage area as a result of climatic changes in tropical forests would suggest that no significant reductions in species richness are to be expected, given the area–diversity relationship of these forests. The exception would be the cloud forests, which appear to be sensitive to climatic changes and which contain high proportions of species of plants and animals with very restricted ranges (see Terborgh and Winter 1980; Gentry 1992). In addition, montane forests are critical components for the survival of the numerous species of animals that track their appropriate environmental conditions by means of altitudinal migrations.

Another appealing approach to predicting future biodiversity scenarios in the light of climatic changes is that of Rochefort and Woodward (1992), in which known tolerances of vascular plant families to low temperatures and precipitation were used to model changes in familial diversity. The models assumed a 3°C increase in temperature, a 10% increase in precipitation, and a doubling of CO_2 concentration. Overall, they predicted that the combination of increased precipitation and a doubling of CO_2 offsets the deleterious effects of global warming and that familial diversities would be globally similar to those predicted in present-day vegetation. The exceptions would be "the tropics" (i.e., Amazon Basin and West African rain forests), where the diversity is even predicted to increase. This is an interesting approach, but its resolution is clearly very coarse in spatial scale.

In addition to these predictions based on models, other authors speculate about possible qualitative scenarios of climate change in tropical forests. Hartshorn (1992) suggests that it is unlikely that the higher temperatures expected under global warming per se will have negative effects on tropical forests. High temperatures are modulated by cloud cover in tropical rain forest areas; if degree of cloudiness increases with global warming, this will ameliorate the effects of increased temperatures. In addition, the impressive and common interannual variation in precipitation, so characteristic of tropical forests, augurs that changes in total precipitation may not have a major effect on the biological diversity of these forests. In contrast, if changes in seasonality occur as a consequence of global warming (Phillips 1997), noticeable effects are to be expected in the biodiversity and ecological processes, particularly via modifications of the phenological rhythms of tropical communities (Foster 1982). In addition, some functional groups, composed by species preadapted to living in the more seasonally dry forests, are likely to be benefited (e.g., reptiles, cacti, epiphytic Bromeliads) at the expense of the more messic-adapted groups (e.g., amphibians, some ferns, and evergreen trees and lianas) (Phillips 1997).

Even if increased temperatures are not a direct threat to tropical biodiversity and ecological processes, increased sea surface temperatures could lead to more frequent and severe tropical cyclones (Hartshorn 1992), which could have devastating effects on tropical forests. Indeed, under the warmer conditions expected for the near future, more frequent and intensive hurricanes have been forecasted by O'Brien et al. (1992). If typhoons are to increase as a result of climatic change, then differential effects on life forms are to be expected (Ewel and Bigelow 1996). In particular, palms, which are remarkably resistant to strong winds, could become predominant at the expense of the more susceptible large trees. Because trees constitute a crucial support to lianas and epiphytes, these life forms and functional groups are likely to decrease in abundance in future tropical forests. Such a shift in the abundance of life forms (with canopy tree reduction) would have devastating effects on animal diversity (see Erwin 1992 and the earlier discussion on animal diversity).

Land Use

Deforestation

In contrast to the predictions of land cover as a result of climatic change, the projections of the IMAGE model for land-use/forest-cover change for tropical forests forecast a dramatic reduction in coverage (see Chap. 3). The most apparent trend is the great conversion to agricultural lands in Africa, where the prediction for year 2100 is almost a complete eradication of this biome. Asia is second in the projected degree of conversion. Latin America shows a much less dramatic conversion; however a closer look at available data on deforestation shows that neotropical forests are also seriously altered in several parts of their distribution in the Americas (e.g., Dodson and Gentry 1991; Dirzo and García 1992).

It is commonly claimed that less than half the original area of tropical moist forest remains in primary forest (Wilson 1992). Janzen (1988) argues that in Central America more than 95% of the tropical dry forest has been converted to agricultural lands. The overall assessment of the IMAGE model and the statistics referred to earlier coincide with the increasingly accepted view that land use is by far the major global environmental threat in tropical forests (Orians et al. 1995). In addition, current rates of deforestation suggest that the pulse of forest conversion is still very high, of the order of 10.5 million ha/year for closed forests (Reid 1992), and is likely to remain so in the forthcoming decades (Laurance and Bierregaard 1997).

Such a dramatic pulse of habitat destruction suggests that a significant wave of species extinction must be underway, and several researchers have been prompted to provide possible scenarios of the magnitude of species loss.

Table 12.1. The predicted extinctions (%) of species from tropical closed forests under three scenarios of deforestation at two future dates

Year	Region	Low scenario		Mid scenario		High scenario	
		z = 0.15	z = 0.35	z = 0.15	z = 0.35	z = 0.15	z = 0.35
2015	Africa	1	3	3	6	4	9
	Asia	2	5	5	11	8	18
	Latin America	2	4	4	8	6	13
	Total	2	4	4	8	6	14
2040	Africa	3	6	6	13	10	21
	Asia	5	11	12	26	28	53
	Latin America	3	8	8	18	15	32
	Total	4	8	9	19	17	35

The three deforestation scenarios correspond to the current estimated rate mid (10 million ha/year), low (5 million ha/year), and high (15 million ha/year). The two values for each column correspond to estimates based on two slopes of the species–area relationship (0.15 and 0.35). From Reid 1992.

A common approach to these forecasts has been the application of the species–area relationship of biogeographical theory (MacArthur and Wilson 1967). One exercise at the global scale (Table 12.1) shows the range of percentage of species decline for three scenarios of deforestation magnitude and two future dates (Reid 1992). The midscenario, which corresponds to the estimated current rate of deforestation of ca. 10.5 million ha/year, forecasts a loss of 4–8% (depending on the slope, z, of the species–area relationship used, 0.15 and 0.35, respectively) of tropical species by year 2015. By year 2040 the corresponding figures are 9–19%. These values are reduced roughly by half under the low deforestation scenario (assuming, for example, that policies are implemented to reduce deforestation by half the current value). Under the high deforestation scenario (i.e., if deforestation rates were to increase from 10 to 15 million ha/year), the percentage of decline in species richness is roughly doubled. Some regional variation is predicted also, with Asia showing the highest values. Other independent exercises yield similar scenarios of reduction in species richness (see Wilson 1992).

An obvious limitation of these forecasts is that, given our ignorance of the number of tropical species, estimates of potential absolute species loss can only be guessed. One exercise for the Lacandon forest (Chiapas, Mexico), where current rates of deforestation have been calculated accurately (Mendoza and Dirzo 1999), and where a relatively complete floristic account has been estimated in 4314 plant species (Martínez et al. 1994), allows the generation of future scenarios of absolute plant species loss (Table 12.2). According to this study, by year 2034, when (at the current rate of average deforestation of 1.6% per year) the area of forest is reduced to 50% of its original coverage, the number of species committed to extinction would be

Table 12.2. The estimated number of plant species committed to extinction on the basis of area loss in the Lacandon forest, Chiapas, Mexico

Year	% remaining area	Number species lost	
		$z = 0.15$	$z = 0.35$
1991	100	—	—
2009	75	187	418
2034	50	831	933
2077	25	814	1662
2135	10	1264	2389

The estimates were derived from the species area relationship at different future dates, when current rate of deforestation in the area (1.6%/year) reduce forest coverage to 75, 50, 25, and 10% of the original area. The two sets of values correspond to estimates based on two slopes of the species–area relationship (0.15 and 0.35) and the original number of species in 1991 is 4314. From Mendoza and Dirzo 1999.

between 831 and 933. Assuming rates of deforestation remain the same, forest area would be reduced to 25% of its original by year 2077, and expected species loss would be between 814 and 1662. Such analyses of species richness decline offer a gloomy picture of future tropical biodiversity, but at the same time they show that significant reductions in the potential species extinction could be achieved if rates of forest destruction can be strongly diminished.

Finally, even if the application of the island biogeographic approach to generate scenarios of species loss remains controversial (see Reid 1992; Turner et al. 1994), such scenarios suggest that, at least, a significant wave of population extinctions must be in motion. This is a major pulse of diversity loss that has received virtually no attention in tropical forests.

Fragmentation and Edges

Reduction of forest coverage is typically accompanied by a significant degree of forest fragmentation, whereby a considerable proportion of the remaining forested areas consists of fragments of habitat, more or less isolated from one another (Skole and Tucker 1993). No statistics are available to provide a synthetic assessment of the degree of global tropical forest fragmentation, although the data for specific regions suggest that this may be a pervasive phenomenon. The time-course of tropical deforestation in conjunction with fragmentation is exemplified in Figure 12.1 for the area of Los Tuxtlas, Veracruz, Mexico (Dirzo and García 1992). This study case arguably represents the predominant features of deforestation and fragmentation in the neotropics. This example suggests the following changes in the future spatial configuration of tropical landscapes: (1) an increase in the number of patches, (2) a reduction of the median size of patches, (3) an increase in the skewness of the size distribution, and (4) a reduction of the median distance edge-center of the patches (R. Dirzo, unpublished data).

In general, fragmented landscapes will be embedded in a matrix of transformed lands, predominantly grasslands for cattle ranching or agricultural fields for cropping, areas for timber products extraction, abandoned areas with secondary growth and secondary forest, and lands at various degrees of degradation. These components of the transformed landscapes will be

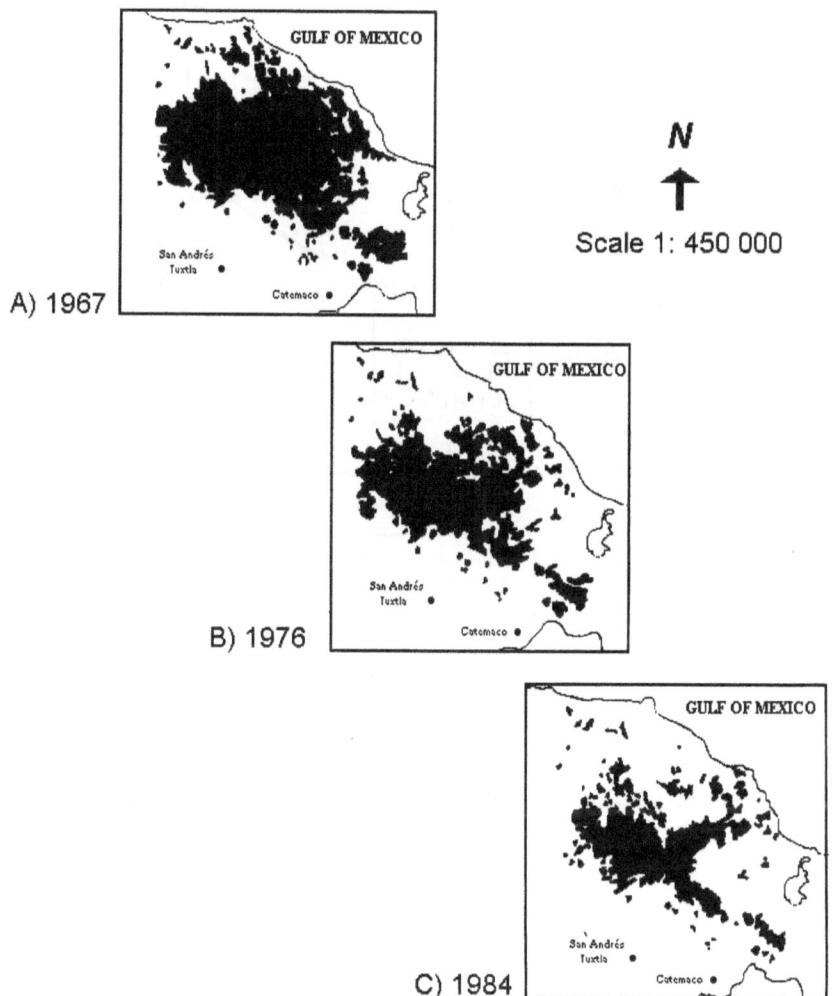

Figure 12.1. Time-course changes in tropical forest coverage in the region of Los Tuxtlas, Veracruz, Mexico. The images show a sequence of forest spatial configuration in 1967 (A), 1976 (B), and 1986 (C). From Dirzo and García 1992.

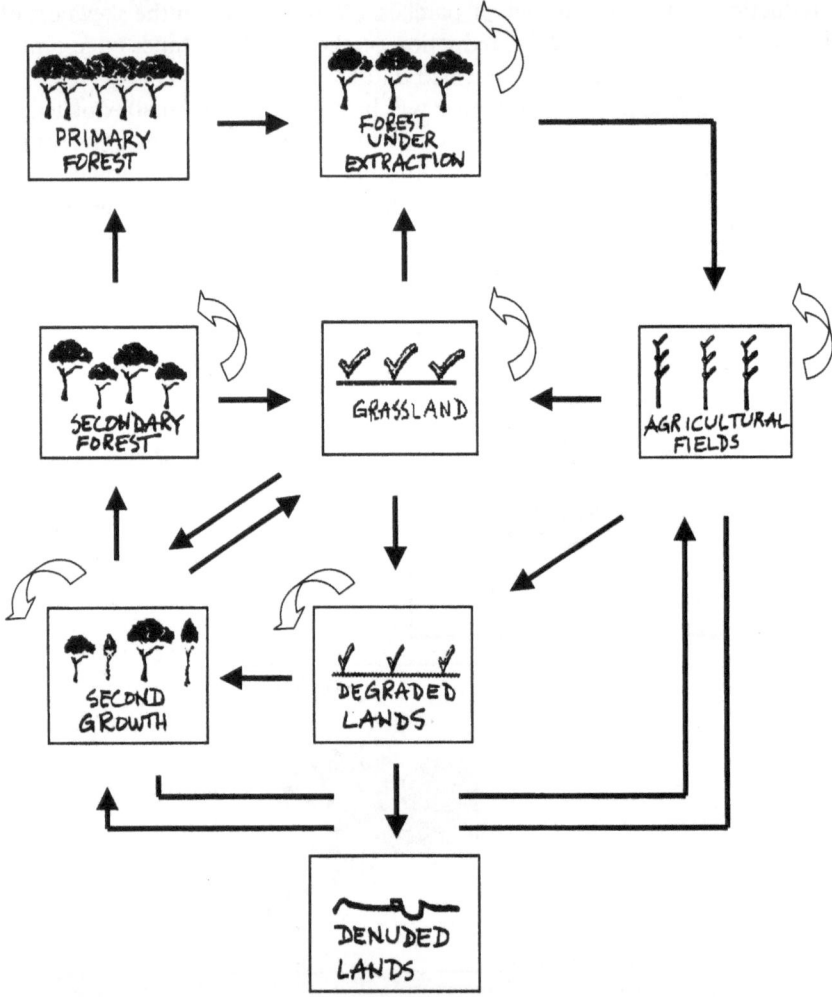

Figure 12.2. The land-use patterns of tropical forests and dynamics of conversion into different land uses. The arrows indicate the different routes of conversion, including the situation in which a given type of patch remains in the same condition (white arrows). From O. Masera and R. Dirzo (unpublished).

dynamic systems with changes and transition paths as shown by the arrows in Figure 12.2. Under a specific set of land-use practices, transition probabilities can be assigned to the paths of change (arrows) to model the dynamics and composition of landscape elements of the tropical forest areas of the future. Such analytical tools and conceptualizations will be useful, both to describe tropical landscapes and their dynamics in future times, and to monitor biodiversity.

It could be argued that, under the current trends of land use, a fragmented area offers a greater diversity at the landscape level (cf. Fig. 12.2). Even if the mosaics of fragmented areas immersed in a matrix of land use practices provides a more diverse landscape, biodiversity at the species and genetic levels is sure to be reduced greatly: The transformed elements of the landscape are very species-poor and also very poor in life-form diversity (e.g., cattle grasslands, monocultural crop areas); in addition, biodiversity loss is likely to be exacerbated according to the natural history knowledge, theory, and ecological principles that are discussed next.

On the basis of island biogeography theory, larger fragments are expected to support more species than smaller ones (Mac Arthur and Wilson 1967), and several empirical studies support this expectation (see Laurance and Bierregaard 1997). This effect is born out by the fact that the smaller fragments are the predominant ones in fragmented landscapes (cf. Fig. 12.1). Species that naturally occur at low densities and species that specialize in restricted habitats will be more susceptible to local or regional extinctions due to sampling error when their habitats are subdivided (Phillips 1997). Erosion of genetic diversity may also endanger species survival due to increased susceptibility to environmental variation. Finally, fragmentation will bring about increased exposure to forest edges, and evidence is accumulating in the literature regarding the plethora of negative effects of forest edges on biodiversity (see Laurance and Bierregaard 1997, and chapters and references therein).

Ecological consequences of land use will be addressed shortly, but a brief discussion of two aspects related to land use (i.e., addition and extraction of organisms) may be in order first.

Invasions by and Extraction of Plants and Animals

Tropical forests unquestionably provide a few of the most infamous examples of species invasions. The example par excellence is that of Africanized bees that, once escaped in South America, have invaded up to the northernmost tropical forests in Mexico (personal observation, and Roubik 1991). Other examples are known, both for plants and animals, but in the most thorough analysis (Rejmánek 1996), compelling evidence is produced to suggest that in comparison to extratropical biomes, tropical ones have accumulated a much-reduced contingent of invasive plant species. In addition, about 50% of the plant species known to have invaded tropical forests did so only on islands, and out of the 21 species known to have invaded tropical primary forests, eight are reported only from forest gaps. Moreover, even tropical islands seem to be resistant to invasions provided that they are not disturbed by other human activities (Rejmáneck 1996, and refs.) Many alien plant species can be seen in tropical regions, but these are generally confined to urban, rural, and disturbed areas. To what extent, even if of rare occurrence, do invasions in tropical forests affect biodiversity negatively? There is simply not enough evidence to address this question, but Phillips (1997 and refs.) cites two cases that suggest that significant disruptions can be expected: (1) introduction of rats into the Hawaiian archipelago may have precipitated the extinction of several endemic birds; (2) rainbow trout introduction in some Andean montane forest streams may have been responsible for the modification of both the diversity and trophic structure of forest streams and rivers. In addition, the introduction of the legume *Myrica faya* (and its nitrogen-fixing actinomycete root symbiont) into Hawaii, where no such biotic associations had previously existed, has changed forest community composition, forest structure, and nitrogen cycling (Vitousek and Walker 1989). This example also illustrates the potential consequences of the introduction of a novel functional group.

In contrast, extractions of plants and animals have taken a considerable toll of several tropical species, both of plants and animals. It is a well-established fact that some precious timber species [e.g., mahogany (*Swietenia macrophylla*)] have disappeared from most of the forests where they were previously well represented. A similar situation is presented by several fruit trees in the neotropics (Vázquez and Gentry 1989). To the extent that some of these species play keystone roles, a cascade of negative effects on biodiversity could be expected, but there does not seem to be any documented case of this type.

Extraction of animals has been so intensive in several tropical forests that Redford (1992) was prompted to describe the current defaunation pulse as leading to the existence of "empty forests." He estimated that the number of individual mammals killed by hunters in Amazonian Brazil was 14 million per year. If the number of killed birds and reptiles is added, he estimated that the number of game animals killed each year in Amazonian Brazil may reach 19 million! Most of this massive contemporary defaunation is concentrated on the medium- to large-sized vertebrates, but because of the nocturnal or secretive habits of these animals the extent of defaunation is very difficult to assess unless detailed studies are carried out. The consequences for tropical biodiversity of this current and future pulse of defaunation are beginning to be studied and understood (Dirzo and Miranda 1990, 1991). This will be discussed in the next section.

Ecological Consequences

No doubt the most conspicuous consequence of the current and future threats to tropical forests is the loss of species and populations per se; however I will explore here the functional consequences of such threats. The ecological consequences of biodiversity loss can be envisaged at different levels: landscapes, life forms, species and populations, and intrapopulational genetic variation. There are obviously interactions among the effects that occur at these levels because there are interactions, feedbacks, and synergisms among the different drivers of biodiversity loss discussed in the previous sections. In addition, the consequences can be manifested in ecological processes at the ecosystem, community, and population levels. A thorough review of this complex interplay of causes and effects is beyond the scope of this chapter; in addition, the critical information for such an assessment is still very limited. Some reviews (Orians et al. 1995, 1996; Phillips 1997) attempt to compile and analyze the relevant information. I will offer here a sketchy panorama of the consequences I view as representative of the expected future functional changes in the tropics that result from biodiversity changes.

Ecosystem Processes

One of the six major ecosystem processes in which biodiversity is considered to play a role is atmospheric properties and feedbacks (see Mooney et al. 1995). Tropical deforestation and burning, and its predominant conversion to grasslands or other simplified systems, is currently a major source of atmospheric CO_2, but this is largely unrelated to species loss per se. As would be expected, however, cutting and burning of different types of forest has different contributions to carbon emissions. In a comparative analysis of the emissions contributed by cutting and burning of the different *types* of forest biomes in Mexico (Masera et al. 1997) it was found that of the 67 million tons of carbon emitted per year, tropical evergreen forests, although accounting for only one fourth of the area of the country affected by deforestation, were responsible for 49% of the total annual carbon balance. Tropical dry forests emitted an additional 32%, and deforestation of the other (temperate) forest biomes contributed with only 19%. Tropical rain forests clearly play the most significant role for this ecosystem function, especially because its conversion to agricultural lands (particularly grasslands for cattle) is currently the most common type of conversion, and is forecast to be the predominant type of land use in the future (see Chap. 3).

Primary productivity seems to be correlated positively with plant species richness only when the number of species is well below that which characterizes most tropical forests (Wright 1996), so future species loss is not expected to affect this ecosystem process. Species richness or intraspecific variation, however, may influence primary productivity if there is much variation in weather or under conditions of frequent and severe disturbances. Such buffering effect has been proposed for species-rich grasslands in comparison to impoverished grasslands (Tilman and Downing 1994), but no evidence exists to support this expectation in tropical forests.

Nutrient cycling and other ecosystem processes could potentially be affected by the loss or addition of species, life forms, or functional groups. Potential effects can be inferred from the known role of these biodiversity components on ecosystem processes; however detailed information is appallingly scarce, and only a few examples hint at the type of effects to be expected. In a unique experiment, Weaver (1972) removed epiphytes from a Puerto Rican dwarf cloud forest and found that even though this treatment did not affect the total amount of water reaching the forest floor, it did produce a significant change in its spatial distribution. In epiphyte-free plots, water was mostly channeled down stems in torrents to the forest floor, whereas water moved more or less evenly through the canopy in control plots. Other ecosystem processes were unfortunately not measured in this remarkable experiment. Judging from the critical role this life-form/functional group plays (e.g., retention of inorganic nitrogen present in cloud water, mist, and precipitation; provision of the physical matrix for the establishment of phytotelmata, augmentation of leaf area, diversion of water from soil to the atmosphere, etc.), plus their sensitivity to expected environmental changes, as discussed earlier, one could expect several critical functional processes of ecosystems to be impaired in the future.

A forest fragmentation mega-project in Central Amazonia (see description in Lovejoy et al. 1986) provides an excellent source of exploration of ecosystem functioning consequences that results from fragmentation and its effects on biodiversity. The following are two notable examples.

Dung beetle (Scarabaeinae) diversity was strongly affected by forest fragmentation (Fig. 12.3). In an elegant experiment Klein (1989) found that the mean number of species sampled with a standard protocol decreased from around 25 in the continuous forest to around 15, 13, and 4 in fragments of 10 and 1 ha, and in cleared pasture, respectively. Such a decrease was accompanied by a considerable decline in the mean rate of dung decomposition,

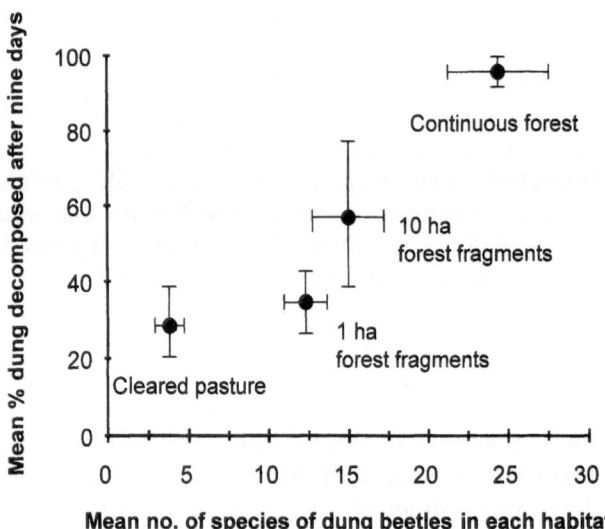

Figure 12.3. The relationship between land use and forest fragment size in Central Amazonia, Brazil, and dung beetle (Scarabaeinae) species diversity and dung decomposition rates. Data points correspond to means (± SE) of three replicates in each forest fragment/type. Redrawn from Klein 1989.

thus experimentally showing that animal species diversity plays a crucial role in ecosystem functioning.

A comparison of tree biomass in forest fragments at the same Central Amazonian site (Laurance et al. 1997) revealed that fragments suffered a considerable loss of above-ground biomass of trees up to 100 m from the edges due to exposure to winds and other changes in microclimate. Such losses were not compensated for by new tree recruitment. The authors of this study suggest that the loss of biomass of recently fragmented forests could be an additional significant source of greenhouse emissions.

Biotic Interactions

A major characteristic of tropical forests is the complexity and profusion of biotic interactions, which are considered to be an important factor in the maintenance and promotion of biodiversity. Land-use patterns that lead to alterations, particularly reductions of species richness or abundance, have profound effects on several biotic interactions. I will discuss several examples of animal species loss and highlight some of their ecological consequences and the future potential repercussions this may have on the structure, functioning, and diversity of tropical forests.

Forest fragmentation reduces the species diversity of several important pollinators (e.g., the specialized Euglossine bees, as shown in a study in the fragmentation megaexperiment at Central Amazonia) (Fig. 12.4). Visitation rates at chemical baits were measured for three species of *Euglossa* in continuous forest and fragments of 100, 10, and 1 ha. For all three species, visitation to the artificial floral attractants was markedly reduced as fragment size diminished. Because these bees are the most important pollinators of several orchids and other aroids, their reduction or eradication in forest fragments is likely to negatively affect the reproductive success and genetic variability of these plants. Reductions in seed set in small fragments compared with relatively continuous forest have been observed with several plant species (e.g., *Costus* spp.) that are easier to monitor at ground level (Dirzo, unpublished data). No single study has fully analyzed the range of potential consequences of decline in pollinator services as a result of land-use patterns, including visitation patterns, plant reproductive success, and genetic variability. Nevertheless, partial studies suggest this is a likely functional alteration of future tropical forests (see also Aizen and Feinsinger 1994).

The combined effects of habitat reduction (deforestation), fragmentation, and contemporary hunting have caused a significant reduction or outright eradication of several species of medium- to large-sized mammals (i.e., tapirs, peccaries, deer, and agouties) from tropical forests. Although the magnitude of such tropical defaunation is difficult to quantify [but, see my reference to Redford's work (1992) earlier], and the consequences of it are difficult to document in the absence of detailed field work, work is beginning to unravel the plethora of disruptions this brings about (Dirzo and Miranda 1991; Asquith

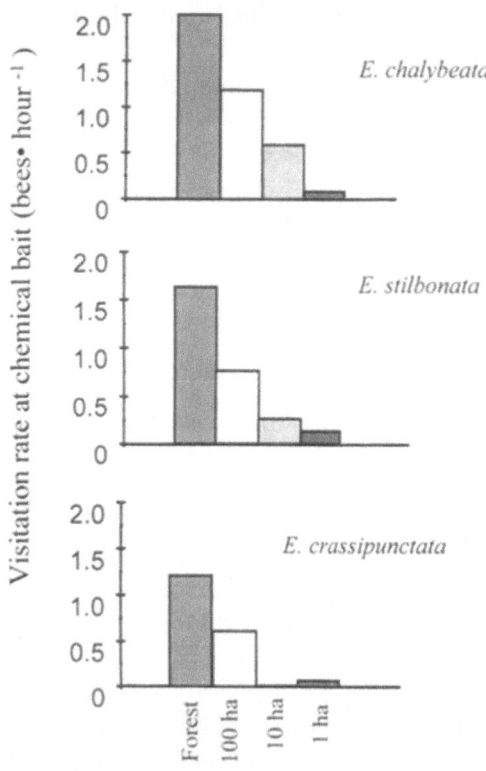

Figure 12.4. The decline in visitation rates at chemical baits, by three species of *Euglossa*, in continuous forests and fragments of three different sizes in Central Amazonia, Brazil. From Powell and Powell 1987.

et al. 1997). In Mexican tropical forests Dirzo and Miranda (1991) compared two sites of contrasting level of conservation of the understory mammalian herbivores, and some of the relevant results are summarized in Table 12.3. In the heavily defaunated site, Los Tuxtlas, most of the herbivory on the understory plants was due to insect damage; no damage by vertebrates was detected. In the well-preserved site, Montes Azules, vertebrate herbivory on seedlings and saplings was around 30% (Table 12.3A). That is, a critical eco-logical process (i.e., mammalian herbivory) in a defaunated site can be heavily reduced or even eradicated. Moreover, foliage herbivory is a surro-gate descriptor of a host of other effects attributable to understory mammals (e.g., seed predation, trampling, and dung deposition). Therefore, additional consequences could be expected. In the same study Dirzo and Miranda found that the understory of Los Tuxtlas could be characterized by a "defaunation syndrome" (Table 12.3B) in which carpets of seedlings are present at high density, and with a significant numerical dominance of a few species. This contrasts with the understory at Montes Azules (the well-preserved site), in which seedling density is about 50% and diversity (species richness and diversity index) is at least twice that of the defaunated site. Experimental work in progress suggests that such alterations of the structure and diversity of defaunated are caused by the loss of fauna and not by other factors (Dirzo and Frías, unpublished).

The study of defaunation in Mexico illustrates an additional point: Individual threats to biodiversity do not act in isolation. Defaunation is caused by absolute habitat area loss, and is exacerbated by fragmentation, hunting, and illegal trading of vertebrates (Dirzo and Miranda 1991). This may in turn precipitate the seemingly innocuous effect of other threats (e.g., species invasions). In addition, feedbacks are likely to occur. For example, the dense and monotonous carpets of seedlings of defaunated understories may promote insect herbivory and pathogenic infection, or this may attract increased numbers of the natural enemies of the phytophagous insects.

In a more general sense, as Phillips (1997) aptly put it, there are three basic kinds of interaction between threats and forest ecology: direct causation, where a threat leads to another threat (e.g., fragmentation leading to increased edge effects); combinative causation, where a threat combines with another (e.g., fragmentation/deforestation and hunting lead synergistically to major defaunation) and feedbacks (dense seedling carpets affecting other trophic interactions). Study and documentation of these three interactions is needed in order to assess the future biodiversity scenarios and functional properties of tropical forests.

Table 12.3. Consequences of defaunation on (A) forest understory levels of herbivory on seedlings and saplings (% plants or leaves damaged) by mammals, and (B) on the structure and diversity of the understory plant community

(A) Mammalian herbivory

Site	Seedlings		Saplings		Overall	
	Plants	Leaves	Plants	Leaves	Plants	Leaves
Montes Azules	29	31	30	24	29	27
Los Tuxtlas	0	0	0	0	0	0

(B) Structure and diversity of the understory

	Los Tuxtlas	Montes Azules
Mean density (range)	52.8 (11–204)	22.6 (10–62)
Number plots with one dominant species	14	2
Mean number different species	2.3	6.7
Mean diversity index (H)	1.07	3.71

Data were derived from a comparison of two sites with contrasting conservation of the fauna: Montes Azules (intact) and Los Tuxtlas (heavily defaunated). See text for details. From Dirzo and Miranda 1991.

Conclusions

The great biodiversity of tropical forests and the magnitude and nature of the current and future threats to it due to human activity suggest that significant losses of tropical biodiversity can be expected.

Of the five major drivers of biodiversity change addressed in this book, land use is unquestionably the most significant one for tropical forests. The abundance of species, life forms, and functional groups can be dramatically reduced by land use due to area loss and forest insularization (i.e., deforestation and fragmentation), accompanied by extraction of organisms. Such a large change in this driver is expected to impact biodiversity directly, as well as several critical ecological processes in a very significant way.

This review suggests that climatic change will not have major effects on tropical diversity, other than in midelevation localities and due to potential increase in incidence of typhoons or changes in seasonality. The consequences of the expected changes in this driver could be significant in some aspects, such as life-form relative representation in cloud forests, or phenological patterns and the functional repercussions of these changes.

Changes in atmospheric CO_2 are not expected to be particularly significant in tropical ecosystems given that CO_2 mixes globally rather rapidly. In addition, most of the model simulations available and discussed in this chapter indicate that either coverage or species richness will not be negatively impacted (coverage may even be increased slightly). Uncertainties about effects on ecological processes, such as changes in plant chemistry (e.g., modifications of C–N ratios and nature of defensive compounds) leading to changes in biotic interactions are still significant and should not be ignored.

Information regarding nitrogen deposition rates and consequences is still very scarce for tropical ecosystems. Given that N deposition is currently greatest in northern localities and close to major urban centers, and that the impact is expected to be largest on biomes which are N-limited, overall impact in tropical forests could be low.

Biotic invasions as a driver are likely to be of relatively minor impact in the future, largely because evidence suggests a low probability of successful establishment. In addition, the purposeful or inadvertent movement of biotic entities (i.e., seeds, bulbs, plants, animals) does not seem abnormally high toward tropical regions, although several important trading routes are likely to be developed in the future. Nevertheless, the establishment of alien biotic components in tropical systems takes place in disturbed areas or on islands, particularly disturbed islands. Thus, the major disturbances brought about by land use may be an important facilitating agent of biodiversity change due to biotic introductions.

The previous assessment does not take another crucial aspect into consideration: Interactions among changes in the analyzed drivers. The information discussed in this chapter suggests that this is likely to be a major force in shaping the biodiversity scenarios of the future. Thus the qualitative assessment provided in this section should be considered as a conservative evaluation.

Acknowledgments. I thank Osvaldo Sala and Terry Chapin for the invitation to participate in the workshop that led to this volume, and for their patience and encouragement to write this chapter. Their editorial assistance is very much appreciated. Eduardo Mendoza, Ek del Val, and Elizabeth Huber-Sannwald read a previous draft and offered useful suggestions. Raúl I. Martínez helped in various aspects in the preparation of this chapter. This chapter was prepared with support from DGAPA (UNAM) and CONACYT, and with the facilities provided by Northern Arizona University, Department of Biological Sciences.

References

Aizen M, Feinsinger P (1994) Forest fragmentation, pollination and plant reproduction in a Chaco dry forest, Argentina. Ecology 75:320–341.

Ashton PS (1977) A contribution of rain forest research to evolutionary theory. Annals of the Missouri Botanical Garden 64:694–705.

Ashton PS (1988) Dipterocarp biology as a window to the understanding of tropical forest structure. Annual Review of Ecology and Systematics 19:347–370.

Ashton PS (1993) Species richness in plant communities. In: Fiedler PL, Jain SK (eds) *Conservation Biology*, pp. 4–22. Chapman and Hall, New York.

Asquith N, Wright SJ, Clauss MJ (1997) Does mammal community composition control recruitment in neotropical forests? Evidence from Panama. Ecology 78: 941–946.

Bocher TW (1977) Convergence as an evolutionary process. Bot J Linn Soc 75:1–19.

Box EO (1981) *Macroclimate and Plant Forms: An Introduction to Predictive Modeling in Phytogeography.* Dr. DW Junk Publishers, The Hague.

Clark DA, Clark DB (1992) Life history diversity of canopy and emergent trees in a neotropical rain forest. Ecological Monographs 62:315–334.

Denslow JS (1987) Tropical tree fall gaps and tree species diversity. Annual Review of Ecology and Systematics 18:431–451.

Denslow JS (1996) Functional group diversity and responses to disturbance. In: Orians GH, Dirzo R, Cushman JH (eds) *Biodiversity and Ecosystem Processes in Tropical Forests*, pp. 127–151. Springer-Verlag, Berlin.

Didham RK, Ghazou J, Stork NE, Davis AJ (1996) Insects in fragmented forests: a functional approach. Trends in Ecology and Evolution 11:255–260.

Dirzo R, Miranda A (1991) Altered patterns of herbivory and diversity in the forest understory: a case study of the possible consequences of contemporary defaunation. In: Price PW, Lewinsohn TM, Fernandes GW, Benson WW (eds) *Plant-Animal Interactions: Evolutionary Ecology in Tropical and Temperate Regions*, pp. 273–287. John Wiley and Sons, New York.

Dirzo R, García MC (1992) Rates of deforestation in Los Tuxtlas, a neotropical area in Southeast Mexico. Conservation Biology 6:84–90.

Erwin TL (1992) Tropical forests: their richness in Coleoptera and other arthropod species Coleopterists' Bulletin 36:74–75.

Ewel JJ, Bigelow SW (1996) Plant life-forms and tropical ecosystem functioning. In: Orians GH, Dirzo R, Cushman JH (eds) *Biodiversity and Ecosystem Processes in Tropical Forests*, pp. 101–126. Springer-Verlag, Berlin.

Foster RB (1982) The season rhythm of fruitfall on Barro Colorado Island. In: Leigh EG, Rand AS, Windor DM (eds) *The Ecology of a tropical rain forest: Seasonal*

Rhythms and Long-Term Changes, pp. 151–172. Smithsonian Institution Press Washington, DC.

Gaston KJ (1991) The magnitude of global insect species richness. Conservation Biology 5:283–296.

Gentry AH (1988) Changes in plant community diversity and floristic composition on environmental and geographical gradients. Annals of the Missouri Botanical Garden 75:1–34.

Gentry AH (1992) Tropical forest biodiversity: distributional patterns and their conservational significance. Oikos 63:19–28.

Gentry AH, Dodson CH (1987) Contribution of nontrees to species richness of a tropical rain forest. Biotropica 19:149–156.

Gentry AH, Emmons LH (1987) Geographical variation in fertility, phenology and composition of the understory of Neotropical forests. Biotropica 19:216–227.

Gómez-Laurito J, Gómez LD (1989) *Ticodendron*: a new tree from Central America. Annals of the Missouri Botanical Garden 76:1148–1151.

Grubb PJ, Lloyd JR, Pennington TD, Whitmore TC (1964) A comparison of montane and lowland rain forest in Ecuador. I. The forest structure, physiognomy, and floristics. Journal of Ecology 51:567–601.

Hall JB, Swaine MD (1981) *Distribution and Ecology of Vascular Plants in a Tropical Rain Forest: Forest Vegetation in Ghana*. Junk Publishers, The Hague.

Hallé F, Oldeman RAA, Tomlinson PB (1978) *Tropical Trees and Frests: An Architectural Analysis*. Springer-Verlag, Berlin.

Hartshorn, GS (1992) Possible effects of global warming on the biological diversity of tropical forests. In: Peters RL, Lovejoy TE (eds) *Global Warming and Biological Diversity*, pp. 137–146. Yale University Press, New Haven.

Holdridge LR (1967) *Life Zone Ecology*. Tropical Science Center, San José, Costa Rica.

Huston MA (1994) *Biodiversity*. Cambridge University Press, Cambridge, UK.

Ibarra-Manriquez G, Martínez-Ramos M, Dirzo R, Núñez-Farfán J (1997) La vegetación. In: González E, Dirzo R, Vogt RC (eds) *Historia natural de Los Tuxtlas*, pp. 61–85. CONABIO-UNAM, Mexico City.

Janzen DH (1987) Insect diversity of a Costa Rican dry forest: why keep it and how? Biological Journal of the Linnean Society 30:343–356.

Janzen DH (1988) Tropical dry forests: the most endangered tropical ecosystem. In: Wilson EO, Peters FM (eds) *Biodiversity*, pp. 130–137. National Academy Press Washington, DC.

Klein DR (1989) The effects of forest fragmentation on dung and carrion beetle (Scarabaeinae) communities in Central Amazonia. Ecology 70:1715–1725.

Laurance WF, Bierregaard RO (eds) (1997) *Tropical Forests Remnants: Ecology, Management and Conservation of Fragmented Forests*. University of Chicago Press, Chicago.

Laurance WF, Laurance SG, Ferreira LV, Rankin-de Merona JM, Gascon C, Lovejoy TE (1997) Biomass collapse in Amazonian forest fragments. Science 278:1117–1118.

Lovejoy TER, Bierregaard O, Rylands AB, Malcolm JR, Quintela CE, Harper LH, et al. (1986) Edge and other effects of isolation on Amazon forest fragments. In: Soulé M (ed) *Conservation Biology: The Science of Scarcity and Diversity*, pp. 257–285. Sinauer, Sunderland, MA.

MacArthur RH, Wilson EO (1967) *The Theory of Island Biogeography*. Princeton University Press, Princeton.

Martínez E, Ramos CH (1989) Lacandoniaceae (Triuridales): una nueva familia de México. Annals of the Missouri Botanical Garden 76:128–135.

Martínez E, Ramos CH, Chiang F (1994) Lista florística de la selva Lacandona, Chiapas. Boletín de la Sociedad Botánica de México 54:99–177.

Masera O, Ordóñez MJ, Dirzo R (1997) Carbon emissions from Mexican forests: current situation and long-term scenarios. Climatic Change 35:265–295.

Mendoza E, Dirzo R (1999) Deforestation in Lacandonia (Southeast Mexico): evidence for the declaration of the northernmost tropical hot-spot. Biodiversity and Conservation Boletín de la Sociedad Botánica de México 8:1621–1641.

Miranda F, Hernández-X E (1963) Los tipos de vegetación de México y su clasificación. Bol Soc Bot Mex 28:29–178.

Mooney HA, Lubchenco J, Dirzo R, Sala O (1995) Introduction. Section 6. In: Watson RT, Heywood VH (eds) Global Biodiversity Assessment, pp. 333–452. Cambridge University Press, Cambridge, UK.

Moore HE (1973) Palms in the tropical forest ecosystems of Africa and South America. In: Meggers B, Ayensu E, Duckworth E (eds) Tropical Forest Ecosystems in Africa and South America: A comparative Review, pp. 63–68. Smithsonian Institution Press. Washington, DC.

O'Brien ST, Hayden BP, Shugart H (1992) Global climatic change, hurricanes and a tropical forest. Climatic Change 22:175–190.

Orians GH, Dirzo R, Cushman JH, Medina E, Wright SJ (1995) Tropical forests. In: Watson RT, Heywood VH (eds) Global Biodiversity Assessment, pp. 339–344. Cambridge University Press, Cambridge, UK.

Orians GH, Dirzo R, Cushman JH (eds) (1996) Synthesis. In: Biodiversity and Ecosystem Processes in Tropical Forests, pp. 195–220. Springer-Verlag, Berlin.

Phillips O (1997). The changing ecology of tropical forests. Biodiversity and Conservation 6:291–311.

Pimm SL (1984) The complexity and stability of ecosystems. Nature 307:321–326.

Powell A, Powell GV (1987) Population dynamics of male Euglossine bees in Amazonian forest fragments. Biotropica 19:176–179.

Redford KH (1992) The empty forest. BioScience 42:412–426.

Reid WV (1992) How many species will be there? In: Whitmore TC, Sayer JA (eds) Tropical Deforestation and Species Extinction, pp. 55–73. Chapman and Hall, London.

Rejmánek M (1996) Species richness and resistance to invasions. In: Orians GH, Dirzo R, Cushman JH (eds) Biodiversity and Ecosystem Processes in Tropical Forests, pp. 153–172. Springer-Verlag, Berlin.

Rochefort L, Woodward FI (1992) Effects of climate change and doubling of CO_2 in vegetation diversity. Journal of Experimental Botany 43:1169–1180.

Roubik DW (1991) Aspects of Africanized bee colonization in tropical America. In: Spivak M, Fletcher DJC, Breed MD (eds) The African Honeybee, pp. 259–281. Westview Press, Boulder, CO.

Rzedowski J (1978) Vegetación de México. Editorial LIMUSA, Mexico City.

Silver WL, Brown S, Lugo A (1996) Biodiversity and biogeochemical cycles. In: Orians GH, Dirzo R, Cushman JH (eds) Biodiversity and Ecosystem Processes in Tropical Forests, pp. 49–67. Springer-Verlag, Berlin.

Skole D, Tucker C (1993) Tropical deforestation and habitat fragmentation in the Amazon: satellite data from 1978 to 1988. Science 260:1905–1910.

Terborgh J (1977) Bird species diversity along an Andean elevational gradient. Ecology 56:562–576.

Terborgh J, Winter B (1980) Some causes of extinction. In: Soulé ME, Wilcox BA (eds) Conservation Biology: An Evolutionary Perspective, pp. 110–134. Sinauer, Sunderland, MA.

Tilman D, Downing JA (1994) Biodiversity and stability in grasslands. Nature 367: 363–365.

Trejo I (1998) Distribución y diversidad de selvas bajas de Mexico: relaciones con el clima y el suelo Ph. D. Dissertation. Universidad Nacional Autónoma de México, Mexico City.

Turner IM, Tan HTW, Wee YC, Ibrahim AB, Chew PT, Corlett RT (1994) A study of plant species extinction in Singapore: lessons for the conservation of tropical diversity. Conservation Biology 8:705–712.

Valencia R, Valslev H, Paz y Miño GC (1994) High tree alpha-diversity in Amazonian Ecuador. Biodiversity and Conservation 3:21–28.

Vázquez R, Gentry AH (1989) Use and misuse of forest-harvested fruits in the Iquitos area. Conservation Biology 3:350–361.

Villers L, Trejo I (1997) Assessment of the vulnerability of forest ecosystems to climate change in Mexico. Climate Research 9:87–93.

Vitousek PM, Walker LR (1989) Biological invasion by *Myrica faya* in Hawaii: plant demography, nitrogen fixation and ecosystem effects. Ecological Monographs 59:247–265.

Walter H (1973) *Vegetation of the Earth in Relation to Climate and the Ecophysiological Conditions*. Springer-Verlag, Berlin.

Weaver PL (1972) Cloud moisture interception in the Luquillo mountains of Puerto Rico. Caribbean Journal Science 12:129–144.

Wilson EO (1992) *The Diversity of Life*. W.W. Norton, New York.

Wright SJ (1996) Plant species diversity and ecosystem functioning in tropical forests. In: Orians GH, Dirzo R, Cushman JH (eds) *Biodiversity and Ecosystem Processes in Tropical Forests*, pp. 11–32. Springer-Verlag, Berlin.

13. Lakes

David M. Lodge

This chapter will develop scenarios of how lake biodiversity and ecosystem function will be affected by future global changes (Fig. 13.1). As a guide to that goal, much of the chapter will be devoted to an examination of the responses of lake biodiversity to past and ongoing global changes. Related reviews have emphasized the response of freshwaters to changes in climate alone (Carpenter et al. 1992; Firth and Fisher 1992; Arnell et al. 1996; Cushing 1997). This chapter will examine responses of biodiversity in lakes to a broader range of ongoing global changes, including land-use changes and attendant irrigation, eutrophication, and chemical pollution; introduction of nonindigenous species (NIS); overexploitation of fisheries; increases of UV-B radiation; and climate change. Unlike terrestrial ecosystems, lakes are unlikely to respond directly to changes in atmospheric CO_2 concentration because most lakes are supersaturated with CO_2 (Cole et al. 1994).

In this chapter, the multitude of global changes will be viewed as drivers of change in the physical and chemical structure of lake ecosystems, which in turn greatly affect biodiversity in lakes (Schindler 1990; Fig. 13.1). Most research on these topics has focused on how physical, chemical, and aggregated biotic parameters (e.g., primary production of phytoplankton communities) respond to these global changes. Changes in ecosystem parameters, therefore, can be more confidently predicted than changes in biodiversity. Each global change driver, however, also directly affects lake biota. Through-

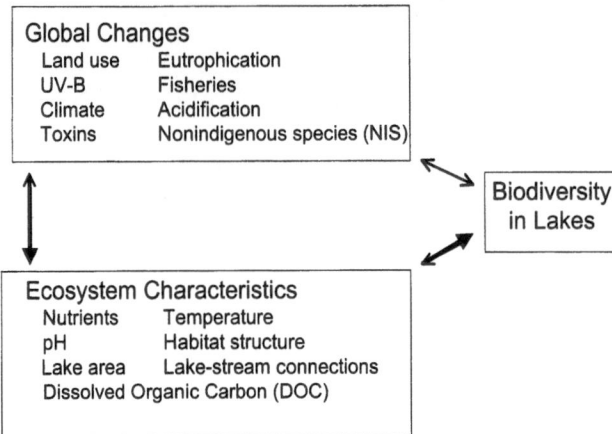

Figure 13.1. Hypothesized direct affects among global changes, lake ecosystem characteristics, and lake biodiversity. Arrow thickness indicates the hypothesized relative importance of affects.

out the chapter, examples of strong feedbacks among global changes, biodiversity, and lake ecosystem function will be emphasized (Fig. 13.1).

The chapter's primary focus is on species biodiversity because species are unique genetic units of nature, are highly valued by humans, are critical to ecosystem function, and contribute to the services that ecosystems provide to humans. In this chapter, I use *biodiversity* to mean species diversity (Wilson 1988), acknowledging that this definition does not encompass the genetic, population, community, ecosystem, and landscape diversity that are critical to nature's function (Angermeier and Karr 1994; Moyle and Yoshiyama 1994; Humphries et al. 1995; Nielsen 1995; Turner et al. 1995; Wheeler 1995; Hughes et al. 1997; Nee and May 1997; Callicott et al. 1999). Examples focus heavily on fishes because they have been much more intensively studied than other taxonomic groups. Consistent with conventional ecological usage, I include the relative abundance of species within "species diversity." For example, if the abundance of a naturally common fish species has been reduced by overexploitation, I consider that as a reduction in species diversity. The focus is on species diversity because most of the available information pertains to diversity at the species level. An important intellectual challenge—with major resource management implications—is clearly to understand how interactions between global changes and species diversity affect function at other levels of organization (Carpenter et al. 1996).

Importance of Lakes to Human Welfare

Inland lakes range from very fresh to hypersaline, with saline lakes reaching their greatest abundance in subtropical climates, where they are of economic and ecological importance (Williams 1993). I will give some attention to saline lakes in the chapter, but my focus is on freshwater lakes. Even though freshwaters cover only about 0.5% of global surface area, they contribute at least 5%, or 1.7×10^{12}, of total global flow value of ecosystem goods and services (Costanza et al. 1997). About 90% of the freshwater value derives from water supply and regulation, with waste treatment, food, and recreation being the only other freshwater ecosystem services for which global value estimates exist (Costanza et al. 1997). The public also values biodiversity of freshwaters (e.g., many endangered fishes are highly valued by humans) (Loomis and White 1996), but no aggregate statistics on monetary value are available. Because freshwater has no substitutes for most of its uses, the marginal value of freshwater is probably underestimated more severely than that of many other biomes (Postel and Carpenter 1997).

Furthermore, on a global basis, much of available freshwater is already co-opted by humans: Of the total global renewable freshwater supply, 23% is used by humans, with the majority of it as irrigation water or as evapotranspiration through crops (both irrigated and unirrigated) (Postel et al. 1996). Considering that much of the world's runoff is likely to remain inaccessible to humans (e.g., much of the Amazon, Zaire/Congo, and many arctic and subarctic rivers), moderate estimates of human popuation growth suggest that freshwater could become very limiting in the first few decades of the twenty-first century (Postel et al. 1996). Before it limits population growth, it will become much more costly, and could easily be the focus of international warfare, especially in the Mideast (Postel 1992; Gleick 1993a; McCaffrey 1993). As a function of population growth, and perhaps climate warming, the number of water-scarce countries (currently 20–80; Postel 1992; Pimentel et al. 1997a) will likely double by 2025 (Kaiser 1997). On an annual basis, nine countries, all in the Mideast, already consume greater than 100% of their renewable freshwater supplies (Gleick 1993a). Such consumption is only the most direct and obvious of the many anthropogenic impacts on lake ecosystems worldwide. This chapter examines the impact of water use and many other anthropogenic effects on lake biodiversity and ecosystem function.

Species Diversity and Ecosystem Function

Directly and indirectly, biodiversity is invaluable to humans (Pimentel et al. 1997b). The old hypotheses that species diversity increases the productivity, sustainability, and stability of ecosystems have gained recent empirical support from terrestrial field studies on plants (Chapin et al. 1998). Positive relationships exist between biodiversity and terrestrial ecosystem responses at low species numbers, but the relationships asymptote at higher species numbers (Tilman et al. 1996; Tilman et al. 1997). The asymptotic nature of the relationships have emphasized the need to examine how functional diversity relates to species number (Chapin et al. 1997; Grime 1997; Hooper and Vitousek 1997; Tilman et al. 1997; Wardle et al. 1997). Even though terrestrial experiments testing the links between species number and ecosystem function have focused on primary producers, freshwater studies have rarely explicitly tested the importance of species number (Naeem and Li 1997). These studies have instead emphasized the functional differences among consumer species, and on the large impacts some consumers have on ecosystem function.

In freshwaters, the addition or deletion of single consumer species often causes large changes in ecosystem function. Because it is difficult to identify such species a priori (Hurlbert 1997), preserving native biodiversity is an important hedge against adverse effects of nonindigenous species on ecosystems. Because experimental results for impact of single species on lakes have been reviewed (Lodge 1993a; Carpenter et al. 1996), I will only briefly describe some examples here and develop other examples in the case studies.

Beaver (*Castor canadensis*) were historically widespread and abundant throughout North America. They were virtually extinct in much of their former range by 1900. The dramatic ecological impacts of beaver, therefore, have only now been studied (Snodgrass and Meffe 1998). They impose long-lasting increases in landscape diversity by converting large areas from terrestrial ecosystems to lakes, creating larger riparian zones, and changing carbon and nutrient storage (Pollock et al. 1995).

Single species of invertebrates can also make a large difference in ecosystem function, especially when they are functionally different than resident species. The introduction of mysid shrimp to Flathead Lake (Montana) reduced abundance of large zooplankton on which salmon had previously relied (Spencer et al. 1991). The salmon population collapsed, resulting in the decline of eagles and bears, which had relied on salmon. Angling, ecotourism, and the regional economy also suffered (Spencer et al. 1991). Many other introductions of mysids have also caused major changes in lake biodiversity and ecosystem function (Nesler and Bergersen 1991).

Finally, cascading trophic interactions (e.g., the impact on lower trophic levels of changes in a higher trophic level) have been convincingly demonstrated in lakes. Additions of a piscivore to a piscivore-poor lake often cause reductions in zooplanktivory and, consequently, increases in algal standing crop and changes in algal productivity (Carpenter et al. 1987; Carpenter and Kitchell 1993). Many different macrophyte species, which as an aggregate increase and then decline along the lake trophic gradient, can modify such cascading trophic interactions, and strongly influence lake nutrient dynamics and cycling (Jeppesen et al. 1998).

Overall, where functional redundancy exists, ecosystem parameters change little as one species is added or deleted because of complementary responses by other functionally similar species (Carpenter et al. 1993; Frost et al. 1995; Schindler 1995). Strong changes in ecosystem function in response to the introduction or deletion of a single species occur when a lack of functional redundancy exists. Thus, there is increasing recognition, fueled largely by freshwater research, that strong feedbacks exist between biodiversity, the structure of food webs, and ecosystem function (Polis and Winemiller 1996).

I will provide additional examples in the case studies (see later), including zebra mussels (*Dreissena polymorpha*) in the Great Lakes, of freshwater species–ecosystem links. As in terrestrial ecosystems, the best evidence suggests that conservation of lake ecosystem function depends to a large degree on conservation of species functional diversity, if not on biodiversity per se. Given that species are being lost from many lakes at a high rate, a priority for research must be better testing the relationship between biodiversity and ecosystem function.

Impact of the Rate of Global Changes on Lake Biodiversity

Over a period of months about 7150 years ago, the freshwater Black Sea (Ukraine, Russia, Georgia, Turkey) expanded and became brackish when Mediterranean Sea water flooded over the Bosporus isthmus at the rate of up to $50\,km^3$/day (Ryan et al. 1997). This was an abrupt culmination of gradual global warming over thousands of years, subsequent deglaciation, and rising sea levels. Thus, slow, long-term changes can result in abrupt and extreme environmental impacts, that result in wholesale changes in biodiversity. For a natural event, the rate of the Black Sea transformation was unusually high. Freshwater responses to climate change in other areas have been slower. For example, over the last 5000 years, frequency of high lake levels in the North American southwest changed over years to centuries in response to regional climate changes, large-scale atmospheric circulation, and ENSO events (Ely et al. 1993). Many anthropogenic effects on lakes, however, including those caused by climate, are likely to be manifest over years to decades, not millenia (Schindler 1998). Since 1970, for example, the volume of the Aral Sea (Kazakhstan, Uzbekistan) has shrunk by 72% from withdrawal of irrigation water, salinity has increased, and many species have disappeared (see case study later). These changes will often outpace the ecological and evolutionary mechanisms that might otherwise allow organisms to disperse to other lakes or adapt (Kareiva et al. 1993; Lodge 1993a).

Global Distribution of Lakes, Biodiversity, and Human Impact

Relative to most terrestrial biomes, lakes are like islands, and lake biota thus have inherent barriers to migration (Magnuson 1976; Naiman et al. 1995). This is especially true for lakes without stream connections to other lakes. Such isolation, especially in ancient lakes, has allowed the evolutionary radiation of endemic species flocks. Connecting these endemic biotas to other biotas (e.g., through the introduction of nonindigenous species) can cause devastating losses of global biodiversity. On the other hand, natural migration among lakes that are connected by streams often occurs (Hinch et al.

Table 13.1. Major lake types of the world that are important for global biodiversity

Lake type	Typical age (years)	Typical latitude	Typical physical characteristics		Typical biological characteristics	
			Lake area	Lake depth (m)	Species number	Endemism
1. Glacial	10^3–10^4	>40°	Small-Large	10^1–10^2	Low	Low
2. Floodplain	<10^0–10^2	All	Small-Med.	<10^0–10^1	Low-High	Low
3. Tectonic	10^2–10^7	All	Large	10^2–10^3	High	High
4. Volcanic	10^2–10^3	All	Small	10^2	Low	High

Simplified from Wetzel 1983.
Lake types are defined by geological origin. Lake types are listed in probable decreasing order of global abundance, although no complete census of lakes exists (Meybeck 1995). Glacial, tectonic, and volcanic lakes are emphasized in this chapter; although floodplain lakes are numerous, their physical and biological characteristics are heavily affected by the rivers and the riverine biotas with which they are associated (see Chapter 14).

1991; Horvath et al. 1996). It is important for restoring local biodiversity in glacial lakes that have had winterkill or some other natural loss of species (Tonn et al. 1990; Magnuson et al. 1997). In these situations, loss of stream connections, as often results from anthropogenic water withdrawal or climate warming, may result in loss of local (but not regional) biodiversity. Thus, the biodiversity consequences of connecting or disconnecting lakes are system specific. In general, lakes are very vulnerable to changes in the surrounding terrestrial biome (See Chaps. 2 and 3), including changes in evapotranspiration, nutrient loading, and other watershed influences. Except for those of glacial origin, lakes have no strong association with any climatic zone (Meybeck 1995; Lewis 1996) and are therefore unlike the terrestrial biomes of the world (covered in most other chapters of this volume) (Table 13.1, Fig. 13.2).

Geological origin determines local geomorphology and lake morphometry, which in turn interacts strongly with climate to determine many aspects of physical, chemical, and biological dynamics of the lake ecosystem (Wetzel 1983; Naiman et al. 1995; Fee et al. 1996; Riera et al. 2000). Four major types of geological origin are responsible for most lakes on earth. Glacial lakes are concentrated in lake districts of several large and thousands of small lakes in northern North America and Eurasia because glaciers and land masses intersected there. Ongoing postglacial colonization (Hinch et al. 1991), stream connections among many lakes, and young age have allowed little isolation of populations (Tonn et al. 1990) and, consequently, little speciation. Nevertheless, these lakes are the most numerous on the globe (Herdendorf 1990; Meybeck 1995), they are important biologically, and they are central to many human activities in the north temperate zone. Thus, the first case study in this chapter will focus on north temperate and boreal glacial lakes.

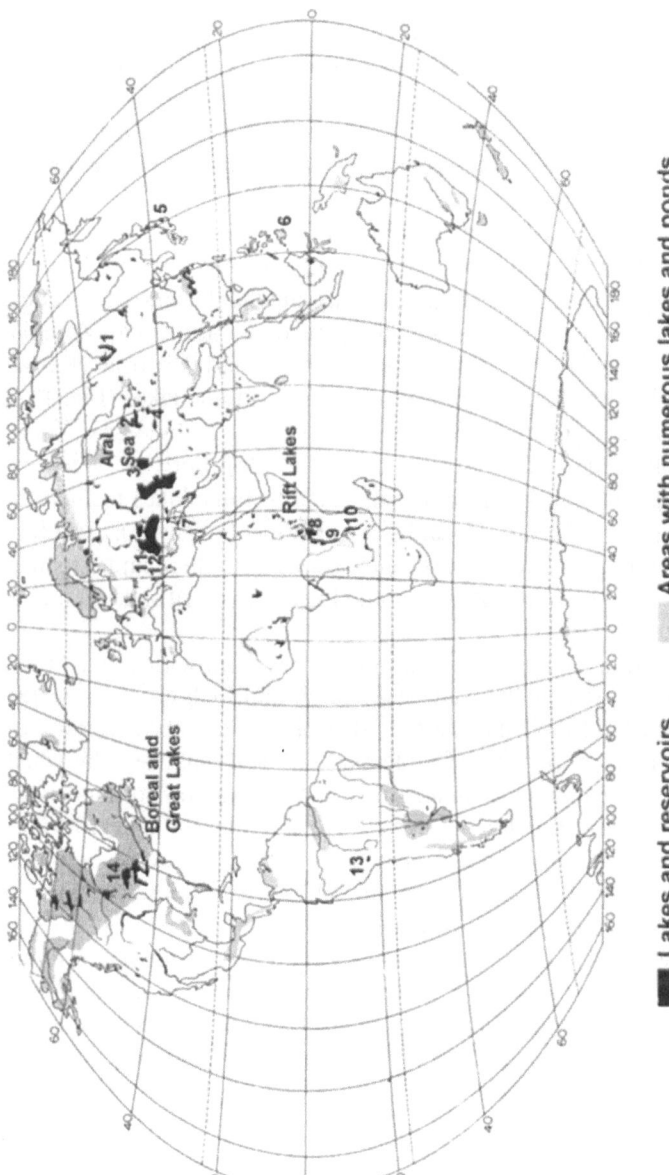

Figure 13.2. Distribution of large lakes (in black), including brackish and saline lakes, and concentrations of smaller lakes (shaded areas) on earth. Labeled lakes or lake districts are used as case studies in text. Numbers refer to lakes listed by the same numbers in Table 13.2; most of these have high endemism. This map was extensively modified (by study of 1:5,000,000 to 1:2,500,000 projections in the *The Times Atlas of the World* [1980]) from the one appearing inside the cover of Taub 1984.

■ Lakes and reservoirs   ▨ Areas with numerous lakes and ponds

Rivers create lakes in several ways, including the isolation of a former segment of river channel (e.g., oxbow lakes), or seasonal filling of flood-plain depressions (e.g., Australian billabongs; Hillman 1986), or extensive and long-lasting flooding that creates unique lake and wetland communities [Naiman et al. 1995; e.g., the Sudd of the Nile River (Melack 1992), the lakes and varzea swamps of the Amazonian (Forsberg et al. 1988; Sippel et al. 1992; Mertes et al. 1995; Junk and Weber 1996; Melack 1996) and Orinoco floodplains (Hamilton and Lewis 1990); the Pantanal of Bolivia and Brazil (Eckstrom 1996; Van der Beck et al. 1996)]. These floodplain lakes are strongly influenced by their associated rivers (see Chap. 14), are poorly studied, and are beyond the scope of this chapter.

Tectonic lakes are formed from faults (grabens) or streams dammed by uplift. They are typically species-rich, with high endemism, because they are often large, much older than glacial lakes, and isolated in the landscape (Table 13.1). Lakes Baikal (Russia), Tanganyika (Burundi, Tanzania, Zambia, Zaire), Malawi (Tanzania, Mozambique, Malawi), and Victoria (Kenya, Tanzania, Uganda) are examples of such speciose tectonic lakes (Fig. 13.2, Table 13.2). Geologically similar lakes in North America include Lake Tahoe (Nevada and California) and Pyramid Lake (Nevada), both of which are much less speciose. Closed basin, brackish-saline tectonic lakes include the Caspian Sea (Iran, Azerbaijan, Kazakhstan, Turkmenia, Russia), Aral Sea, and the Great Salt Lake (Utah). The second case study in this chapter focuses on the Aral Sea, and the third focuses on the East African rift lakes, empha-sizing Lake Victoria.

A small proportion of lakes are volcanic craters filled with water (Table 13.1). Because they are usually small, with small and high elevation water-sheds, and are isolated from other water bodies, they tend to be species-poor, but they have a high proportion of endemics. Other mechanisms by which natural lakes are created include stream damming by lava flows or landslides; dissolution of soluble bedrock (e.g., karst regions of the Balkans, Florida); wind-formed dune lakes and playas; polygonal, cryogenic lakes (e.g., millions of small lakes in arctic Alaska; Milner et al. 1997); and damming by beavers (reviewed earlier). Although dam construction by people has created many reservoirs on the globe, it has simultaneously reduced or eliminated flood-plain lakes along many of the streams and rivers of the world. The impact of damming and reservoirs is discussed more fully in relation to streams and rivers elsewhere in this volume (see Chap. 14), but will not be considered further in this chapter.

Lake Biodiversity Hotspots and Human Impact

Concentrations of freshwater biodiversity exist (Dobson et al. 1997a; Table 13.2), especially in ancient lakes that are isolated from other water bodies. Isolation is often the result of the aridity of the surrounding landscape.

Because geographic isolation enhances speciation, many of these speciose lakes also have highly endemic biotas (Table 13.2). Freshwater fish endemism in the arid U.S. southwest is often greater than 70% for individual lakes and drainages (Warren and Burr 1994). For taxonomic groups other than fishes, molluscs, and crustaceans, data on diversity and endemism are few (Table 13.2). Smaller organisms in particular (e.g., algae, most invertebrates) are very poorly studied, and even the number of species remains unknown (Michaelis 1986; Tyler 1996; Palmer et al. 1997).

In North America high proportions of freshwater species (including stream and lake species), especially bivalves (73%), crayfishes (65%), and fishes (34%), are declining (Master 1990; Bogan 1993; Williams et al. 1993; Taylor et al. 1996; Cummings et al. 1997; Lodge et al. 1998a). High proportions of the stream and lake fish faunas are also imperiled in Europe (42%), Iran (22%), South Africa (63%), Sri Lanka (28%), Australia (26%), and Latin America (9%) (Moyle and Leidy 1992; Kirchhofer and Hefti 1996). Twenty percent of freshwater fish globally are extinct or in serious decline (Moyle and Leidy 1992). Lake Victoria is an extreme example, where perhaps half of the vast flock of endemic cichlid fishes has been dramatically reduced or driven extinct (Table 13.2; see case study). Biodiversity in the world's lakes is in decline (Pimm et al. 1995), and in lakes with high rates of endemism the local loss of a species means global extinction.

For lakes, an unfortunate positive correlation seems to exist between biodiversity (either number of species or level of endemism) and probability of severe human impact. Arid-land lakes, many of which have high biodiversity and high levels of endemism, are subject to the greatest levels of human influence because water is a limiting resource in these arid regions, despite low per capita use relative to western countries (Gleick 1993b). Fast-growing human populations in many of these regions translate into intense impact on water resources and the high biodiversity they contain (e.g., the U.S. southwest, the Mideast, north and east Africa) (Postel 1992; Falkenmark and Lindh 1993; Dobson et al. 1997b). In contrast, biodiversity and endemism are lower where lakes are most abundant and people are in lower abundance (e.g., boreal and north temperate lakes) (Table 13.2). Thus, these lakes suffer lower anthropogenic impacts on biodiversity.

From the perspective of impacts on ecosystem function, however, the situation may be reversed. Functional redundancy in more speciose lakes may mean that ecosystem responses are small, even as many species are driven extinct. In contrast, in species-poor boreal lakes, the extirpation of one species (e.g., the one piscivorous fish species present) may cause large ecosystem responses (Schindler 1988). These contrasting possible relationships between probability of human impact on biodiversity, on one hand, and on ecosystem function, on the other hand, pose serious challenges to freshwater research and conservation biology.

Table 13.2. Percentage of endemic species for a selection of the world's lakes with the greatest endemism, and for the lakes (African Rift Lakes, Aral Sea, North American boreal and Great Lakes) used as case studies in this Chapter. Lake numbers are used to locate lakes on Figure 13.1. At the bottom of the table, sources are listed by lake number.

Lake name (Country)	Lat./long.	% Endemic species (Total species, excluding NIS where possible)									
		Nonvasc.	Vasc.	Platyhel.	Annelida	Rotifera	Mollusca	Chelicerata	Crustacea	Insecta	Osteichthyes
Asia											
1 Lake Baikal (Russia)	57°N, 110°E			77% (106)	72% (166)	16% (206)	76% (170)	50% (6)	82% (537)	16% (231)	48% (56)
2 Lake Balkhash (Kazakhstan)	45°N, 76°E										60% (5)
3 Aral Sea (Kazakhstan), Uzbekistan	45°N, 60°E										5% (19)
4 Lake Issyk-kul (Kyrgyzstan)	42°N, 77°E										100% (10)
5 Lake Biwa (Japan)	35°N, 135°E	<1% (424)	3% (67)	13% (15)	10% (20)	0% (103)	45% (44)	0% (7)	8% (38)	7% (169)	
6 Lakes Lanao and Dapao	8°N, 125°E	0% (70)				0% (7)			30% (6)		
7 Lake Kinneret (Israel)	32°N, 32°E			0% (20)	0% (14)		0% (8)		7% (42)	0% (58)	26% (27)
Africa											
8 Lake Victoria (Kenya, Tanzania, Uganda)	1°S, 33°E										74% (285)
9 Lake Tanganyika (Burundi, Tanzania, Zambia, Zaire)	6°S, 30°E			64% (11)	61% (28)	7% (70)	61% (75)	37% (46)	58% (219)	69% (155)	73% (330)

Lake	Location								
10 Lake Malawi (Tanzania, Mozambique, Malawi)	12°S, 35°E								42% (646)
Europe									
11 Lake Ohrid (Albania, Macedonia)	41°N, 20°E			64%			69%		60%
12 Lakes Megali Prespa and Mikri Prespa (Albania, FYROM, Greece)	41°N, 21°E								6% (23)
South America									
13 Lake Titicaca (Peru, Bolivia)	15°S, 69°W	5% (158)	4% (23)	0% (15)	0% (7)	21% (24)	43% (39)	3% (36)	79% (29)
North America									
14 Great Lakes (United States, Canada)	45°N, 85°W								<1% (168)

[1] Martin 1994, Martin 1996; [2] Petr 1992; [3] Bortnik et al. 1992, Petr and Mitrofanov 1995; [4] Savvaitova and Petr 1992; [5] Nakajima 1994, Martin 1996; [6] Lewis 1979 [for Nonvasc., Rotifera, and Crustacea (planktonic species only for all three groups], Frey 1969 [for fishes (cyprinids only)]; [7] Gophen and Nishri 1994, Martin 1996; [8] Greenwood 1994, Martin 1996; [9] Coulter 1994, Martin 1996; [10] Ribbink 1994, Martin 1996; [11] Salemaa 1994, Martin 1996; [12] Crivelli et al. 1997; [13] Dejoux 1994, Martin 1996; [14] Burr and Mayden 1992. "Nonvasc." and "Vasc." refer to nonvascular plants (algae) and vascular plants (macrophytes), respectively.

Table 13.3. For five global drivers (rows), the degree of expected change in the driver, the degree of impact on biodiversity per unit change in the driver, and an index of the global impact on biodiversity of the changes in each driver (= product of the first two columns)

	Change in driver	Impact on biodiversity per unit change in driver	Index of global impact on lake biodiversity
Land use	5	5	25
NIS	5	5	25
Harvesting	5	3	15
Climate	3	3	9
CO_2	2	0	0

1 = small; 5 = large. Biodiversity refers to the number of native species at the landscape scale. Values are approximations of global averages, weighted more by the number of lake regions than by the total number of lakes (to avoid scores being skewed by the numerous boreal and north temperate lakes). Because lakes are scattered across the earth in different climatic zones and are subject to different drivers in different locations, these scores may not apply well to some lake districts. Land use includes habitat loss by drying, salinization, siltation; nutrient loading (N, P); acidification by atmospheric deposition of N, S; and pollution by metals, pesticides, and herbicides. NIS refers to the introduction of nonindigenous species (= alien species = exotic species). Harvesting includes exploitation of fishes and waterfowl. Climate refers to temperature, precipitation, and UV radiation. CO_2 refers to atmospheric CO_2 concentrations.

Global Changes

Many of the largest anthropogenic impacts on freshwaters parallel the impacts on terrestrial ecosystems that have been well illustrated by Vitousek (1994) (Vitousek et al. 1996). Table 13.3 summarizes the broad consensus on the major global threats to lake biodiversity, ecosystem function, and ecosystem services (National Research Council 1992, 1996; Naiman et al. 1995). In order of global importance, these threats are changing land use, introduction of NIS, overexploitation of fishes and other species, and changes in climate and the atmosphere (Table 13.3). I will describe each of the important global change drivers more fully. For each case study I will then describe the relative importance of different drivers and responses.

Land Use

Conversion of natural habitats is a major global change that affects biodiversity and ecosystem function (Dobson et al. 1997b), and will likely increase dramatically in many of the lake districts of the world (Sykes, this volume). The terrestrial vegetation and climate in a watershed have large impacts on the solute budgets of aquatic ecosystems, lake productivity, the balance between autochthonous versus allocthonous resources, and consequently on aquatic biodiversity (Likens 1985; National Research Council 1992, 1996; Kling 1995; Naiman et al. 1995; Wallace et al. 1997). Including terrestrial vegetation responses to climate change produces qualitatively different predictions of aquatic ecosystem parameters relative to predictions generated assuming only a direct climate–aquatic ecosystem link (Band et al. 1996).

For example, conversion from traditional polyculture to intensive agriculture to increase production for a growing human population results in strong environmental side effects on aquatic ecosystems (Matson et al. 1997). Forty to 60% of N and much P that is applied as fertilizer is lost to the intended crop, much of it to runoff (Matson et al. 1997). This causes eutrophication, which is the most serious and widespread global change affecting lakes (National Research Council 1992, 1996; Naiman et al. 1995). The extremely well-studied impacts of eutrophication include increased productivity of phytoplankton, often of noxious cyanobacteria; decreased light penetration and altered thermal structure; increases in hypoxia; and shifts in species composition of many taxa, including fishes (National Research Council 1992, 1996; Naiman et al. 1995; Carpenter et al. 1996).

Increased loading of sediments and toxins into lakes is inextricably associated with runoff of nutrients. About 60% of eroded soil is deposited in freshwater ecosystems (Pimentel et al. 1995), and associated pesticides and herbicides cause fish kills (Nash 1993; Matson et al. 1997) and reduce marketability of fishes (Pimentel et al. 1992). Bioaccumulation of organic toxins results in many government advisories about safe rates of fish consumption (National Research Council 1996). The loss of wetlands simultaneously destroys spawning and nursery habitat for fishes and other species (Magnuson and Lathrop 1992), and decreases removal rates of particles, nutrients, and chemical pollutants from waters entering lakes (National Research Council 1996). Furthermore, irrigation for agriculture or water withdrawal for urban use lowers the water table, decreasing lake area and volume (National Research Council 1992; Pimentel et al. 1997a). Many of these agriculture-related drivers will increase disproportionately as intensification of agriculture brings more steeply sloping land under cultivation (Pimentel et al 1995).

Lake acidification from precipitation-containing fossil fuel–derived oxides of S and N is a major driver of biodiversity losses in the boreal lakes of North America, Europe, and Scandanavia because these lakes are naturally low in buffering capacity (Schindler 1988). Although S emissions in North America have decreased in recent decades, global acidification will continue and perhaps increase (Schindler 1998), especially in regions where poorly buffered lakes overlap with air masses from regions currently undergoing economic development and industrialization. In other regions acidification results from acid mine drainage (National Research Council 1992). Finally, acidification with gaseous NO_X emissions from soils may add proportionately more to this problem in future as fertilizer use in agriculture increases (Matson et al. 1997). Contamination with heavy metals and organic toxins result from a combination of aerial deposition and runoff that affect even the most remote lakes on earth; important chronic biological effects are now being recognized (Naiman et al. 1995; National Research Council 1996).

Many of these land-use changes involve substantial time lags and complex temporal dynamics (Lodge et al. 1998b). For example, eutrophication proceeds slowly, but it may induce abrupt switches in alternate lake states. A vegetated, clearwater lake may become a turbid, algal dominated ecosystem within a year (Jeppesen et al. 1998). In acidified lakes reproduction of fishes may cease years before the existing fishes die and their food web impacts disappear (Schindler 1988). Thus, global changes that produce what initially seem to be modest biodiversity responses may, in fact, have major long-term consequences. Overall, then, changing land use is probably the most important driver of changes in lake biodiversity and ecosystem function (Table 13.3).

Nonindigenous Species (NIS)

In studies that span multiple biomes (Ruesink et al. 1995; Vitousek et al. 1996; Williamson 1996; Czech and Krausman 1997), and especially in freshwater studies (National Rearch Council 1992, 1996; Lodge 1993a,b; U.S. Congress 1993; MacKaye et al. 1995; Naiman et al. 1995; Morton 1997; Foin et al. 1998; Lodge et al. 1998a), there is a consensus that NIS are among the top threats to global biodiversity and ecosystem function. Among freshwater taxa, this is particularly true for fishes (Fletcher 1986; Pollard and Burchmore 1986; Moyle and Leidy 1992; Warren and Burr 1994; cf. Dill and Cordone 1997) and crayfishes (Light et al. 1995; Merrick 1995; Taylor et al. 1996). It is likely to be increasingly true for bivalves as zebra mussels colonize North America (Hunter et al. 1997; Lodge et al. 1998a). Within less than a decade after zebra mussel (*Dreissena polymorpha*) invaded Lake St. Clair (between lakes Huron and Erie), abundance and biodiversity of native unionids declined by 98% and 72%, respectively (Nalepa et al. 1996).

Effects of NIS on native species include parasitism, predation, competition, and fouling, but some NIS also change abiotic features and habitat structure in more general ways. Water hyacinth (*Eichhornia crassipes*) and common carp (*Cyprinus carpio*) provide two examples. Water hyacinth is native to South America, but in North America and Africa it forms a floating mat that covers large portions of the surface area of subtropical and tropical lakes, occluding light, decreasing dissolved oxygen, and dramatically shifting composition of native plant and animal communities (Kaufman et al. 1996). In many temperate areas of the world outside their native Asia, common carp decrease macrophyte abundance and biodiversity, increase turbidity, increase water temperature, and alter pH and dissolved oxygen dynamics (Roberts et al. 1995). Carp, water hyacinth, and many other NIS, especially fish species, were and are often intentionally stocked outside their native range. Additional examples of NIS-caused losses of biodiversity and alteration in ecosystem function are provided in the case studies.

Overexploitation of Species

As reviewed earlier, the extirpation of beaver from much of North America caused the loss of many lakes and associated local biodiversity, but this was hardly documented (Snodgrass and Meffe 1998). In contrast, reductions by commercial and recreational fishing of populations of many native fish species in all lake regions of the world are well documented (Lodge 1993a; Naiman et al. 1995; Reckahn 1995; National Research Council 1996; see case studies). Managing, studying, and exploiting fish has been a major activity of government agencies and scientists for decades (Magnuson 1991). Because top-down food web effects are so strong in lakes, fishing (usually targeting piscivorous species) has also indirectly resulted in shifts in species composition in other fishes, benthic invertebrates, zooplankton, and phytoplankton, as well as changes in overall lake productivity (Carpenter and Kitchell 1993; Brett and Goldman 1997).

Harvesting of waterfowl has had a major impact on waterfowl abundance and biodiversity in Europe, Asia, and North America, with potentially large but mostly untested consequences for the biodiversity of the consumed plants, and for the function of lake ecosystems (Lodge et al. 1998b). In many areas of North America, waterfowl populations are now increasing again because of regulations on hunting (Lodge et al. 1998b). At the same time, lake and wetland habitat area is declining from conversion to agriculture and other human development. Herbivory by increasing concentrations of waterfowl reduces aquatic plant biomass, sometimes increasing and sometimes decreasing plant biodiversity (Lodge 1991; Froelich and Lodge, 2000).

Climatic Changes and Increases in UV-B

Likely consequences of climate change on lakes were summarized by Arnell et al. (1996): Increases in temperature will increase evapotranspiration, which will often produce dryer conditions, even when precipitation increases; increases in water temperature and changes in thermal structure of lakes will affect abundance and species composition of individual lakes and regional biota; changes in runoff will alter nutrient and dissolved organic carbon (DOC) loading to lakes; ranges of many organisms will move poleward where water connections allow or go extinct where they do not; and climatic effects, including extinctions, will be greater at higher latitudes.

Altered hydrologic budgets will change stream flow (see Chap. 14) and groundwater inputs to lakes (Kratz et al. 1991). Lake responses could be extreme at high latitudes or in already arid habitats. For example, the numerous floodplain lakes of the Mackenzie River (Canada) delta may dry (Marsh and Lesack 1996), and permanent inflows to many boreal lakes may become ephemeral or disappear (Schindler et al. 1996a). Changes in mean water level or in the seasonality or magnitude of fluctuations in water level will affect shoreline vegetation and spawning in many fish species (Arnell et al. 1996), and affect the many biotic and abiotic interactions that depend on littoral–pelagic links (Lodge et al. 1988). Basins that become closed will be subject to salinization (Laird et al. 1996). More subtle changes in lake area, volume, or DOC loading could also result in large changes in thermal structure and the biological interactions shaped by it (Fee et al. 1996). Because vegetation of watershed wetlands and terrestrial communities often determines inputs of DOC and other solutes into aquatic ecosystems (Dillon and Molot 1997), the substantial lags that occur in vegetation responses to climate will also be expressed in aquatic responses (Likens et al. 1996; Wallace et al. 1997; Schindler 1998).

More direct effects of temperature will likely reduce biodiversity. Poleward movements of species will occur where water connections provide the necessary conduits for migration (Shuter and Post 1990), but many extinctions and extirpations will occur at the lower latitude boundaries of species distributions (Lodge 1993a; Arnell et al. 1996). At the highest latitudes, extinctions or extirpations will occur where habitat for coldwater species disappears; in addition, reduction in ice cover (and therefore reduction in magnitude and frequency of winterkill) will remove one of the filters maintaining regional fish biodiversity (see boreal lakes case study). At lower latitudes biodiversity will decline with increasing temperature if thermal maxima are exceeded, or if species cannot migrate as quickly as climate changes (Arnell et al. 1996). In some regions, however, biodiversity may increase as warmer temperatures make lakes tolerable for more species.

As a result of stratospheric ozone depletion, UV-B radiation is increasing, particularly at high latitudes (Kerr and McElroy 1993). Even current levels of UV-B, however, have strong biological effects, depending on the UV transparency of water (which is primarily determined by DOC). In clearwaters, ambient UV-B reduces zooplankton survival, and lowers abundance and production of phytoplankton and periphyton (Schindler et al. 1996b; Vinebrook and Leavitt 1996; Williamson et al. 1996). Increased UV increases photolysis of DOC, creating a positive feedback loop that clarifies the water (Schindler 1998). These effects and their interactions with other drivers are elaborated in the next section.

Interactions Among Global Change Drivers

Acidification, UV-B, and Climate Warming

Important interactions among different global changes are increasingly recognized. The best documented are the three-way interactions among lake acidification, UV-B radiation, and climate warming (Fig. 13.3; Schindler et al. 1996b; Yan et al. 1996; Schindler 1998). For lakes of naturally low buffering capacity (like most boreal and many north temperate lakes), acidic deposition reduces lake pH; changes algal composition and productivity; reduces abundance and biodiversity of molluscs, crustaceans, fishes, and other taxa; and reduces DOC and increases water clarity (Schindler 1988, 1998).

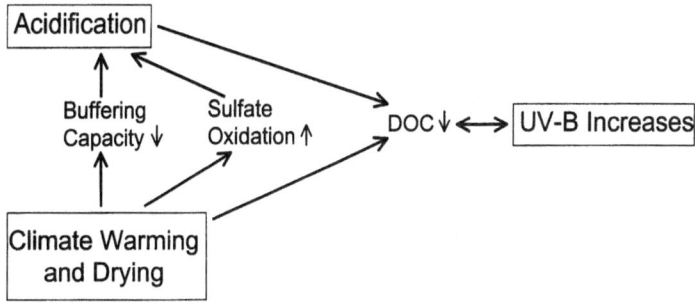

Figure 13.3. Interactions, including positive feedbacks, among three global changes of major importance for many north temperate and boreal lakes (from text in Schindler 1998). All three global changes drive many significant biological responses (see text).

Climate warming also increases UV-B (Fig. 13.3; Schindler 1998). Warming and drying increase fire frequency in the terrestrial landscape, which changes vegetation and soil characteristics. Reduced runoff decreases inputs of DOC to lakes (Schindler et al. 1997). Although increased evaporation of lakewater concentrates solutes, biologically active ones (nutrients, DOC) decrease because increased water residence time exposes them longer to biological uptake and/or breakdown. Thus, algal biomass declines and lakes become clearer, which increases thermocline depth, thus decreasing habitat for cold stenotherm fishes (e.g., lake trout). UV-B penetration, with associated biological effects, thus increases further as DOC and algal productivity decline (Fig. 13.3; Schindler 1998).

Climate warming also interacts strongly with acidification (Fig. 13.3; Schindler 1998). When wetlands or littoral sediments dry, S previously stored in reduced form is exposed to air and oxidizes (Yan et al. 1996). During rain storms, strong acids of S are then pulsed into lakes. During warmer climate conditions, lower runoff reduces base cation inputs to lakes; however, cation concentrations and buffering capacity in the lake increase from several mechanisms, including increased water residence time (Schindler et al. 1996a). Thus, decreases in cation inputs and pulses of acids result in less buffered lakes (Schindler 1998). Thus, both climate warming and acidification increase UV penetration, even in the absence of increasing incident UV (Fig. 13.3).

Climate warming and UV also interact with terrestrial–aquatic links. For example, clearcut logging decreases soil C storage. This reduces DOC inputs to lakes (thus increasing UV penetration), but it may also cause a net loss to the atmosphere of C—another positive feedback of land use to climate warming (Likens et al. 1996; Schindler 1998). Freshwater studies are leading the way in unraveling the strong interactions among global changes, but most empirical and modeling studies still focus on one or two drivers and largely ignore interactions.

Interactions Among Other Global Change Drivers

A few of many other possible interactions among drivers of global change include eutrophication enhancing the invasibility of ecosystems by NIS (Carpenter et al. 1996); eutrophication, and fish management switching lakes from C sources to C sinks (Schindler et al. 1997); positive feedbacks between climate drying and reduction in lake area (see case study of Aral Sea); extirpation of predatory trophic levels (e.g., lake trout in arctic lakes, where functional redundancy is low) resulting from starvation as increased temperature raises predator metabolic rates more than the production rate of food organisms (McDonald et al. 1996); increased bioaccumulation of heavy metals and pesticides at higher temperature (Arnell et al. 1996); and increased bioaccumulation of hydrocarbons by exotic relative to native species (Bruner et al. 1994a,b). Interactions among global change drivers are probably much more common and important than now recognized.

Case Studies

North Temperate and Boreal Lakes

The glaciated temperate and boreal regions of North America, Scandinavia, and Siberia contain vast numbers of small lakes (perhaps as many as 4 million) that collectively cover about 10% of the total boreal area (Schindler 1998), and some of the largest lakes of the world, including the Laurentian Great Lakes (Table 13.1, Fig. 13.2). The Great Lakes are particularly important in commerce (i.e., shipping), as a water source for the many large cities on their shores, and as a focus of commercial and recreational fisheries. They and smaller north temperate and boreal lakes have perhaps the most extensive history of limnological research in the world (Frey 1963). In addition, these lakes have been the focus of regional projections of climate change impact (McKnight et al. 1996; Magnuson et al. 1997; Schindler 1998).

The biggest threats to aquatic biodiversity in this region have been eutrophication (in the more heavily populated southern parts of this region), NIS, fish stocking and harvesting, acid precipitation, stratospheric ozone depletion, wetland destruction, forest clearcutting, and climate warming. Even though the threat from eutrophication has declined because of effective regulation of external nutrient loading, threats from the other drivers continue or increase.

Eutrophication

Eutrophication impact and recovery have been summarized for lakes Erie (National Research Council 1996), Michigan (Schelske and Carpenter 1992), and Huron (Reckahn 1995). Eutrophication was extreme, especially in Lake Erie, such that species composition of many taxa changed, productivity of algae increased, water clarity declined, macrophytes declined, and biodiversity declined, but is now recovering as a result of nutrient abatement (National Research Council 1996; Kolar et al. 1997).

NIS

Introductions of NIS in the Great Lakes may have declined with implementation of ballast water controls in the early 1990s, but other vectors remain ineffectively regulated (Ludwig and Leitch 1996). At least 139 NIS now inhabit the Great Lakes, and established exotics are constantly being discovered (Mills et al. 1993). NIS have had a major impact on lake biodiversity and ecosytem function (Mills et al. 1994).

One of the most widespread NIS in the north temperate and boreal lakes of North America is the rusty crayfish (*Orconectes rusticus*), which is a native of the lower midwest (Indiana). Its range has expanded via the bait buckets of fisherman, intentional stocking by crayfish harvesters, and natural dispersal (Lodge and Hill 1994). It now occurs in most, if not all, of the Great Lakes, and in many inland lakes and streams of Michigan, Illinois, Wisconsin, Minnesota. Range expansion continues in all directions in the United States and northward through Ontario. As the rusty crayfish expands its range, resident crayfishes are extirpated (Lodge et al. 1986; Olsen et al. 1991) by an interaction of competition among crayfishes and predation by fishes (DiDonato and Lodge 1993; Hill and Lodge 1994; Hill and Lodge 1999). Hybridization also occurs with some resident crayfishes and may contribute to species replacement (Perry et al. 2001). Other ecological impacts of rusty crayfish include clearcutting of littoral zones by herbivory (Lodge et al. 1994; Lodge et al. 1998b), elimination of snails (Lodge et al. 1994; Lodge et al. 1998a), reductions in other invertebrates (Perry et al. 1997), and increases in density of periphyton (Weber and Lodge 1990; Luttenton et al. 1998). Predatory fishes reduce, but do not eliminate, these food-web effects of the rusty crayfish (Hill and Lodge 1995; Kershner and Lodge 1995).

Other recently arrived NIS may compound the many impacts of past and current NIS in north temperate lakes. Recent nuisance species still in a phase of rapid range expansion and population growth include the round goby (*Neogobius melanostomus*), and the Eurasion river ruffe [*Gymnocephalus cernuus*, a percid fish (Busiahn 1997; Fullerton et al. 1998)]. The NIS of the past decade with the most dramatic impacts is the zebra mussel (see the section on Lake Huron).

Climate

At least some parts of the north temperate and boreal region have warmed about 2°C since 1975 (Anderson et al. 1996; Schindler 1998), which is a trend that is likely to continue. Projections of the direction of change in future precipitation differ among models, and differ on small spatial scales within models. There is general agreement among scenarios, though, that whether precipitation shows modest increases or decreases, the ratio of precipitation:

evapotranspiration will decrease because of increased temperatures (Clair and Ehrman 1996; Mortsch and Quinn 1996). Great Lakes water levels may drop by 2.5 m, with inflows declining by up to 54% (Magnuson et al. 1997). The net climatic effect on watershed vegetation is likely to be a shift northward of temperate and boreal forest, a shift from forest to grassland, and decreasing agricultural land use (see Chaps. 2 and 3).

In this region, fish species richness is strongly and positively correlated to temperature, and many species are expected to expand their range 500–600 km northward as temperatures increase, if migration routes are available (Mandrak 1989; Minns and Moore 1995; Magnuson et al. 1997). Changes in temperature interact strongly with lake morphometry and trophic status to affect thermal stratification and dissolved oxygen (DO) (Fee et al. 1996). Overall, coldwater species [e.g., lake trout (*Salvelinus namaycush*), cisco (*Coregonus artedii*)] will experience large habitat reductions, except possibly in the Great Lakes (Magnuson et al. 1990; Schindler et al. 1990; De Stasio et al. 1996; Magnuson et al. 1997). Cool- and warmwater fishes will gain habitat, with the magnitude of changes positively related to latitude (Stefan et al. 1995, 1996). In some lakes, however, increases in epilimnetic temperatures (75–90% of air temperature increases) may exceed the upper lethal limit for even warmwater fishes (De Stasio et al. 1996).

Epilimnetic temperature increases, similar or colder hypolimnetic temperatures, and longer ice-free, stratified seasons (McCormick 1990; Assel et al. 1995) will probably increase summertime hypolimnetic hypoxia (De Stasio et al. 1996). Increased hypoxia would increase nutrient recycling from the hypolimnion (Magnuson et al. 1997). On the other hand, decreased duration of ice-cover would reduce wintertime hypoxia and reduce winterkill, which is a major structuring force of fish communities in small inland lakes (Magnuson 1991). Thus, regional biodiversity might decline as the fish communities in different lakes converge. Because fish biodiversity so strongly affects the entire food web and lake productivity, indirect effects of such convergence of fish communities on zooplankton, other invertebrates, and algae would be large (De Stasio et al. 1996).

Warmer temperatures, a longer growing season, less light limitation (because of less ice cover), and greater internal nutrient loading all suggest that annual primary production will increase (Magnuson et al. 1997), but Schindler et al. (1996a) showed that climate warming was associated with lower lake chlorophyll. In addition, reductions in agriculture (see Chap. 3) might reduce external nutrient loading and counter any increase in internal loading. Thus, confidence is low for predictions of the direction of change of annual production.

Multiple Drivers in Lake Huron

The history of Lake Huron (Canada, United States) during the last 100 years illustrates the typical impact of multiple drivers (Reckahn 1995). Beginning in the late nineteenth century, and continuing into the first three decades of the twentieth century, human overexploitation in Lake Huron reduced populations of whitefish (*Coregonus clupeaformis*) dramatically. During the earlier portion of this time, destruction of spawning habitat from forestry practices (bark and wood deposition over spawning grounds) and obstruction of spawning runs by dams also reduced populations. The construction of the Welland Canals (1824–1932) subsequently allowed the invasion of Lake Huron (and other Great Lakes) by the sea lamprey (*Petromyzon marinus*) and alewife (*Alosa pseudoharengus*), both of which negatively affected whitefish. Beginning in 1906, rainbow smelt (*Osmerus mordax*) were intentionally introduced, with further negative impact on native whitefish. Subsequent recovery of lake whitefish has resulted from a combination of factors, including reversal of eutrophication, chemical (lamprey) and predator (alewife) control of harmful NIS, and die-offs of NIS (smelt and alewife) (Reckahn 1995).

Since 1990 another NIS, the filter-feeding zebra mussel (*Dreissena polymorpha*), has had a major impact on the biodiversity and ecosystem function of Lake Huron. Nalepa and Fahnenstiel (1995) summarize results of a large number of on-going coordinated studies on Saginaw Bay (Lake Huron) during a zebra mussel invasion. Zebra mussel reduced phytoplankton chlorophyll (59%) and production (38%), reduced total P (42%), speeded rates of nutrient recycling, and increased water transparency (60%). The impacts of zebra mussel on water quality were thus of the same magnitude as the very effective nutrient abatement programs of the 1970s and 1980s. In response to increased water transparency, abundance and productivity of macrophytes and benthic algae increased.

Thus, the benthic filter feeder changed the biotic and abiotic characteristics of the ecosystem dramatically, changed the morphology of the benthic substrate, and shifted energy flow from the pelagic to the benthic habitat. These conclusions are consistent with impacts of zebra mussel on Lake Erie and other North American lakes (MacIsaac 1996). In addition to the effects noted in Saginaw Bay, zebra mussels in other North American lakes have reduced native bivalves and shifted species composition of other benthic macroinvertebrates (Nalepa and Schloesser 1993; Mellina et al. 1995; Arnott and Vanni 1996; James et al. 1997). Many native bivalves will be extirpated by zebra mussels (Schloesser et al. 1998); other taxonomic groups are experiencing large shifts in species abundance.

Conclusions for North Temperate and Boreal Lakes

Impacts on lake biodiversity of eutrophication, NIS, overexploitation, acidification, and UV-B are already large, and some of these drivers interact with changing climate in positive feedbacks (Fig. 13.3). The extent of on-going global changes is large on the Great Lakes, the shorelines of which are heavily populated. It is more surprising and cautionary that many of the remote, small boreal lakes are heavily affected by NIS, changes in aerial deposition of acids, pollutants and nutrients, UV-B increases, and climate change, even though they have little or no direct contact with humans. As large as these anthropogenic impacts on the Great Lakes and boreal lakes are, they are small relative to ongoing and likely future global change impacts on the Africa rift lakes and the Aral Sea.

African Rift Lakes

Most of the rift lakes of east Africa occupy closed or nearly closed basins, including the three largest, best known lakes: Victoria, Malawi, and Tanganyika (Fig. 13.2). Biodiversity is extremely high in the larger rift lakes, as is endemism (Martens et al. 1994; Kaufman et al. 1996; Table 13.2), but there have been many species losses (Witte et al. 1995). According to Witte et al. (1995), there were more than 300 species of cichlids in Lake Victoria, 99% of which were endemic (but estimates differ; Table 13.2). A large proportion of these species have now been extirpated or persist at population levels too low to detect (Witte et al. 1995). I will review the drivers that have contributed to this dramatic loss of biodiversity, focusing on Lake Victoria because it is the most studied of the rift lakes, and has been affected most severely. Its watershed is the most heavily populated of the large lakes. It is also the shallowest of the large lakes, and thus least able to dilute incoming nutrients and pollutants (Hecky 1993).

Land Use

The overriding impact on the rift valley terrestrial and aquatic landscape will continue to come from the wholesale shift of land use toward agriculture as a result of human population increases (see Chap. 3). Human population growth rate in the Lake Victoria basin is currently 3–4% per year (about twice the global average), with most of the population engaged in subsistence agriculture and animal husbandry. Annual deforestation rates in the Lake Victoria watershed are currently about 1–2% (Kaufman et al. 1996). As a result of increased human land use over many decades, Lake Victoria began a marked eutrophication in the early decades of the twentieth century (Hecky 1993). Eutrophication caused a shift in phytoplankton species composition, and increases in phytoplanktonic production (Hecky 1993). As a result, anoxia has crept upward at least 10 m since 1970, so that the water overlying at least 35% of the lake's bottom is now often anoxic; in 1970, the entire water column was usually oxygenated (Hecky et al. 1994). Thus, the deepwater habitat for fishes and other organisms has been lost. Higher turbidity from siltation and phytoplankton has also reduced color discrimination by cichlids, reducing sexual selection, reducing reproductive isolation, and therefore reducing one of the evolutionary mechanisms that produces and maintains cichlid biodiversity in Lake Victoria (Seehausen et al. 1997). Thus, eutrophication has certainly contributed to the loss of biodiversity.

Overexploitation

In addition, overfishing has also reduced abundance of many species, and shifted species composition (Pitcher and Hart 1995; Kitchell et al. 1997). Constantly increasing numbers of people have employed traditional gear, increased use of mechanized gear (included trawlers), and shifted to smaller mesh sizes. Overfishing has clearly depressed populations of many endemic fishes in Lake Victoria and in other lakes (Cohen et al. 1996).

NIS

Introduced fishes, especially the large, predaceous Nile perch (*Lates nilotica*) and some tilapia, have driven several species of native fishes to extinction in several rift valley lakes (Pitcher and Hart 1995; Cohen et al. 1996). Nile perch were first introduced to Lake Victoria in the 1950s, but they did not become abundant and important until the 1980s, after eutrophication was already well underway (Hecky 1993). Water hyacinth has more recently overgrown much of the littoral zone of the lake, and will surely have large effects on biodiversity and ecosystem function (Kaufman et al. 1996).

Climate Change

East African terrestrial and freshwater ecosystems have been very responsive to past climate change, and thus very inconstant (Arnell et al. 1996; Livingstone 1996). Lake Victoria was dry about 12,500 years ago (Johnson et al. 1996). Future climate scenarios for the rift valley region of East Africa suggest temperatures will change little or increase, with most models showing 2–4°C increases (Cohen et al. 1996). Lake Victoria temperatures have increased by about 1°C since 1970, increasing thermal stability and decreasing mixing (Hecky 1993).

Scenarios are mixed for directions and magnitude for precipitation change, depending on which lake and which model (Cohen et al. 1996). For Lake Victoria, most models suggest decreases in precipitation (Cohen et al. 1996), which would produce a shift of natural vegetation to that typical of dryer climates (see Chapter 2). Ten to 20% declines in annual rainfall would drop lake levels and cause basin closure, even if temperatures remained unchanged.

Given scenarios of both decreasing rainfall and increasing temperature, the levels of these lakes are likely to decline (Arnell et al. 1996). Water level of Lake Victoria, however, increased in early 1960s after a few seasons of above-average rainfall, and has remained high (Arnell et al. 1996). Knowledge of the hydrology of these basins, which is critical for developing more detailed climate change scenarios, is poor.

Conclusions for African Rift Lakes

Overall, eutrophication has interacted with increasing lake temperature and predation by exotic fishes to reduce the biodiversity of endemic fishes in Lake

Victoria dramatically. The thermal and DO stratification pattern that has developed in Lake Victoria in the last decades leaves little safe habitat for endemic fishes. If they stay in deep waters, they are killed by anoxia; in shallow, well-oxygenated waters, they are concentrated with exotic, predatory Nile perch (Hecky 1993). Thus, the endemic flock of haplochrome cichlids in Lake Victoria has been reduced roughly by half (Witte et al. 1995). Anthropogenic drivers and biodiversity losses are likely to increase.

Aral Sea

The Aral Sea is one of several large closed-basin, oligotrophic, fresh-saline lakes in the arid regions of central Asia (Fig. 13.2; Mandych 1995). Other similar lakes include Lake Balkhash (Kazakhstan) (Petr 1992), and Lake Issyk-kul (Kyrgyzstan) (Savvaitova and Petr 1992). In general, even though these lakes are depauperate in species, they are high in endemicity (Table 13.2). Their ecology is very poorly known. Severe regional water management problems have substantially reduced inflows to most of these lakes (Petr and Mitrofanov 1994). Because of the aridity of the regional climate, the hydrology and biology of these lakes are very responsive to diversion or damming of inflowing streams. The greatest anthropogenic drivers for the Aral Sea all result from agriculture: water withdrawal for irrigation, salinization, pollution of water with pesticides, and eutrophication. Nonindigenous species have also been major drivers of biodiversity loss. Climate change in the recent past has been more of a response to the agriculture-related impacts than an independent driver.

Land Use

The Aral Sea once had the world's fourth largest lake surface area, but it is now disappearing (Smith 1995). Beginning in the 1960s, both major inflows to the lake were diverted, primarily for irrigation of cotton and rice. Between 1974 and 1986, one of the major rivers (Syr Darya) did not even reach the lake (Smith 1995). Now, 90% of the inflowing water is diverted (Golubev 1996). Between 1960 and 1993 the lake level dropped 31%, the lake split into two basins (around 1990), with a total decline in area of 50%, and a decline in lake volume of 72%. Over the same period, salinity of the lake increased from 10 g/L (when the lake was populated predominantly by freshwater species) to 37 g/L (more concentrated than seawater, and intolerable to freshwater species). By the year 2000, salinity was projected to be 65–70 g/L (Micklin 1992; Raskin et al. 1992). The small proportion of water that emerges from agriculture as surface water or groundwater contains dangerously high concentrations of nitrates and pesticides, including high concentrations of DDT (Smith 1995). Poor water quality, from both biological and chemical contamination, is causing many human health problems (Smith 1995).

NIS

Like most other central Asian lakes (Petr and Mitrofanov 1994), many species have been introduced into the Aral Sea. Approximately 19 fish species were native to the Aral Sea (Table 13.2). At least 17 fishes have been introduced (Glazovsky 1995). Many invertebrates have also been introduced, including mysids (Glazovsky 1995). In 1950, a commercial fishery, focused on both native and introduced species, directly or indirectly employed about 60,000 people. By the 1980s, the fishery was destroyed (Micklin 1992). Many, if not all, native and introduced species of fishes, invertebrates, and other taxa have since declined or been extirpated. Attributing blame for this wholesale loss of biodiversity is impossible, though, because of the enormous physical and chemical changes in the lake coincident with the introduction of NIS. It is likely that interactions between drivers were important as they were in the North American Great Lakes.

Climate Change

Some climate scenarios for the Aral Sea region suggest increasing temperatures, decreasing precipitation, and conversion of natural vegetation from grassland and steppe to desert (see Chap. 2). Others suggest, however, that precipitation and runoff might increase up to 47% (Smith 1994). Thus, it is unclear even what direction climate change will push water availability in the future. Land-use scenarios suggest no dramatic shift (see Chap. 3). Large changes in local climate have already occurred in the Aral Sea basin since 1970, largely as feedbacks from the reduction in lake size (Smith 1994). The climate has become more continental, with greater daily and seasonal (1–3°C) ranges of air temperature. The growing season has shortened by about 3 weeks (Glazovsky 1995). The number of days with substantial dust and salt storms has increased 50%, and desertification of the Aral Sea basin is well underway (Smith 1994; Glazovsky 1995). Thus, the conversion of land to irrigated cotton and rice agriculture ultimately produces an even dryer climate, as well as loss of terrestrial and freshwater biodiversity.

Conclusions for Aral Sea

Human impacts are literally obliterating the Aral Sea. Independent of any future climate change, the Aral Sea is drying up and is now very saline because of current water diversions. Although few biological data exist, freshwater biodiversity, including endemic species, has apparently already been largely exterminated. In turn, the disappearance of the lake has changed regional climate and started desertification of the surrounding terrestrial ecosystem.

Conclusions

Biodiversity of the world's lakes is already declining at a rapid rate. Impact of some anthropogenic drivers has been reduced in recent decades in some parts of the world (e.g., eutrophication in the North American Great Lakes). Most drivers are increasing, however, driven by increasing human population and increasing per capita resource consumption. Land use (and attendant eutrophication, irrigation, siltation, etc.), NIS, overexploitation of fisheries, and pollution by acidic deposition and other toxins all pose large and certain threats to the biodiversity and ecosystem function of the world's lakes (Table 13.3). Climate changes pose smaller and less certain, although still potentially very serious, threats (Table 13.3).

Impacts of all these drivers, including climate change, on lake ecosystem processes are already rigorously documented in many parts of the globe, providing a strong basis for predicting future responses. In addition, interactions among drivers, including positive feedbacks, are common, and will likely increase in severity with time (Fig. 13.3). Although it is more difficult to predict biodiversity responses than abiotic ecosystem responses (Fig. 13.1), it is clear that biodiversity in the world's lakes will continue to decline as a result of anthropogenic drivers. The ecosystem services provided by lakes to humankind will also be reduced, partly because of changes in biodiversity.

Acknowledgments. For references, reprints and preprints, email exchanges, and conversations that shaped many parts of this chapter, I thank Steve Carpenter, Andy Cohen, Alan Covich, Keely Lange, Bill Lewis, John Magnuson, Peter Moyle, and Roger Pielke. I also benefitted greatly from participation in the numerous email exchanges among the co-authors of Chapter 14 in this volume: LeRoy Poff, Paul Angermeier, Scott Cooper, Sam Lake, Kurt Fausch, Kirk Winemiller, Leal Mertes, Frank Rahel, Mark Oswood, and Jim Reynolds. Laura Eidietis and Molly McCracken brought order to the bibliography. Molly McCracken and Bill Perry made the figures and tables. The chapter was much improved by timely, thorough, and constructive reviews by Cindy Kolar, Bill Lewis, John Magnuson, Bill Perry, Daniel Schindler, Dave Schindler, and Yvonne Vadeboncoeur.

References

Anderson WL, Robertson DM, Magnuson JJ (1996) Evidence of recent warming and El Nino-related variations in ice breakup of Wisconsin lakes. Limnology and Oceanography 41:815–821.

Angermeier PL, Karr JR (1994) Biological integrity versus biological diversity as policy directives. BioScience 44:690–697.

Arnell N, Bates B, Lang H, Magnuson JJ, Mulholland P (1996) Hydrology and freshwater ecology. In: Watson RT, Zinyowera MC, Moss RH, Dokken DJ (eds) *Climate Change 1995. Impacts, Adaptations, and Mitigation of Climate Change: Scientific-Technical Analyses*, pp. 325–364. Cambridge University Press, Cambridge, UK.

Arnott DL, Vanni MJ (1996) Nitrogen and phosphorus recycling by the zebra mussel (Dreissena polymorpha) in the western basin of Lake Erie. Canadian Journal of Fisheries and Aquatic Sciences 53:646–659.

Assel RA, Robertson DM, Hoff MH, Selgeby JH (1995) Climatic-change implications from long-term (1823–1994) ice records for the Laurentian Great Lakes. Annals of Glaciology 21:383–386.

Band LE, Mackay DS, Creed IF, Semkin R, Jefferies D (1996) Ecosystem processes at the watershed scale: sensitivity to potential climate change. Limnology and Oceanography 41:928–938.

Bogan AE (1993) Freshwater bivalve extinctions (Mollusca: Unionoida): a search for causes. American Zoologist 33:599–609.

Bortnik VN, Kuksa VI, Tsytsarin AG (1992) Present status and possible future of the Aral Sea. Post-Soviet Geography 33:315–333.

Brett MT, Goldman CR (1997) Consumer versus resource control in freshwater pelagic food webs. Science 275:384–386.

Bruner KA, Fisher SW, Landrum PF (1994a) The role of the zebra mussel, Dreissena polymorpha, in contaminant cycling: I. The effect of body size and lipid content on the bioconcentration for PCBs and PAHs. Journal of Great Lakes Research 20:725–734.

Bruner KA, Fisher SW, Landrum PF (1994b) The role of the zebra mussel, Dreissena polymorpha, in contaminant cycling: II. Zebra mussel contaminant accumulation from algae and suspended particles, and transfer to the benthic inverebrate, Gammarus fasciatus. Journal of Great Lakes Research 20:735–750.

Burr BM, Mayden RL (1992) Phylogenetics and North American freshwater fishes. In: Mayden RL (ed) *Systematics, Historical Ecology, and North American Freshwater Fishes*, pp. 18–75. Stanford University Press, Dalo Alto.

Busiahn TR (1997) Ruffe control: a case study of an aquatic nuisance species control program. In: D'Itri FM (ed) *Zebra Mussels and Aquatic Nuisance Species*, pp. 69–86. Ann Arbor Press, Inc., Ann Arbor.

Callicott JB, Crowder LB, Murnford K (1999) Current normative concepts in conservation. Conservation Biology 13:22–35.

Carpenter SR, Fisher SG, Grimm NB, Kitchell JF (1992) Global change and freshwater ecosystems. Annual Review of Ecology and Systematics 23:119–139.

Carpenter SR, Frost TM, Kitchell JF, Kratz TK (1993) Species dynamics and global environmental change: a perspective from ecosystem experiments. In: Kareiva PM, Kingsolver JG, Huey RB (eds) *Biotic Interactions and Global Change*, pp. 267–279. Sinauer Associates, Inc., Sunderland.

Carpenter S, Frost T, Persson L, Power M, Soto D (1996) Freshwater ecosystems: linkages of complexity and processes. In: Mooney HA, Cushman JH, Medina E, Sala OE, Schulze E, (eds) *Scope 55: Functional Roles of Biodiversity: A Global Perspective*, pp. 299–325. John Wiley and Sons, Chinchester.

Carpenter SR, Kitchell JF (ed) (1993) *The Tropic Cascade in Lakes*. Cambridge University Press, Cambridge, UK.

Carpenter SR, Kitchell JF, Hodgson JR, Cochran PA, Elser JR, Elser MM, et al. (1987) Regulation of lake primary productivity by food web structure. Ecology 68:1863–1876.

Chapin FS III, Walker BH, Hobbs RJ, Hooper DU, Lawton JH, Sala OE, et al. (1997) Biotic control over the functioning of ecosysems. Science 277:500–504.

Chapin FS III, Sala OE, Bruke IC, Grime JP, Hooper DU, Lauenroth WK, et al. (1998) Ecosystem consequences of changing biodiveristy. BioScience 48:45–52.

Clair TA, Ehrman JM (1996) Variations in discharge and dissolved organic carbon and nitrogen export from terrestrial basins with changes in climate: a neural network approach. Limnology and Oceanography 41:921–927.

Cohen AS, Kaufman L, Ogutu-Ohwayo R (1996) Anthropogenic threats, impacts and conservation strategies in the African great lakes: a review. In: Johnson T, Odada E (eds) *Limnology, Climatology, and Paleoclimatology of the African Great Lakes*, pp. 575–624. Location: Publisher.

Cole JJ, Caraco NF, Kling GW, Kratz TK (1994) Carbon dioxide supersaturation in the surface waters of lakes. Science 265:1568.

Costanza R, d'Arge R, de Groot R, Farber S, Grasso M, Hannon B, et al. (1997) The value of the world's ecosystem services and natural capital. Nature 387:253–260.

Coulter GW (1994) Lake Tanganyika. Ergebnisse der Limnologie 44:13–18.

Crivelli AJ, Catsadorakis G, Malakou M, Rosecchi E (1997) Fish and fisheries in the Prespa Lakes. Hydrobiologia 351:107–125.

Cummings KS, Buchanan AC, Mayer CA, Naimo TJ (eds) (1997) Conservation and Management of Freshwater Mussels II. Initiatives for the Future. Proceedings of a UMRCC Symposium; October 16–18, 1995; St. Louis, MO. Available from: Upper Mississippi River Conservation Committee, Rock Island, IL.

Cushing (ed) (1997) *Special Issue: Freshwater Ecosystems and Climate Change.* Hydrolic Processes 11:817–1067.

Czech B, Krausman PR (1997) Distribution and causation of species endangerment in the United States. Science 277:1116.

Dejoux C (1994) Lake Titicaca. Ergebnisse der Limnologie 44:55–64.

DeStasio BT Jr., Hill DK, Kleinhans JM, Nibbelink NP, Magnuson JJ (1996) Potential effects of global climate change on small north-temperate lakes: physics, fish, and plankton. Limnology and Oceanography 41:1136–1149.

DiDonato GT, Lodge DM (1993) Species replacements among *Orconectes* crayfishes in northern Wisconsin lakes: the role of predatory fish. Canadian Journal of Fisheries and Aquatic Sciences 50:1484–1488.

Dill WA, Cordone AJ (1997) History and status of introduced fishes in California, 1871–1996: conclusions. Fisheries 22:15–18.

Dillon PJ, Molot LA (1997) Dissolved organic and inorganic carbon mass balances in central Ontario lakes. Biogeochemistry 36:29–42.

Dobson AP, Bradshaw AD, Baker AJM (1997b) Hopes for the future: restoration ecology and conservation biology. Science 277:515–522.

Dobson AP, Rodriguez JP, Roberts WM, Wilcove DS (1997a) Geographic distribution of endangered species in the United States. Science 275:550–553.

Eckstrom CK (1996) A wilderness of water: Pantanal. Audubon 98:52–65.

Ely LL, Enzel Y, Baker VR, Cayan DR (1993) A 5000-year record of extreme floods and climate change in the southwestern United States. Science 262:410–412.

Falkenmark M, Lindh G (1993) Water and economic development. In: Gleick PH (ed) Water in crisis: a guide to the world's fresh water resources, pp. 80–91. Oxford University Press, New York.

Fee EJ, Hecky RE, Kasian SEM, Cruikshank DR (1996) Effects of lake size, water clarity, and climatic variability on mixing depths in Canadian Shield lakes. Limnology and Oceanography 41:912–920.

Firth P, Fisher SG (eds) (1992) *Global Climate Change and Freshwater Ecosystems.* Springer-Verlag, New York.

Fletcher AR (1986) Effects of introduced fish in Australia. In: DeDecker P, Williams WD (eds) *Limnology in Australia*, pp. 231–238. Dr W. Junk Publishers, Boston.

Foin TC, Riley SPD, Pawley AL, Ayres DR, Carlsen TM, Hodum PJ, et al. (1998) Improving recovery planning for threatened and endangered species. BioScience 48:177–184.

Forsberg BR, Devol AH, Richey JE, Martinelli LA, dos Santos H (1988) Factors controlling nutrient concentrations in Amazon floodplain lakes. Limnology and Oceanography 33:41–56.

Frey DG (ed) (1963) *Limnology in North America.* University of Wisconsin Press, Madison, WI.

Frey DG (1969) A limnological reconnaissance of Lake Lanao. Verh Internat Verein Limnol 17:1090–1102.

Froelich AJ, Lodge DM (2000) Waterfowl-induced changes in submerged and emergent macrophyte communities at Lake Mattamuskeet, NC, USA. In: Comin FA, Herrera JA, Ramirez J (eds) *Limnology and Aquatic Birds*, pp. 43–60. Universidad Autonoma de Yucatan, Merida, Mexico.

Frost TM, Carpenter SR, Ives AR, Kratz TK (1995) Species composition and complementarity in ecosystem function. In: Jones CG, Lawton JH (eds) *Linking Species and Ecosystems*, pp. 224–239. Chapman and Hall, New York.

Fullerton AH, Lamberti GA, Lodge DM, Berg MB (1998) Prey preferences of Eurasian Ruffe and Yellow Perch: comparison of laboratory results with composition of Great Lakes benthos. Journal of Great Lakes Research 24:319–328.

Glazovsky NF (1995) Aral Sea. In: Mandych AF (ed) *Enclosed Seas and Large Lakes of Eastern Europe and Middle Asia*, pp. 119–154. SPB Academic Publishing, Amsterdam.

Gleick PH (1993a) Water in the 21st century. In: Gleick PH (ed) *Water in Crisis: A Guide to the World's Fresh Water Resources*, pp. 105–116. Oxford University Press, New York.

Gleick PH (1993b) An introduction to global fresh water issues. In: Gleick PH (ed) *Water in Crisis: A Guide to the World's Fresh Water Resources*, pp. 3–12. Oxford University Press, New York.

Golubev GN (1996) Caspian and Aral Seas: two different paths of environmental degradation. International Association of Theoretical and Applied Limnology 26:159–166.

Gophen M, Nishri A (1994) Lake Kinneret. Ergebnisse der Limnologie 44:43–54.

Greenwood PH (1994) Lake Victoria. Ergebnnisse der Limnologie 44:19–26.

Grime JP (1997) Biodiversity and ecosystem function. Science 277:1260–1261.

Hamilton SK, Lewis WM Jr. (1990) Physical characteristics of the fringing floodplain of the Orinoco River, Venezuela. Interciencia 15:491–499.

Hecky RE (1993) The eutrophication of Lake Victoria. Verh Internat Verein Limnol 25:39–48.

Hecky RE, Bugenyi FWB, Ochumba P, Talling JF, Mugidde R, Gophen M, et al. (1994) Deoxygenation of the deep water of Lake Victoria, East Africa. Limnology and Oceanography 39:1476–1481.

Herdendorf CE (1990) Distribution of the world's largest lakes. In: Tilzer MM, Serruya C (eds) *Large Lakes: Ecological Structure and Function*, pp. 3–38. Springer-Verlag, Berlin.

Hill AM, Lodge DM (1994) Diel changes in resource demand: interaction of competition and predation in species replacements by an exotic crayfish. Ecology 75: 2118–2126.

Hill AM, Lodge DM (1995) Multi-trophic-level impact of sublethal interactions between bass and omnivorous crayfish. Journal of the North American Benthological Society 14:306–314.

Hill AM, Lodge DM (1999) Evaluating competition and predation as mechanisms of crayfish species replacements. Ecological Applications. (In press).

Hillman TJ (1986) Billabongs. In: De Deckker P, Williams WD (eds) *Limnology in Australia*, pp. 457–470. Dr W. Junk Publishers, Dordrecht.

Hinch SG, Collins NC, Harvey HH (1991) Relative abundance of littoral zone fishes: biotic interactions, abiotic factors, and postglacial colonization. Ecology 72:1314–1324.

Hooper DU, Vitousek PM (1997) The effects of plant composition and diversity on ecosystem properties. Science 277:1302–1305.

Horvath TG, Lamberti GA, Lodge DM, Perry WL (1996) Zebra mussel dispersal in lake-stream systems: source-sink dynamics? Journal of the North American Benthological Society 15:564–575.

Hughes JB, Daily GC, Ehrlich PR (1997) Population diversity; its extent and extinction. Science 278:689–692.

Humphries CJ, Williams PH, Vane-Wright RI (1995) Measuring biodiversity value for conservation. Annual Review of Ecology and Systematics 26:93–111.

Hunter RD, Toczylowski SA, Janech MG (1997) Zebra mussels in a small river: impact on unionids. In: D'Itri FM (ed) *Zebra Mussels and Aquatic Nuisance Species*, pp. 161–186. Ann Arbor Press, Inc, Ann Arbor, MI.

Hurlbert SH (1997) Functional importance vs. keystoneness: reformulating some questions in theoretical biocenology. Australian Journal of Ecology 22:369–382.

James WF, Barko JW, Eakin HL (1997) Nutrient regeneration by the zebra mussel *(Dreissena polymorpha)*. Journal of Freshwater Ecology 12:209–216.

Jeppesen E, Sondergaard M, Sondergaard M, Christofferson K (eds) (1998) *The Structuring Role of Submerged Macrophytes in Lakes.* Springer-Verlag, New York.

Johnson TC, Scholz CA, Talbot MR, Kelts K, Ricketts RD, Ngobi G, et al. (1996) Late Pleisocence desiccation of Lake Victoria and rapid evolution of cichlid fishes. Science 273:1091–1093.

Junk WJ, Weber GE (1996) Amazonian floodplains: a limnological perspective. Verh Internat Verein Limnol 26:149–157.

Kaiser J (1997) Helping those most at risk. Science 278:217.

Karieva PM, Kingsolver JG, Huey RB (eds) (1993) Biotic interactions and global change. Sinauer Associates, Inc, Sunderland, MA.

Kaufman L, Chapman LJ, Chapman CA (1996) The Great Lakes. In: McClanahan TR, Young TP (ed) *East African Ecosystems and Their Conservation*, pp. 191–216. Oxford University Press, New York.

Kerr JB, McElroy CT (1993) Evidence for large upward trends of ultraviolet-B radiation linked to ozone depletion. Science 262:1032–1034.

Kershner MW, Lodge DM (1995) Effects of littoral habitat and fish predation on the distribution of an exotic crayfish, Orconectes rusticus. Journal of the North American Benthological Society 14:414–422.

Kirchhofer A, Hefti D (eds) (1996) *Conservation of Endangered Freshwater Fish in Europe.* Birkhauser Verlag, Basel.

Kitchell JF, Schindler DE, Oguto-Ohwayo R, Reinthall PN (1997) The Nile Perch in Lake Victoria: interactions between predation and fisheries. Ecological Applications 7:653–664.

Kling GW (1995) Land-Water Interactions: the influence of terrestrial diversity on aquatic ecosystems. Ecological Studies 113:297–310.

Kolar CS, Hudson PL, Savino JF (1997) Conditions for the return and simulation of the recovery of burrowing mayflies in western Lake Erie. Ecological Applications 7:665–676.

Kratz TK, Benson BJ, Blood ER, Cunningham GL, Dahlgren RA (1991) The influence of landscape position on temporal variability in four North American ecosystems. American Naturalist 138:355–378.

Laird KR, Fritz SC, Grimm EC, Mueller PG (1996) Century-scale paleoclimatic reconstruction from Moon Lake, a closed-basin lake in the northern Great Plains. Limnology and Oceanography 41:890–902.

Lewis WM Jr. (1979) Zooplankton community analysis. Springer-Verlag, Berlin.

Lewis WM Jr. (1996) Tropical lakes: how latitude makes a difference. In: Schiemer F, Boland KT (eds) *Perspectives in Tropical Limnology*, pp. 43–64. SPB Academic Publishing, Amsterdam.

Light T, Erman DC, Myrick C, Clarke J (1995) Decline of the shasta crayfish *(Pacifastacus fortis* Faxon) of northeastern California. Conservation Biology 9:1567–1577.

Likens GE (ed) (1985) *An Ecosystem Approach to Aquatic Ecology.* Springer-Verlag, New York.

Likens GE, Driscoll CT, Buso DC (1996) Long-term effects of acid rain: response and recovery of a forest ecosystem. Science 272:244–246.

Livingstone DA (1996) Historical ecology. In: McClanahan TR, Young TP (ed) *East African Ecosystems and Their Conservation*, pp. 3–17. Oxford University Press, New York.

Lodge DM (1991) Herbivory on freshwater macrophytes. Aquatic Botany 41:195–224.

Lodge DM (1993a) Species invasions and deletions: community effects and responses to climate and habitat change. In: Kareiva PM, Kinsolver JG, Huey RB (eds) *Biotic Interactions and Global Change*, pp. 367–387. Sinauer Associates Inc, Sunderland, MA.

Lodge DM (1993b) Biological invasions: lessons for ecology. Trends in Ecology and Evolution 8:133–137.

Lodge DM, Hill AM (1994) Factors governing species composition, population size, and productivity of cool-water crayfishes. Nordic J Freshw Res 69:111–136.

Lodge DM, Kratz TK, Capelli GM (1986) Long-term dynamics of three crayfish species in Trout Lake, Wisconsin. Canadian Journal of Fisheries and Aquatic Sciences 43:993–998.

Lodge DM, JW, Barko JW, Chair, Strayer D, Melack JM, Mittelbach GG, et al. (1988) Spatial heterogeneity and habitat interactions in lake communities. In: Carpenter SR (ed) *Complex Interactions in Lake Communities*, pp. 181–208. Spinger-Verlag, New York.

Lodge DM, Kershner MW, Aloi JE, Covich AP (1994) Effects of an omnivorous crayfish (*Orconectes rusticus*) on a freshwater littoral food web. Ecology 75:1265–1281.

Lodge DM, Stein RA, Brown KM, Covich AP, Bronmark A, Garvey JE, et al. (1998a) Predicting impact of freshwater exotic species on native biodiversity: challenges in spatial scaling. Australian Journal of Ecology 23:53–67.

Lodge DM, Cronin G, van Donk E, Froelich AJ (1998b) Impact of herbivory on plant standing crop: comparisons among biomes, between vascular and non-vascular plants, and among freshwater herbivore taxa. In: Jeppesen E, Sondergaard M, Sondergaard M, Christofferson K (eds) *The Structuring Role of Submerged Macrophytes in Lakes*, pp. 149–174. Springer-Verlag, New York.

Loomis JB, White DS (1996) Economic values of increasingly rare and endangered species. Fisheries Economics 21:6–10.

Ludwig HR Jr., Leitch JA (1996) Inter-basin transfer of aquatic biota via anglers' bait buckets. Fisheries 21:14–18.

Luttenton MR, Horgan M, Lodge DM (1998) Effects of crayfish on littoral algal communities. Crustaceana. (In press).

MacIsaac HJ (1996) Potential abiotic and biotic impacts of zebra mussels on the inland waters of North America. American Zoologist 36:287–299.

MacKaye KR, Ryan JD, Stauffer JR Jr., Perez LJL, Vega GI, van den Berghe EP (1995) African tilapia in Lake Nicaragua. Ecosystem in transition. BioScience 45:406–411.

Magnuson JJ (1976) Managing with exotics-a game of chance. Transactions of the American Fisheries Society 105:1–9.

Magnuson JJ (1991) Fish and fisheries ecology. Ecological Applications 1:13–26.

Magnuson JJ, Lathrop RC (1992) Historical changes in the fish community. In: Kitchell JF (ed) *Food Web Management: A Case Study of Lake Mendota*, pp. 193–227. Springer-Verlag, Berlin.

Magnuson JJ, Meisner JD, Hill DK (1990) Potential changes in the thermal habitat of Great Lakes fish after gloal climate warming. Transactions of the American Fisheries Society 119:254–262.

Magnuson JJ, Assel RA, Bowser CJ, Dillon PJ, Eaton JG, Evans HE, et al. (1997) Potential effects of climate change on aquatic systems: Laurentian Great Lakes and Precambrian shield region. Hydrologic Processes. (In press).

Magnuson JJ, Tonn WM, Banerjee A, Toivonen J. Isolation versus extinction in the assembly of fishes in small northern lakes. Ecological Monographs. (In press).

Mandrak NE (1989) Potential invasion of the Great Lakes by fish species associated with climatic warming. Journal of Great Lakes Research 15:306–316.

Mandych AF (ed) (1995) *Enclosed Seas and Large Lakes of Eastern Europe and Middle Asia*, pp. 33–70. SPB Academic Publishing, Amsterdam.

Marsh P, Lesack LFW (1996) The hydrologic regime of perched lakes in the Mackenzie Delta: potential responses to climate change. Limnology and Oceanography 41:849–856.

Martens K, Goddeeris B, Coulter G (eds) (1994) *Advances in Limnology: Speciation in Ancient Lakes*. Schweizerbart, Stuttgart.

Martin P (1994) Lake Baikal Ergebnisse der Limnologie 44:3–11.

Martin P (1996) Oligochaeta and Aphanoneura in ancient lakes: a review. Hydrobiologia 334:63–72.

Master L (1990) The imperiled status of North American aquatic animals. Biodiversity Network News 3:5–8.

Matson PA, Parton WJ, Power AG, Swift MJ (1997) Agricultural intensification and ecosystem properties. Science 277:504–509.

McCaffrey SC (1993) Water, politics, and international law. In: Gleick PH (ed) *Water in Crisis: A Guide to the World's Fresh Water Resources*, pp. 92–104. Oxford University Press, New York.

McCormick MJ (1990) Potential changes in thermal structure and cycle of Lake Michigan due to global warming. Transactions of the American Fisheries Society 119:183–194.

McDonald ME, Hershey AE, Miller MC (1996) Global warming impacts on lake trout in arctic lakes. Limnology and Oceanography 41:1102–1108.

McKnight D, Brakke DF, Mulholland PJ (eds) (1996) Freshwater ecosystems and climate change in North America. Limnology and Oceanography 41(5).

Melack JM (1992) Recriprocal interactions among lakes, large rivers, and climate. In: Firth P, Fisher SG (eds) *Global Climate Change and Freshwater Ecosystems*, pp. 68–87. Spring-Verlag, New York.

Melack JM (1996) Recent developments in tropical limnology. Verh Internat Verein Limnol 26:211–217.

Mellina E, Rasmussen JB, Mills EL (1995) Impact of zebra mussel (Dreissena polymorpha) on phosphorus cycling and chlorophyll in lakes. Canadian Journal of Fisheries and Aquatic Sciences 52:2553–2573.

Merrick JR (1995) Diversity, distribution and conservation of freswater crayfishes in the easter highlands of New South Wales. Proceedings of the Linnean Society of New South Wales 115:247–258.

Mertes LAK, Daniel DL, Melack JM, Nelson B, Martinelli LA, Forsberg BR (1995) Spatial patterns of hydrology, geomorphology, and vegetation on the floodplain of the Amazon River in Brazil from a remote sensing perspective. Geomorphology 13:215–232.

Meybeck M (1995) Global distribution of lakes. In: Lerman A, Imboden DM, Gat JR (eds) *Physics and Chemistry in Lakes*, pp. 1–36. Springer-Verlag, New York.

Michaelis FB (1986) Conservation of Australian aquatic fauna. In: DeDecker P, Williams WD (eds) *Limnology in Australia*, pp. 599–613. Dr W. Junk Publishers, Boston.

Micklin PP (1992) The Aral crisis: introduction to the special issue. Post-Soviet Geography 33:269–282.

Mills EL, Leach JH, Carlton JT, Secor CL (1993) Exotic species in the Great Lakes: a history of biotic crises and anthropogenic introductions. Journal of Great Lakes Research 19:1–54.

Mills EL, Leach JH, Carlton JT, Secor CL (1994) Exotic species and the integrity of the Great Lakes: lessons from the past. BioScience 44:666–676.

Milner AM, Irons JG III, Oswood MW (1997) The Alaskan landscape: an introduction for limnologists. In: Milner AM, Oswood MW (eds) *Freshwaters of Alaska: Ecological Syntheses*, pp. 1–44. Springer-Verlag, New York.

Minns CK, Moore JE (1995) Factors limiting the distributions of Ontario's freshwater fishes: the role of climate and other variables, and the potential impacts of climate change. In: Beamish RJ (ed) *Climate change and northern fish populations.* Canadian Special Publications in Fisheries and Aquatic Sciences 121:137–160.

Morton B (1997) The aquatic nuisance species problem: a global perspective and review. In: D'Itri FM (ed) *Zebra Mussels and Aquatic Nuisance Species*, pp. 1–54. Ann Arbor Press, Inc, Ann Arbor.

Mortsch LS, Quinn FH (1996) Climate change scenarios for Great Lakes Basin ecosystem studies. Limnology and Oceanography 41:903–911.

Moyle PB, Leidy RA (1992) Loss of biodiversity in aquatic ecosystems: evidence from fish faunas. In: Fiedler PL, Jain SK (eds) *Conservation Biology: The Theory and Practice of Nature Conservation, Preservation, and Management*, pp. 127–169. Chapman and Hall, New York.

Moyle PB, Yoshiyama RM (1994) Protection of aquatic biodiversity in California: a five-tiered approach. Fisheries 19:6–18.

Naeem S, Li S (1997) Biodiversity enhances ecosystem reliability. Nature 390:507–509.

Naiman RJ, Magnuson JJ, McKnight DM, Stanford JA (eds) (1995) *The Freshwater Imperative: A Research Agenda.* Island Press, Washington, DC.

Nakajima T (1994) Lake Biwa. Ergebnisse der Limnologie 44:43–54.

Nalepa TF, Schloesser DW (1993) *Zebra Mussels Biology Impacts, and Control.* Lewis Publishers, Boca Raton.

Nalepa TF, Fahnenstiel GL (1995) Dreissena polymorpha in the Saginaw Bay, Lake Huron ecosystem: Overview and perspective. Journal of Great Lakes Research 21:411–416.

Nalepa TF, Hartson DJ, Gostenik GW, Fanslow DL, Lang GA (1996) Changes in the freshwater mussel community of Lake St. Clair: from Unionidae to Dreissena polymorpha in eight years. Journal of Great Lakes Research 22:354–369.

Nash L (1993) Water quality and health. In: Gleick PH (ed) *Water in Crisis: A Guide to the World's Fresh Water Resources*, pp. 25–39. Oxford University Press, New York.

National Research Council (1992) *Restoration of Aquatic Ecosystems.* National Academy Press, Washington, DC.

National Research Council (1996) *Freshwater Ecosystems.* National Academy Press, Washington, D.C.

Nee S, May RM (1997) Extinction and the loss of evolutionary history. Science 278:692–694.

Nesler TP, Bergersen EP (1991) Mysids and their impacts on fisheries: an introduction to the 1988 mysid-fisheries symposium. American Fisheries Society Symposium 9:1–4.

Nielsen C (1995) Animal evolution: interrelationships of the living phyla. Oxford University Press, New York.

Olsen TM, Lodge DM, Capelli GM, Houlihan RJ (1991) Mechanisms of impact of three crayfish congeners (*Orconectes* spp.) on littoral benthos. Canadian Journal of Fisheries and Aquatic Sciences 48:1853–1861.

Palmer MA, Covich AP, Finlay BJ, Gilbert J, Hyde KD, Johnson RK, et al. (1997) Biodiversity and ecosystem processes in freshwater sediments. Ambio 26:571–577.

Perry WL, Lodge DM, Lamberti GA (1997) Impact of crayfish predation on exotic zebra mussels and native invertebrates in a lake-outlet stream. Canadian Journal of Fisheries and Aquatic Sciences 54:120–15.

Perry WL, Feder JL, Lodge DM (2001) Hybridization and introgression between introduced and resident *Orconectes* crayfishes in northern Wisconsin. Journal of Conservation Biology (in press).

Petr T (1992) Lake Balkhash, Kazakhstan. Int J Salt Lake Res 1:21–26.

Petr T, Mitrofanov VP (1994) Fisheries in arid countries of central Asia and in Kazakhstan under the impact of irrigated agriculture. FAO Fisheries Report No. 512 Supplement: 40–79.

Pimental D, Acquay H, Biltonen M, Rice P, Silva M, Nelson J, et al. (1992) Environmental costs and economic costs of pesticide use. BioScience 42:750–760.

Pimental D, Harvery C, Resosudarmo P, Sinclair K, Kurz D, McNair M, et al. (1995) Environmental and economic costs of soil erosion and conservation benefits. Science 267:1117–1123.

Pimentel D, Houser J, Preiss E, White O, Fang H, Mesnick L, et al. (1997a) Water resources: agriculture, the environment and society: and assessment of the status of water resources. BioScience 47:97–106.

Pimental D, Wilson C, McCullum C, Huang R, Dwen P, Flack J, et al. (1997b) Economic and environmental benefits of biodiversity. BioScience 47:747–768.

Pimm SL, Russell GJ, Gittleman JL, Brooks TM (1995) The future of biodiversity. Science 269:347–350.

Pitcher TJ, Hart PJB (eds) (1995) *The Impact of Species Changes in African Lakes.* Chapman and Hall, London.

Polis GA, Winemiller KO (eds) (1996) *Food Webs: Integration of Patterns and Dynamics.* Chapman and Hall, New York.

Pollard DA, Burchmore JJ (1986) A possible scenario for the future of Australia's freshwater fish fauna. In: DeDecker P, Williams WD (eds) *Limnology in Australia*, pp. 615–636. Dr W Junk Publishers, Boston.

Pollock MM, Naiman RJ, Erickson HE, Johnston CA, Pastor J, Pinay G (1995) Beavers as engineers; influence on biotic and abiotic characteristics of drainage basins. In: Jones CG, Lawton JH (eds) *Linking Species and Ecosystems*, pp. 117–126. Chapman and Hall, New York.

Postel S (1992) Last oasis: facing water scarcity. W.W. Norton and Company, New York.

Postel S, Carpenter S (1997) Freshwater ecosystem services. In: Daily GC (ed) Nature's Services: Societal dependence on natural ecosystems, pp. 195–214. Island Press, Washington, DC.

Postel SL, Daily GC, Ehrlich PR (1996) Human appropriation of renewable fresh water. Science 271:785–788.

Raskin P, Hansen R, Zhu Z, Stavisky D (1992) Simulation of water supply and demand in the Aral Sea region. Water International 17:55–67.

Reckahn JA (1995) A graphical paradigm for the sequential reduction and spectacular rehabilitation of the Lake Whitefish of Lake Huron. In: Munawar M, Edsall T, Leach J (eds) *The Lake Huron Ecosystem: Ecology Fisheries and Management*, pp. 171–190. SPB Academic Publishing, Amsterdam.

Ribbink AJ (1994) Lake Malawi. Ergebnisse der Limnologie 44:27–33.

Riera JL, Magnuson JJ, Kratz TK Webster KE. A geomorphic template for the analysis of lake districts applied to the Northern Highland Lake District, Wisconsin, USA. Freshwater Biology 43:301–318.

Roberts J, Chick A, Oswald L, Thompson P (1995) Effect of Carp, *Cyprinus carpio* L., an exotic benthivorous fish, on aquatic plants and water quality in experimental ponds. Marine and Freshwater Research 6:1171–1180.

Ruesink JL, Parker IM, Groom MJ, Karieva PM (1995) Reducing the risks of nonindigenous species introductions: guilty until proven innocent. BioScience 45:465–477.

Ryan WBF, Pitman WC III, Major CO, Shimkus K, Mosalenko V, Jones GA, et al. (1997) An abrupt drowning of the Black Sea shelf. Marine Geology 138:119–126.

Salemaa H (1994) Lake Ohrid. Ergebnisse der Limnologie 44:55–64.

Savvaitova K, Petr T (1992) Lake Issyk-kul, Kirgizia. International Journal of Salt Lake Research 1:21–46.

Schelske CL, Carpenter SR (1992) Lake Michigan. In: National Research Council. *Restoration of Aquatic Ecosystems*, pp. 380–392. National Academy Press, Washington, D.C.

Schindler DE, Carpenter SR, Cole JJ, Kitchell JF, Pace ML (1997) Influence of food web structure on carbon exchange between lakes and the atmosphere. Science 277:248–251.

Schindler DW (1988) Effects of acid rain on freshwater ecosystems. Science 239:149–157.

Schindler DW (1990) Natural and anthropogenically imposed limitations to biotic richness in fresh waters. In: Woodwell GM (ed) *The Earth in Transition: Patterns and Processes of Biotic Impoverishment*, pp. 425–462. Cambridge University Press, Cambridge, UK.

Schindler DW (1995) In: *Linking Species and Ecosystems.*

Schindler DW (1998) A dim future for boreal waters and landscapes. BioScience 48:157–164.

Schindler DW, Beaty KG, Fee EJ, Cruikshank DR, DeBruyn ER, Findlay DL, et al. (1990) Effects of climatic warming on lakes of the central boreal forest. Science 250:967–970.

Schindler DW, Bayley SE, Parker BR, Beaty KG, Cruikshank DR, Everett JF, et al. (1996) The effects of climatic warming on the properties of boreal lakes and streams at the Experimental Lakes Area, northwestern Ontario. Limnology and Oceanography 41:1004–1017.

Schindler DW, Curtis PJ, Parker BR, Stainton MP (1996a) Consequences of climate warming and lake acidification for UV-B penetration in North American boreal lakes. Nature 379:705–708.

Schindler DW, Curtis PJ, Bayley SE, Parker BR, Beaty KG, Stainton MP (1997) Climate-induced changes in the dissolved organic carbon budgets of boreal lakes. Biogeochemistry 36:9–28.

Schloesser DW, Kovalak WP, Longton D, Ohnesorg KL, Smithee RD. Impact of zebra and quagga mussels (*Dreissena* spp.) On freshwater unionids (Bivalvia: Unionidae) in the Detroit River of the Great Lakes. American Midland Naturalist 140:299–313.

Seehausen O, van Alphen JJM, Witte F (1997) Cichlid fish diversity threatened by eutrophication that curbs sexual selection. Science 277:1808–1811.

Shutter BJ, Post JR (1990) Climate population viability, and the zoogeography of temperate fishes. Transactions of the American Fisheries Society 119:314–336.

Sippel SJ, Hamilton SK, Melack JM (1992) Inundation area and morphometry of lakes on the Amazon River floodplain, Brazil. Archiv Für Hydrobiologie 123:385–400.

Smith DR (1994) Change and variability in climate and ecosystem decline in Aral Sea basin deltas. Post-Soviet Geography 35:142–165.

Smith DR (1995) Kazakhstan. In: Pryde PR (ed) *Environmental Resources and Constraints in the Former Soviet Republics*, pp. 251–274. Westview Press, Oxford, UK.

Snodgrass JW, Meffe GK (1998) Influence of beavers on stream fish assemblages: effects of pond age and watershed position. Ecology 79:928–942.

Spencer CN, McClelland BR, Standford JA (1991) Shrimp stocking, salmon collapse, and eagle displacement. BioScience 41:14–21.

Stefan HG, Hondzo M, Eaton JG, McCormick JH (1995) Predicted effects of global climate change on fishes in Minnesota lakes. In: Beamish RJ (ed) *Climate Change and Northern Fish Populations.* Canadian Special Publications in Fisheries and Aquatic Sciences 121:57–72.

Stefan HG, Hondzo M, Fang X, Eaton JG, McCormick JH (1996) Simulated long-term temperature and dissolved oxygen characteristics of lakes in the north-central United States and associated fish habitat limits. Limnology and Oceanography 41:1124–1135.

Taub FB (ed) (1984) *Ecosystems of the World: Lake and Reservoirs.* Elsevier, Amsterdam.

Taylor CA, Warren ML Jr., Fitzpatrick JF Jr., Hobbs HH III, Jezerinac RF, Pflieger WL, et al. (1996) Conservation status of crayfishes of the United States and Canada. Fisheries 21:25–38.

Tilman D, Wedin D, Knops J (1996) Productivity and sustainability influenced by biodiversity in grassland ecosystems. Nature 379:718–720.

Tilman D, Knops J, Wedin D, Reich P, Ritchie M, Siemann E (1997) The influence of functional diversity and composition on ecosystem processes. Science 277:1300–1302.

The Times Atlas of the World (1980) Comprehensive Edition, 6th ed. Times Books Limited, London.

Tonn WM, Magnuson JJ, Rask M, Toivonen J (1990) Intercontinental comparison of small-lake fish assemblages: the balance between local and regional processes. American Naturalist 136:345–375.

Turner MG, Gardner RH, O'Neill RV (1995) Ecological dynamics at broad scales: Ecosystems and landscapes. BioScience Supplement: S-29–35.

Tyler PA (1996) Endemism in freshwater algae. Hydrobiologia 336:127–135.

U.S. Congress (1993) Harmful non-indigenous species in the United States. Office of Technology Assessment, OTA-F565. U.S. Government Printing Office, Washington, D.C. September.

Van der Beck I, Thornton JA, Griesinger B (1996) Some thoughts on the Pantanal—the world's largest wetland. Lakeline Nov 1996:20–21, 39–42.

Vinebrook RD, Leavitt PR (1996) Effects of ultraviolet radiation on periphyton in an alpine lake. Limnology and Oceanography 41:1035–1040.

Vitousek PM (1994) Beyond global warming: ecology and global change. Ecology 75:1861–1876.

Vitousek PM, D'Antonio CM, Loope LL, Westbrooks R (1996) Biological invasions as global environmental change. American Scientist 84:468–477.

Vitousek PM, Mooney HA, Lubchenco J, Melillo JM (1997) Human domination of earth's ecosystems. Science 277:494–500.

Wallace JB, Eggert SL, Meyer JL, Webster JR (1997) Multiple trophic levels of a forest stream linked to terrestrial litter inputs. Science 277:102–104.

Wardle DA, Zackrisson O, Hornberg G, Gallet C (1997) The influence of island area on ecosystem properties. Science 277:1296–1299.

Warren ML Jr., Burr BM (1994) Status of freshwater fishes of the United States: overview of an imperiled fauna. Fisheries 19:6–18.

Weber LM, Lodge DM (1990) Periphyton food and crayfish predators: relative roles in determining snail distributions. Oecologia 82:33–39.

Wetzel RG (1983) *Limnology,* second ed. Sanders College Publishing, Philadelphia.

Wheeler QD (1995) Systematics and biodiversity. BioScience Supplement: S-21–28.

Williams JD, Warren ML Jr., Cummings KS, Harris JL, Neves RJ (1993) Conservation status of freshwater mussels of the United States and Canada. Fisheries 18:6–22.

Williams WD (1993) Conservation of salt lakes. Hydrobiologia 267:291–306.

Williamson CE, Sternberger RS, Morris DP, Frost TM, Paulsen SG (1996) Ultraviolet radiation in North American lakes: attenuation estimates from DOC measurements and implications for plankton communities. Limnology and Oceanography 41:1024–1034.

Williamson M (1996) Biological invasions. Chapman and Hall, New York.

Wilson EO (ed) (1988) *Biodiversity.* National Academy Press, Washington, D.C.

Witte ZF, Goldschmidt T, Wanink JH (1995) Dynamics of the haplochromine cichlid fauna and other ecological changes in the Mwanza Gulf of Lake Victoria. In: Pitcher TJ, Hart PJB (eds) *The Impact of Species Changes in African Lakes*, pp. 83–110. Chapman and Hall, London.

Yan ND, Keller W, Scully NM, Lean DRS, Dillon PJ (1996) Increased UV-B penetration in a lake owing to drought-induced acidification. Nature 381:141–143.

14. Fish Diversity in Streams and Rivers

N. LeRoy Poff, Paul L. Angermeier, Scott D. Cooper,
P.S. Lake, Kurt D. Fausch, Kirk O. Winemiller,
Leal A.K. Mertes, Mark W. Oswood,
James Reynolds, and Frank J. Rahel

Introduction

Species Diversity in Fluvial Ecosystems

Flowing-water (fluvial) ecosystems are valuable to human societies for many reasons, including aesthetics, recreation, food production, water supply, and waste disposal. Their value derives, to some degree, from their biological diversity, including genetic, species, and community diversity (Angermeier and Schlosser 1995). Of the many components of biodiversity, species diversity is the best documented and provides the focus for this chapter.

Species diversity varies dramatically among regions and among localities within regions. Fluvial ecosystems often support high regional diversity because they are geologically persistent and encompass a wide variety of habitat types, ranging from those that are permanently to intermittently wet, and including transitional habitats at aquatic-terrestrial ecotones. Further, natural environmental variation, which occurs over a range of spatiotemporal scales, maintains this regional habitat diversity (Poff et al. 1997; Cooper et al. 1998). Because global changes in climate and land use are often expressed at regional scales, we will focus primarily on responses of regional species diversity to these changes.

Most of the wide variety of organisms in fluvial ecosystems will respond to global change. Among the common taxa, fishes are probably the best known in terms of distribution and ecological sensitivity. Although we will focus on fishes in our discussion of biotic responses to climate and land-use changes, we emphasize that changes in fish species diversity are a small part of community and ecosystem responses to environmental perturbations.

The most important factors that regulate species diversity from local to regional and continental scales include history, temperature, hydrology, and geomorphology. Local diversity is strongly affected by regional diversity (Moyle and Herbold 1987; Hugueny and Paugy 1995; Angermeier and Winston 1998), which, in turn, reflects phylogenetic history (Tonn et al. 1990; Brooks and McLennan 1993). Because speciation typically requires geographic isolation for long periods, diversity is higher in regions that consistently have provided appropriate habitat through previous cooling and warming periods. For example, high diversity in the southern Appalachian region of North America and the Guyana Shield region of South America reflect the occurrence of persistent refugia during glacial advances or associated dry periods. Some of the regional variation in the diversity of riverine fishes can be explained by the accessibility of thermal or hydrological refugia. For example, fish diversity is greater in temperate North America than it is in Europe in part because north–south dispersal corridors were available to fishes in the Mississippi River basin, but not to fishes in European basins (Briggs 1986).

Thermal regimes have strong effects on species diversity in fluvial ecosystems. Relatively few species tolerate low temperature extremes coupled with a wide annual range of temperatures, resulting in the well-known inverse relationship between diversity and latitude or elevation (Winemiller 1991; Oswood et al. 1995). As thermal regimes shift with latitude and altitude in response to climate change, so will biotic distributions. Because dispersal through fluvial systems is often constrained by natural or artificial barriers (e.g., waterfalls, rapids, dams), accessibility to suitable thermal habitats during global warming may be limited. Thus, climate change generally will reduce regional species diversity over the short term.

The hydrological regime (i.e., the spatial and temporal distribution of runoff and groundwater) regulates the availability of suitable habitat and influences species diversity. For example, rivers and streams that experience very low (or no) flows on an annual (or nearly annual) basis typically support fewer species than perennial rivers and streams in the same biogeographic region (Cross and Moss 1987). At regional scales fish diversity is strongly positively related to amounts of precipitation (McAllister et al. 1986) and, perhaps, inversely related to wide variability in flow conditions that create habitat bottlenecks (Lake, personal observation). Geographic shifts in hydrological regimes caused by climate change will induce shifts in the distribution of fluvial species, most likely resulting in reductions of the regional diversity of species adapted to local hydrological conditions.

The geologic setting of fluvial ecosystems interacts with the hydrological regime to determine the geomorphology of the valley and channel which, in turn, controls the range of habitats available to the biota (see Dunne and Leopold 1978). Rivers flowing through unconsolidated materials in unconstrained valleys create and maintain alluvial floodplains, which provide a variety of productive wetland habitats that enhance the diversity of fluvial species. In general, alluvial habitat, channel size, and discharge increase in a downstream direction, resulting in increasing fish diversity as stream size increases (Sheldon 1968; Lake 1982; Welcomme and de Merona 1988).

In combination, temperature, hydrology, geomorphology, and associated riparian vegetation form a habitat "template" (sensu Southwood 1977) that controls the persistence and diversity of species at local and regional scales (Poff and Ward 1990). Maximum regional diversity is further regulated by historical constraints, including previous climatic bottlenecks and barriers to dispersal. Global changes in climate and land use will modify these regional templates, thereby altering ecosystem processes and patterns of species diversity (Grimm 1992; Poff 1992; Meyer et al. 1999; Lake et al. 2000).

The climatic, geologic, and vegetative features used to delineate biomes on a global scale clearly influence the structure and function of fluvial ecosystems that flow through them (see Minshall 1988). Biome classification, however, may not adequately reflect variation in aquatic biodiversity because streams and rivers are largely linear systems that span more than one biome and provide corridors for species movement (if barriers are absent) across biome boundaries. Environmental factors that directly regulate species diversity can also vary independently of biomes. For example, river hydrological regime, considered to be a "master variable" that limits the distribution and abundance of fluvial species (Power et al. 1995), shows considerable variation within biomes, such that streams in different biomes may be hydrologically more similar than streams within the same biome (e.g., see Poff 1996) due to natural interstream variation in gradient, groundwater inputs, vegetative cover, and so forth. Thus, a narrow focus on the biome scale as the unit of fluvial diversity response to global change is not adequate. We will organize our discussion here along two major template axes: a temperature gradient (tropical to temperate to polar) and a precipitation gradient (wet to dry regions). We will use selected biomes as examples along these axes where appropriate.

Humans Modify the Factors that Regulate Species Diversity in Rivers

Rivers and their floodplains represent areas of concentrated human economic and cultural activities, which pose significant threats to riverine diversity on local, regional, and global scales. Agricultural and urban development, pollution from industrial and domestic wastes, and dam construction have dramatically altered natural templates in rivers throughout the world, with documented reductions in diversity (Nehlsen et al. 1991; Allan and Flecker 1993). The natural connectivity of river corridors has been interrupted, preventing migration and dispersal of many species and blocking access to regional refugia when local conditions deteriorate (Benke 1990). Further, the widespread, intentional introduction of nonnative species into sites outside their historical biogeographic ranges has disrupted relationships among native riverine species and threatens native species diversity in many fluvial ecosystems (e.g., Moyle 1986; Minckley and Deacon 1991). Projected global changes in climate and land use over the next century pose significant new threats to natural ecosystems that are already under stress (see Firth and Fisher 1992).

Hydrologic–Geomorphic Responses to Global Changes in Climate

River flow varies seasonally, laterally, and longitudinally in river corridors. The amount and timing of flow is determined by the hydroclimatology of the region and by the character of the catchment and its local water sources. The meteorologic processes that produce flood events can vary in spatial scale from mesoscale thunderstorms (10^2 km^2) to large tropical and extratropical cyclones (10^4–10^6 km^2) (Hirschboeck 1988). Further, the frequency, timing, and magnitude of these processes vary across biomes (e.g., Hayden 1988). Thus, in a particular basin, precipitation patterns defined by regional hydroclimatology and local meteorology will interact with catchment characteristics (e.g., underlying lithology and vegetative cover) to control the spatial and temporal variability in runoff, as well as patterns of floodplain inundation, although the latter will also reflect local geomorphology (Mertes 1997).

In order to examine the global patterns of flooding and associated biomes, Figure 14.1 (see color insert) was constructed from a flood hydroclimatological map (after Hayden 1988) and the BIOME 3 map. According to Hayden, the atmosphere can generally be divided into two types: baroclinic and barotropic. A baroclinic atmospheric condition is generally characterized by the intersection of different gradients of temperature and pressure. In contrast, baroclinic conditions have gradients of temperature and pressure that are nearly parallel and with the absence of vertical wind shear, convective clouds can grow to immense vertical heights. For example, the Ob'–Irtysh Rivers (CSs**) of Siberia are subjected to perennially baroclinic conditions characterized by moderate rainfall due to an inadequate supply of atmospheric moisture (Hayden 1988). Flooding occurs during snowmelt in the spring. By contrast, the predominantly summer-barotropic Missouri–Mississippi system (TsuCpSe*) is characterized by a potential for flooding throughout the year due to its diverse sources of precipitation, including summer fronts and frontal cyclones, occasional tropical storms, and spring snowmelt. The Amazon River (Tpz) is perennially barotropic with the movement of the intertropical convergence zone (ITCZ) from north to south controlling the seasonal patterns of often intense rainfall. The ITCZ is the convergence zone of equatorial winds. California (TsoCs) rivers experience barotropic

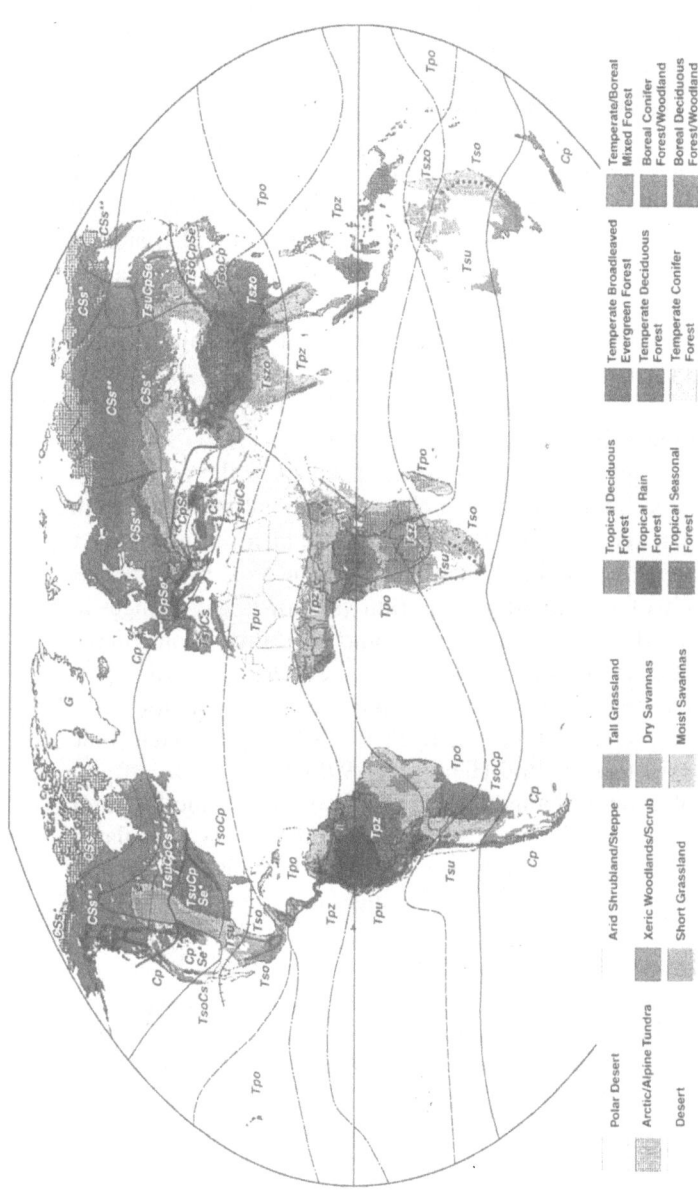

Figure 14.1. Hydroclimatology of floods (after Hayden, 1988, Fig. 9, p. 23) combined with BIOME 3 map. After Hayden, the symbols for the flood zones include: T, barotropic; C, barocline; p, perennial; s, seasonal; z, Inter-Tropical Convergence Zone (ITCZ); o, organized convection; u, unorganized convection; S, snow cover; G, glacial; s, seasonal snow cover; e, ephemeral snow cover; *, snow cover 10 to 50 days; **, snow cover 50 cm or more. The solid and dashed lines are the poleward limit of barotropic conditions in summer and winter, respectively. The average position of the ITCZ is shown by the dash–dot line for January and July. The cross-hatched line indicates the equatorward limit of frontal cyclones for North America. The double-dot–dash line indicates regions with the highest snowfall or duration of snow cover. The thick, solid line indicates the equatorward limit of snow cover for 50 days or more. Mountain regions are shown in solid gray shading. (Cartography kindly provided by D. Lawson.)

conditions in the summer with occasional tropical storms; however, flooding is typically restricted to the winter as the result of cyclones from the northern Pacific.

Several studies have documented how past climate change altered stream and river hydrology. For example, higher frequencies of episodic superfloods in the southwestern United States coincided with historical global temperature changes (Ely et al. 1993). Knox (1985, 1993) showed that a decline in regional precipitation induced by climatic changes during the Holocene affected flood frequency patterns for streams in the Upper Mississippi Valley and that these changes had a dramatic impact on channel morphology (and thus potentially physical habitat characteristics for the biota). Other studies have documented pronounced impacts of El Niño/Southern Oscillation (ENSO) events on the flooding patterns of the Amazon River (Richey et al. 1989), as well as on floods and droughts in rivers of Australia, India–Pakistan, and Africa (Whetton et al. 1990). Burn and Arnell (1993) reported several distinct historical periods showing increased flooding associated with ENSO events across many hydroclimatological zones.

It is therefore important to consider possible changes in the magnitude, duration, frequency, and timing of flows in channels and onto floodplains in response to climate change. The hydrological response of a particular river reflects modifications in the dominant hydroclimatologic regime, alterations in the dominant mode of precipitation (e.g., rain vs. snow), and changes in factors that affect runoff (e.g., vegetation and land use). Predictions of the effects of projected climate changes on global river runoff and flooding patterns are unfortunately too few to allow regional predictions of how river templates will likely respond to climate change (Arnell 1996), although some regional exceptions exist (Fowler and Hennessey 1995).

Although effects of climate change on seasonal patterns in runoff cannot be generally examined in detail, studies of climate change have simulated impacts on average runoff rates for particular river systems. Changes in average precipitation may also be related to changes in extreme flows, and a change in precipitation variation may have a marked effect on the temporal patterns of discharge (Arnell 1996). For example, the Ob'–Irtysh Basin may experience a modest increase in discharge in its lower river reaches (Neilson and Marks 1994). If this effect is combined with increased rain (vs. snow) due to higher temperatures and an earlier spring (e.g., Myneni et al. 1997), then peak runoff may be reduced and may occur earlier in the year. In the Missouri–Mississippi system, runoff is not predicted to change significantly with climate change; however, because snowmelt contributes importantly to early spring flooding, elevated spring temperatures could reduce flood peaks and increase the duration of spring flooding, or cause an earlier snowmelt and reduce late summer baseflows. In the Amazon Basin, regional runoff is predicted to increase by approximately 10%, potentially increasing the levels and duration of flood peaks (Neilson and Marks 1994). Precipitation is predicted to decrease in California, thus exacerbating current semi-arid conditions. As a result, longer dry periods could result in the drying of normally perennial streams.

Variation in ecological responses to changes in discharge and thermal regimes within large hydroclimatological regions can be expected, due to sub-biome scale features that regulate biological responses. For example, local geological conditions that regulate channel slope, floodplain development, and groundwater storage can greatly affect the availability of local refugia from such environmental extremes as floods and droughts. Many species have limited distributions within biomes or hydroclimatological regions; therefore, it is important to consider how rivers and streams within larger regions may respond to global climate and land-use changes. All these potential hydrological modifications can be exacerbated by human activities that alter the hydrologic cycle.

Fish Species Responses to Global Climate and Land Use Changes

Aquatic species, when faced with changing environmental conditions, may respond in one of three ways. First, they may adapt to the new conditions. Assuming the gene pool contains adequate variability, adaptive changes must occur at a rate sufficient to keep pace with environmental change. Global climate change and land-use conversions are occurring at a rate comparable to the greatest rates documented for previous periods of natural climatic change, when mass extinctions occurred. It is therefore highly unlikely that most fluvial species will be able to adapt quickly to such rapidly changing conditions. The second response would be shifts in the present distributions of species. This kind of response was probably very important in earlier periods of climate change (e.g., glacial and interglacial periods in the Pleistocene) (e.g., Briggs 1986). Many present patterns of species diversity and distribution can be understood in light of the historical availability of refugia from harsh conditions imposed by a changing climate. If species are able to redistribute themselves along river corridors as global change occurs, then extensive extinction could be averted. Thus, the regional response of fish species to global climate and land-use change will largely depend on the availability and accessibility of suitable *refugia* from changing environmental conditions. Third, if climate and land use changes produce environmental conditions beyond the tolerance limits of extant species, and refugia are not available or accessible, then extinctions are very likely.

Tropical Systems

Wet Tropical Regions

Tropical freshwaters are a major reservoir of global fish diversity. In South American fresh waters alone, more than 2400 fish species occur (Lowe-

McConnell 1987), and a medium-sized river in South America may contain more species (i.e., >200) than the entire European continent.

In South America three general classes of tropical rivers have been recognized (Sioli 1984): black, clear, and white waters. Black-water rivers are acidic (pH 3.5–6.5) due to dissolved organic matter (mostly tannic compounds) leached from terrestrial vegetation. Clear-water rivers, which originate in high-gradient regions that contain granitic bedrock (e.g., Guyana highlands, Brazilian Plateau), carry few suspended sediments, have low concentrations of dissolved organic compounds, and tend to have circumneutral pHs. White-water rivers carry high sediment loads and are relatively rich in nutrients, and these rivers support rich growths of aquatic macrophytes and algae, especially in floodplain habitats. Major faunal differences occur among black, clear, and white waters in the same geographical region (Goulding et al. 1988; Rodriguez and Lewis 1994). Species may either be restricted to one habitat type due to narrow physiological limits, or they may be generalists that occur in all three habitats. Some species appear to be excluded from physiologically suitable habitats due to biological interactions (e.g., competition, predation) (Goulding et al. 1988).

Tropical fishes can be grouped into functional ecological groups on the basis of life history strategies, trophic specialization, and habitat affinities at various spatial scales (e.g., black vs. white water, upland vs. lowland, and forest vs. savanna). Life history strategies and degree of specialization will influence population responses to environmental disturbances caused by global climate and land-use changes. Among fishes, a continuum of life-history strategies are associated with patterns of environmental variation in critical resources and mortality agents (Winemiller 1992; Winemiller and Rose 1992; Van Winkle et al. 1993). Three distinct strategies can be defined (Winemiller and Rose 1992): equilibrium species (with delayed maturation, brood guarding or hiding, large eggs, and small clutches), opportunistic species (with rapid maturation, short lifespan, frequent spawning, small eggs, and small clutches), and periodic species (with delayed maturation, long lifespan, short spawning season, small eggs, and large clutches). Many commercially-important fish species are periodic.

Under a global warming scenario, widespread opportunistic fishes adapted to high juvenile mortality rates and should be able to withstand frequent, major environmental changes compared with equilibrium and periodic-type fishes, of which the latter often dominate fish communities in tropical rivers (Winemiller 1989, 1992). Gradual irreversible changes to aquatic habitats on a large spatial scale will be detrimental even for resilient opportunistic species. For example, annual killifishes inhabit ephemeral pools, where they coexist with few or no other fish species. If, in response to global warming, the hydrology of these habitats becomes more stable, then annual killifishes will be eliminated by superior competitors or predators. Fishes with equilibrium-type life histories should have the lowest resiliency to disturbances; thus, they should be vulnerable to habitat alterations.

Recruitment in periodic-type fishes depends on large-scale fish movement associated with large-scale variation in environmental conditions, so that climatic or land-use changes (e.g., dams) that obliterate linkages among landscape elements exploited by migratory species could result in their demise. Changes in seasonal cycles of precipitation in tropical regions will alter the predictable dynamics of river flow, flooding, and desiccation on which many periodic fishes depend. For example, the majority of fluvial species depend on seasonal changes in water quality and flooding to provide physiological cues for reproduction in seasonal floodplain habitats.

Tropical fish assemblages show much more trophic specialization compared with temperate assemblages in similar habitats (Winemiller 1991). If other factors are equal, then trophic specialists (e.g., fin nippers, blood parasites, wood-eaters) should be affected more than generalists by rapid changes to the quality and quantity of aquatic habitats; however, additional factors (e.g., life history strategies and the degree of habitat generalization) interact with trophic ecology to affect the population sizes of these species.

The relatively small thermal changes predicted for the tropics should not directly affect tropical freshwater fish diversity. Many fishes in the wet tropics can tolerate temperatures of 37°C or higher for prolonged periods. High-elevation species with narrower tolerance limits may respond to increased temperatures by moving upstream, provided there are no barriers to movement. A few lowland fishes restricted to stream habitats in unbroken tracts of rainforest have lower thermal limits than habitat generalists and savanna species, and would thus be affected to a greater degree by a shift from rainforest to savanna. Perhaps a greater issue than slight increases in temperature is the threat of aquatic hypoxia during periods of reduced instream flow and the recession of seasonal floodwaters. Many tropical fishes possess respiratory adaptations that allow survival in waters essentially devoid of dissolved oxygen (Kramer et al. 1978); however, many do not, and those can suffer high mortality from hypoxia during the dry season. Climate changes that eliminate dry season refugia (or access routes to them) for species intolerant of hypoxia would result in the extirpation of numerous local populations throughout the tropics.

At present, it is not possible to make quantitative estimates of changes in fish diversity in tropical freshwaters in response to changes in climate and land use. Direct impacts of changes in thermal regimes will be minimal, but shifts in biomes and land-use changes will affect species densities and distributions. At present, we predict that changes in land use associated with agriculture will be the major factor driving declines in diversity in fluvial habitats of wet tropical regions. For example, changes in land use will lead to increased erosion and turbidity, which, in turn, will have large and immediate effects on species compostion in clear and black-water rivers because most black-water species are absent from turbid waters. Dramatic drops in local diversity (from 65 to 4 species) have been documented in a small stream in northern Amazonas over an 8-year period when cattle ranching became established in the catchment and stream waters became turbid (Winemiller, personal observation). Projected regional-scale conversions of tropical woodland to agriculture, extensive drainage of wetlands, and other human development impacts (e.g., nonpoint source pollution, dams, and water extraction schemes) clearly pose a severe and immediate threat to the survival of species and ecosystems in the Tropics, which is where such a large fraction of the earth's freshwater fish diversity resides.

Dry Tropical Regions

Savannas typically dominate landscapes in semi-arid tropical climates. We will use the riverine fish communities of African savannas as an example of global change threats to tropical fishes in semi-arid regions.

Faunal assemblages in the rivers of African savannas are dominated by widespread and opportunistic taxa with generalized food habits, as well as high growth and colonization rates. Fish faunas are dominated by cyprinids with important contributions by the Cichlidae, Characidae, Mormyridae, and various catfish families (Greenwood 1958; Lowe-McConnell 1975; Skelton 1988). Most of these fish are generalized invertivores or omnivores. Many riverine fish species undergo upstream spawning migrations during rising water periods, and their young often grow and develop in upstream or floodplain habitats (van Someren 1962; Welcomme 1985). Riverine environments also support diverse amphibian, reptile, bird, and mammal species (Cooper 1996).

With the exception of rivers draining lakes and reservoirs (e.g., the Victorian Nile), discharge in most rivers and streams varies tremendously (by an order of magnitude) throughout the year, reflecting seasonal changes in rainfall (Cooper 1996). Many arid and semi-arid areas have intermittent or ephemeral streams and rivers, and even moderately large rivers may dry up during droughts. Temperatures at elevations less than 1500m are usually greater than 20°C (Vanden Bossche et al. 1990), and many rivers in arid or semi-arid areas are turbid, largely because of the removal of native vegetation by human activities (Dunne 1979).

The IMAGE projections indicate that average temperature changes will be small over the next century in tropical Africa (less than 1°C). Other climatologists, however, have projected temperature increases of up to 2–4°C for this region in response to a doubling of atmospheric carbon dioxide (Mkanda 1996; Unganai 1996). Because temperature conditions determine sex ratios in the developing eggs of some reptiles (e.g., crocodilians, turtles), even small temperature changes may affect the sex ratios and population sizes of reptile populations (Bull 1983). Crocodiles are apex predators in some savanna rivers; therefore, these temperature effects may have repercussions for the rest of the community. Because many of East Africa's rivers flow west and east, fish species cannot migrate latitudinally in response to temperature changes. Many African rivers, however, have their headwaters in high mountains and show large elevational changes as they flow to the ocean. As a consequence, many species can shift their longitudinal (elevational) distributions to match temperature conditions, as long as there are no barriers to movement.

Average precipitation along the equator in Africa is projected to remain the same; however, both drier and wetter conditions are projected to occur away from the equator (IMAGE projections). Parts of the Sahel, for example, are projected to be much wetter, whereas parts of the horn of Africa will be much drier. Stream flows and habitat will expand or contract accordingly. Although many aquatic invertebrates and fish in savanna rivers have adaptations for dealing with high temperatures, low dissolved oxygen, and desiccation, even these capacities may be exceeded if the extent and duration of desiccation is extensive (Cooper 1996).

In the monsoonal savanna systems of East Africa, seasonal and interannual variation in precipitation is likely to increase. As a consequence, streams and rivers may be subjected to an increased frequency and intensity of hydrological disturbance (i.e., droughts, floods). Species richness and biomass of fish assemblages are often directly proportional to discharge levels and/or the extent of floodplain (where present) inundation (Welcomme 1976, 1979, 1985; Welcomme and de Merona 1988). As a consequence, increased periods of desiccation would probably reduce the species diversity of fish assemblages, and possibly eliminate species that cannot withstand desiccation (Welcomme 1985). In addition, increases in the intensity and frequency of flooding may decrease fish populations in headwater areas, but increase the population sizes of fish using floodplain habitats in downstream areas (van Someren 1952; Welcomme 1985).

These projected changes in climatic conditions, however, are completely dwarfed by predicted changes in land-use patterns that ultimately result from human population growth. Human population growth rates in Africa are among the highest in the world, placing increasing demands and stresses on natural resources. African savannas are projected to be either eliminated or reduced to 16% of their present extent in the next 100 years (IMAGE projections A1 or A2). This level of habitat alteration implies the wholesale drainage of wetlands and floodplains, with the attendant loss of species that use these habitats, including resident fish species that live on floodplains and migratory fish species that use floodplains as spawning and nursery habitat. Because cropland development in semi-arid regions implies the attendant development of water resources, it is probable that levels of dam construction, groundwater pumping, and water diversions will also increase, resulting in further losses of species as habitat, resources, and migratory corridors are lost. In short, the extent of projected land-use changes and associated water development forecasts severe, widespread reductions in native species diversity in the rivers of the African savanna.

Tropical Islands

Faunal assemblages in streams on tropical islands are low in richness and dominated by insects, decapods, snails, and gobioid fishes, with insects reaching their highest diversity in headwater areas, and snails, decapods, and fishes becoming increasingly dominant toward river mouths (Resh et al. 1990, 1992). Most of the non-insects are derived from marine ancestors and possess an amphidromous life cycle (i.e., they spend their larval lives in the ocean, but ascend streams to develop, mature, and breed) (Kinzie 1988; Radtke et al. 1988; Fitzsimons et al. 1990; Bell and Brown 1995; Radtke and Kinzie 1996).

Perennial streams on tropical and subtropical islands are usually warm (more than 20°C), short, steep, and low in discharge (Maciolek and Ford 1987; Resh and De Szalay 1995). On the windward sides of high islands rainfall is often high and variable and riparian vegetation is lush. More arid insular areas tend to be dominated by savanna, grassland, and scrubland, and the native vegetation of many islands has been replaced by coconut groves and agricultural land (e.g., sugar cane fields). Because these islands lie in the cyclone belt, frequent flooding occurs (Maciolek and Ford 1987).

Human populations on many tropical islands have grown at a rapid pace, increasing human pressures on stream environments. Water storage dams have been built on many streams, resulting in blocked migration routes for amphidromous species (Stone 1989; Allan and Flecker 1993). In addition, pollution, destruction, diversions, and sedimentation of lower stream reaches have created barriers to the migration of these species (e.g., atyid shrimps, gobies), which can no longer return to upper stream reaches (Resh et al. 1990, 1992; Kenny 1995; Resh and De Szalay 1995; Radtke and Kinzie 1996).

Exotic species have been widely introduced to tropical insular waters (Erdman 1984; Maciolek 1984). Human-altered environments (e.g., artificial reservoirs) have favored many exotics (Erdman 1984; Maciolek 1984) and harmed native species, which are adapted to shaded and fast-flowing stream systems.

The characteristics of habitats and organisms in streams on tropical islands make them particularly vulnerable to climate and land-use changes. Many insular streams are low in discharge and short in length; thus, they support only small animal populations that are vulnerable to local extinction (Stone 1989; Resh and De Szalay 1995). Sea level rises associated with climate change will cause the inundation of lower stream reaches by seawater and further shorten streams. High endemism in streams on some isolated islands makes entire species susceptible to extinction. Most aquatic species occur near their upper thermal and lower oxygen limits, so even small increases in temperature can threaten populations and species (Myers 1992).

Although temperature increases in the Tropics over the next century will be small on average (0–1°C according to IMAGE projections), they may show considerable geographic variability. Since 1941, for example, mean annual temperatures in the insular Caribbean have increased twice as fast as the global average, and they are projected to increase 1–2.3°C over the next 50 years (Hanson and Maul 1989; Granger 1991). Some scientists have documented longitudinal changes in insular faunal assemblages that they attributed, at least partly, to temperature changes (Hynes 1971; Harrison and

Rankin 1976). The availability and accessibility of cool headwater refugia may be critical for those species already living near the upper limit of their thermal tolerances.

Rainfall along the equator is predicted to increase; however, changes in precipitation away from the equator will be quite variable, depending on local topographic and atmospheric conditions (Granger 1991; Pittock et al. 1996; Whetton et al. 1996). It also appears that global warming will enhance monsoonal systems, increase seasonal and interannual variability in rainfall, and increase the magnitude and frequency of storms (Peters 1992). This increased variability, along with land-use changes, will likely exacerbate the frequency and duration of hydrological disturbances, including desiccation and floods. For example, stream drying is likely to increase in response to water development projects (e.g., diversions for drinking water), which have their greatest impacts in downstream areas (Erdman 1984; Stone 1989). The desiccation of stream systems will both destroy habitat and erect migratory barriers for some species; however, even infrequent flows may be sufficient to permit colonization and the maintenance of some upstream populations (Radtke and Kinzie 1996). By contrast, a combination of land-use changes (e.g., deforestation) and increased intensity and frequency of storms will likely also increase the magnitude and frequency of floods. These effects, combined with the likely expansion of harmful exotic species into degraded habitats, will most likely result in the local extinction of many native insular species.

Temperate Systems

Wet Temperate Regions

Temperate deciduous forest is the predominant biome in the eastern United States, western Europe, and northern China. Due to extensive agricultural development, natural vegetation cover is extremely fragmented and reduced in area. The largest remnants of this biome occur in the United States.

Projected climate change will not appreciably affect the global distribution of deciduous forest, but it will exacerbate fragmentation, especially in China. Temperatures in the North American deciduous forest are projected to increase the most (2–4 °C, for a doubling of CO_2). Projected changes in annual precipitation are highly variable, ranging from a 5–10-cm increase in North America to a 0–30-cm decrease in China.

We will focus on the North American temperate deciduous forest because the biota is well known, this region supports the richest temperate freshwater fish fauna in the world (532 species; Page and Burr 1991), and this region will experience relatively large changes in temperature. We expect the region's high species diversity to be especially sensitive to climate change and continuing human land-use impacts.

Many previous studies indicate that some fish species are more vulnerable than others to environmental changes induced by human uses of land and water. For example, a comparison of the historical and current distributions of Virginia fishes showed that species with small geographic ranges or narrow habitat or food requirements were especially prone to extinction (Angermeier 1995). These and other species traits are reasonably well known for many fish in the North American Temperate Deciduous Forest, and they can be used to assess the proportion of the fish fauna vulnerable to projected changes in climate and land or water use.

We identified five traits that will probably affect the sensitivity of fish species to climate change: geographic range, age at first reproduction, number of stream sizes inhabited, number of food types eaten, and flow requirements. Geographic range, which is inversely related to extinction risk (Gaston and Blackburn 1996), was estimated for all 532 fish species in the biome using Page and Burr (1991). The total area represented by the range map for each species was determined relative to a standard areal unit of ca. $250,000 \, km^2$. The latter four traits were examined for species native to Wisconsin or Virginia, which represent northern and southern regions of the biome, respectively. These four traits were assigned to species using Becker (1983) for Wisconsin fishes and Jenkins and Burkhead (1994) for Virginia fishes. Age at first reproduction (which is correlated with body size) is presumably directly related to extinction risk, based on previous observations that long-lived species are the first species to become extinct in chronically stressed ecosystems (Rapport et al. 1985). Large fish are generally more mobile, and they may be more effective colonizers. Specialization on a particular waterbody size or food predisposes species to extinction, especially in variable environments. We recognized three waterbody sizes (small, medium, large) and four general food types (detritus, vegetation, invertebrates, fish). Species that require flowing water are more vulnerable to extinction than those that can live in either flowing or standing water.

Many Temperate Deciduous Forest fishes are vulnerable to climate change due to their small geographic ranges. For example, 47% of the 532 species occurring in the biome have ranges of less than $100,000 \, km^2$, and 19% have ranges of less than $20,000 \, km^2$ (an area ca. 140 km on a side). Range sizes vary substantially among families. Over one third of the 143 darter species have ranges of less than $20,000 \, km^2$, whereas most suckers and sunfishes (35 and 28 total species, respectively) have ranges greater than $600,000 \, km^2$. Species with small geographic ranges are also likely to have limited dispersal abilities.

Most fishes have ecological traits that make them vulnerable to climate change. For example, 74 and 65% of the fish species in Wisconsin and Virginia, respectively, eat only one type of food (Table 14.1). The sensitivity of the fish fauna to environmental change, however, varies among regions. The proportion of fishes that require flowing water is much greater in Virginia than in Wisconsin (68% versus 27%), indicating that Virginia fishes have more climate-sensitive traits than Wisconsin fishes (medians = 2 vs. 1; Table 14.2). Sensitivity also varies among families: Many minnows and sunfishes lack

Table 14.1. Proportion of native fish species in Wisconsin (WI) and Virginia (VA) that use a single waterbody size (NWS) or food type (NFT), require flowing water (FLO), or take more than 2 years to mature (AGE)

	NWS	NFT	FLO	AGE
WI	0.19	0.74	0.27	0.33
VA	0.30	0.65	0.68	0.22

Proportions are based on 146 species in Wisconsin and 197 species in Virginia.

sensitive traits, whereas lampreys and darters often have three or four. Given that the center of fish diversity in North America is in southern Appalachia, the sensitivity of the fauna to climate change for the entire biome is better represented by Virginia fishes than by Wisconsin fishes.

Sensitivity to environmental change and range size are interrelated. For the 197 native Virginia fishes, cumulative ecological sensitivity (indicated by the number of individual traits tallied in Table 14.2) is inversely related to range size (Kendall's tau = -0.38; $p < 0.0001$). In other words, ecologically sensitive fishes also tend to have small ranges. These patterns in sensitivity to environmental change are more obvious in some families than others. For example, the range of ecological sensitivity for Virginia minnows (63 species) is similar to those for all Virginia species combined, but Virginia darters (42 species) are uniformly small, ecologically specialized, and narrowly distributed; thus, they are uniformly vulnerable to environmental change.

Overall losses of fish species from the Temperate Deciduous Forest due to climate and land-use changes cannot be predicted precisely. A substantial proportion of the fish species (25–50%) appear to be acutely sensitive to projected environmental effects of climate and land use changes, based on their range sizes and ecological traits. Some groups (e.g., darters) are especially sensitive, whereas others (e.g., sunfishes) are probably relatively tolerant of such effects. Climate change itself is unlikely to cause complete elimination of many species from the Temperate Deciduous Forest; however, many

Table 14.2. Proportions of native fish species in Wisconsin (WI) and Virginia (VA) that have 0, 1, 2, 3, or 4 ecological traits likely to result in sensitivity to global warming

	0	1	2	3	4
WI	0.12	0.43	0.34	0.09	0.02
VA	0.09	0.25	0.42	0.23	0.02

The four traits analyzed are given in Table 14.1. Proportions are based on 146 species in Wisconsin and 197 species in Virginia.

species have already been reduced due to other human impacts (e.g., habitat loss, degraded water quality and introduced species) (Williams et al. 1989), and climate change may push additional populations to extinction. Even if species are not completely extirpated from the biome, dramatic declines in the distribution and abundance of sensitive species will result in reduced genetic variation, reduced variability among communities, and a loss of biotic diversity at many, specific localities.

Dry Temperate Regions

Almost all that is known of the aquatic biota of grassland/steppe streams comes from the Great Plains of North America, with little or no data on streams in the plains of Europe, Asia, or Africa. In North America, the current diversity of fishes is characterized by a progressive westward decline in the Mississippi River basin, due largely to limits imposed by harsh ecological conditions rather than drainage divides or other factors that limited postglacial dispersal (Cross et al. 1986). For example, the Great Plains physiographic region supports only 77 native fish species, only 33% of the 235 in the western Mississippi River basin, and 8% of the 950 in North America. Moreover, few are endemic species because the fauna is largely a subset of the most tolerant forms of the Central Lowland ecoregion to the east (Cross et al. 1986).

Physicochemical conditions in plains streams are typically harsh, but they are predictable. For example, temperatures typically vary from 0°C to 39°C in streams of the southwestern Great Plains (Matthews and Zimmerman 1990; Labbe and Fausch, unpublished data) and flow regimes may be both intermittent and flashy (cf. Poff and Ward 1989), but these conditions are usually seasonally predictable (Fausch and Bestgen 1997). Great Plains rivers with headwaters in the Rocky Mountains have a predictable early summer discharge peaks from melting snow. In contrast, tributaries originating on the plains flood from spring through fall in response to thunderstorms that vary markedly from year to year and across the landscape (Fausch and Bramblett 1991). Both types of systems were originally intermittent in late summer and fall, had primarily shifting sand substrate, and generally had only small amounts of patchily distributed riparian vegetation (Cross and Moss 1987; Fausch and Bestgen 1997).

Native aquatic biota are well-adapted to this harsh but predictable environment (Matthews 1988). Thermal tolerances of plains stream fishes typically extend as high as 36–40°C for short periods (i.e., critical thermal maxima; Matthews and Zimmerman 1990; Smale and Rabeni 1995), and somewhat lower (ca. 30–34°C) for extended periods (i.e., upper incipient lethal levels; see Smith and Fausch 1997). Fish also have life history adaptations to large natural flow fluctuations. For example, predictable early summer floods from melting snow in southwestern Great Plains rivers selected for a guild of fishes that lay buoyant eggs that develop rapidly while drifting downstream (Moore 1944; Taylor and Miller 1990; Fausch and

Bestgen 1997), an optimal strategy for avoiding the grinding action of shifting sand substrate. Other species attach eggs to the few hard substrates available, and guard and clean their eggs to ameliorate the effects of silt. Most plains stream fishes are small-bodied, mature early, have prolonged breeding, spawn repeatedly, and are highly vagile, all of which adaptations are to counteract the fluctuating physicochemical conditions and intermittency of plains streams (Fausch and Bestgen 1997). In addition to these, there is a suite of sensitive, often endemic fishes that occur only in unique habitats (e.g., spring streams and coolwater streams in the foothills of adjacent mountains). These species typically make up substantial proportions of regional species diversity (e.g., 27% of the species in the two major plains basins of eastern Colorado), but many have already been locally extirpated or severely reduced in range and abundance due to changes in land-use and water management (Fausch and Bestgen 1997).

Arid plains are often irrigated to support intensive agriculture, so in most regions of the world this ecosystem type has already undergone extensive land-use changes (Rabeni 1996). Systems of diversion dams and canals for irrigation fragment lotic systems by hampering the downstream dispersal of fish eggs and larvae, as well as the recolonization of upstream or rewatered reaches by juveniles and adults. Because most plains fishes are small, even these small barriers to upstream dispersal are effective in extirpating species from entire catchments (Winston et al. 1991; Fausch and Bestgen 1997).

Changes in flow regime due to irrigation have both direct and indirect effects on plains stream fishes. Very large hourly or daily fluctuations in flow due to irrigation activities are common (Poff et al. 1997; Strange et al. MS), and likely have direct effects on spawning and recruitment (cf. Schlosser 1985), although this has been largely unstudied. Changes in the seasonal timing of flow (e.g., increased summer baseflow from irrigation return flows) have indirect effects by fostering the growth of riparian vegetation, which subsequently invades the former floodplain and changes braided systems to single narrow channels (Johnson 1994; Scott et al. 1996, 1997), thereby favoing species adapted to such habitat. In contrast, increases in turbidity from inorganic silt due to channel modification or land-use change may have little effect on species tolerant to naturally silty conditions, although reduced turbidity and temperature downstream from reservoirs would likely reduce or extirpate many plains stream species (Bestgen and Platania 1990, 1991). As in other basins, however, added organic matter degrades aquatic systems, primarily via biochemical oxygen demand.

Some predictions can be made about the effects of projected land-use and climate changes on the biodiversity of plains fishes. Matthews and Zimmerman (1990) predicted that temperature increases of 2–4°C would drive many species to extinction because most are living near their upper incipient lethal limits. Even if they survived, fish would likely be concentrated in the few refugia where groundwater upwells into streams, thereby reducing and fragmenting local populations and decreasing the probability of regional persistence. Effects of reduced streamflow and increased temperatures would likely be strongest for sensitive species that occupy unique habitats such as springs or coolwater streams. In contrast, other plains fishes would be most affected by strong increases in flow variation or marked changes in the seasonality of floods or droughts, via effects on reproductive success and the survival of larval fish. Reduced flow may also reduce water quality in reaches receiving waste effluent.

Perhaps the most pervasive effects, however, would be increased fragmentation of habitats through further irrigation diversions or increased intermittency because these would reduce the ability of fish populations to withstand harsh periods in refugium habitats and later recolonize rewatered reaches. On a regional and global scale, the effects of a warming climate would be most severe in basins where drainages tend east to west, rather than in north–south drainages that allow colonization of newly created habitats at higher latitudes. This process would have effects similar to Pleistocene glaciation, when a colder climate extirpated fish in east–west drainages (e.g., the intermountain basins of the western United States and eastern Europe), but not north–south drainages (e.g., the Mississippi Basin) (Briggs 1986).

Invasions of nonnative fish species to plains stream systems have been rare, except in reservoirs and stream reaches immediately downstream where temperatures have become cooler, flows more stable, and water less turbid (Cross and Moss 1987). Thus, if more reservoirs are built to store water as the climate warms and dries in certain regions, native species could also be reduced or extirpated, either because habitat becomes unsuitable, nonnative species exclude them, or both. Relatively few large migratory fish species inhabit plains streams, but these would be strongly affected by dispersal barriers (e.g., dams).

Projections from the IMAGE scenario A1, which incorporate the largest population increases and energy use, indicate that although the area of grassland/steppe ecosystem will decline only 12% worldwide from 1990 to 2100, much larger increases and decreases of more than 50% will occur in 7 of 11 grassland/steppe regions. This suggests that distributions of aquatic biota will need to expand and contract rapidly to adapt to the changing landscape. As climate and land use change in plains ecosystems, therefore, it will be important to maintain the dispersal routes and the integrity of unique aquatic habitats if we want to maintain the diversity of fish biota. This likely holds true for other regions, and other aquatic biota as well, because of similar life history adaptations to harsh environments.

Cold Regions

Temperate Montane

The biota of montane streams is characterized by low diversity and dominance by species requiring cold water (Ward 1994). Only a few families of fishes are represented in temperate montane streams globally. The lower elevational limits of coldwater fishes are closely associated with a critical maximum water temperature (Fausch et al. 1994; Rahel et al. 1996). These systems generally experience a large pulse of meltwater in the spring, and discharge typically drops to low levels during fall and winter (Poff and Ward 1989). Extended periods of ice cover often limit available habitat for fish (Swanston 1991), and the development of frazil or anchor ice increases winter mortality (Reiser and Wesche 1975; Brown et al. 1993).

Climate change is likely to alter both the thermal and hydrological regimes of montane streams. Montane streams near the tops of very high mountains may exhibit decreases in stream temperature due to increased snowpacks in winter and decreased solar radiation in summer (Williams et al. 1996). Very few aquatic species, however, currently reside in these harsh environments.

Of greater concern are streams at middle to lower elevations within the montane zone, including those streams that do not drain very high mountains. Here, both increased water temperature and altered flow regimes have been predicted. Because of their requirements for cold water, montane species are vulnerable to increased maximum temperatures. In North America, summer water temperatures in coldwater streams have been projected to increase by up to 2°C (Eaton and Scheller 1996).

Increases in water temperature are expected to reduce habitat availability and further restrict montane fishes to even higher elevations. In the United States, climate-change studies predict habitat loss for trout in the Rocky Mountains (Keleher and Rahel 1996; Rahel et al. 1996), for various cold-water species in the upper American Midwest (Eaton and Scheller 1996), and for brook trout in the Appalachian Mountains (Meisner 1990). In northern Europe, climate change is predicted to cause a contraction in the geographic ranges of 11 fish species and expansion in the geographic ranges of 16 fish species (Lehtonen 1996). In the Japanese archipelago, the geographic ranges of coldwater species are predicted to shrink substantially (Nakano et al. 1996). In New Zealand, nonnative brown and rainbow trout are predicted to be eliminated from warmer, northern latitudes as mean annual temperatures increase by 3°C (Scott and Poynter 1991).

Species diversity may actually increase in montane streams that undergo warming as fishes from lower elevations find thermal conditions more favorable (Baltz et al. 1982; Lehtonen 1996). Some species of trout and sculpins may be displaced by warm-water species that are introduced or invade upstream sections that become warmer (Larson and Moore 1985; Williams et al. 1989; De Staso and Rahel 1994; Thompson and Rahel 1996). Several subspecies of native trout are also subject to displacement by nonnative salmonids that can tolerate warmer temperatures (e.g., Behnke 1992; Thompson and Rahel 1996). Further, the very steep gradients typifying the uppermost reaches of montane streams can inhibit the dispersal of native coldwater species (Fausch 1989; Kruse et al. 1997), preventing an upstream retreat from nonnative competitors.

Little is known about how alterations in stream discharge might affect montane fish populations. A common projected scenario for flow regimes in response to climate change is increased discharge in winter and spring (due to increased winter precipitation) and reduced discharge in summer and autumn (due to increased evapotranspiration at warmer temperatures) (Houghton et al. 1996; Lehtonen 1996). In some cases, the timing of peak flows may shift from spring runoff to rain-on-snow precipitation events in winter (Lettenmaier and Gan 1990). Reduced streamflow during late summer and autumn would have a negative impact on fish populations because low-flow habitat volume often is an important determinant of fish abundance and species richness (Chapman 1966; Oberdorff et al. 1995). Changes in the timing of runoff events (e.g., from spring to winter) could result in changes in community composition by favoring fishes with different spawning seasons (Schlosser 1985). For example, in a California stream, native spring-spawning species are favored in years with large winter floods, whereas nonnative fall-spawning species are favored in years when high flows do not occur until the spring (Strange et al. 1992). Increased streamflow and associated increases in water velocities during winter and spring could affect anadromous fishes that rely on accumulated energy reserves to reach upstream spawning areas (Hinch et al. 1995).

The loss of coldwater fishes (e.g., trout) from their current lower elevation limits should be one of the first indications of climate change effects in streams. Although local diversity may increase as warmwater species colonize upstream areas, global diversity could be reduced by the loss of species that have been extirpated from most of their historic range by introduced species and which currently persist in high-elevation, coldwater refugia. Although the number of montane fish species that would be affected by climate change is not large, many of the salmon and trout are highly valued as food and sport fish; thus, changes in their distribution will be important to the general public.

High Latitudes

Climate changes in northern high latitude regions are likely to result in a marked physical and biological restructuring of ecosystems because modifications in temperature and precipitation regimes are projected to be substantial. Alaska is a useful example of high latitude, circumboreal ecosystems because of its physiographic diversity (from coastal rain forest to the taiga forest of interior Alaska to the arctic tundra of the North Slope).

Of the fishes of Alaska (Morrow 1980), 45 species have life cycles with a substantial freshwater (running water) component. Running waters are critical to northern fishes as pathways for seasonal migrations as water conditions change. Nearly half of these species (22) are salmonids (i.e., salmon, char, whitefish), which are known for their anadromy, lengthy migrations, and complex life histories. Most of the remaining species are marine-derived, coastal species (e.g., sculpins, smelts, and sturgeons), occupying the lower reaches of rivers at certain times.

The distribution of Alaskan freshwater fishes varies by climatic and hydrological region, with a maximum diversity in southwest Alaska. Decreasing temperatures with increasing latitude likely have diverse effects on fish distributions in Alaska. Cold water temperatures limit the growth and production of both fish and their prey species (Oswood 1997). Streams in very cold areas (i.e., arctic tundra; north-facing watersheds in the zone of intermittent permafrost of interior Alaska) may have little or no winter flow and surface ice extending into the substrate, limiting the fish fauna to those forms capable of seasonal movements to overwintering habitats (Reynolds 1997). The present freshwater fish fauna of Alaska is derived from species that survived Pleistocene glaciation in the Beringian refugium and from species that recolonized from southern glacial refugia in Pacific Coast and Mississippi drainages following glaciation (McPhail and Lindsey 1970).

Observed and projected air temperature increases in Alaska are substantial. Over the period 1960–1990, air temperatures increased 0.4°C per decade in the Arctic and 0.6°C in the interior of Alaska. Model predictions and extrapolations of current trends ($2\times CO_2$; 100 years from now) show 2–3°C increases in summer and 7–11°C increases in winter in the arctic and the interior of Alaska (the greatest temperature increase for any region) as well as higher annual precipitation (Weller et al. 1996).

The potential consequences of climate change for the biota of Alaskan running waters include the obvious increases in water temperature and predictable losses of permafrost, with subsequent changes in hydrothermal conditions and carbon dynamics in streams (Irons and Oswood 1997; MacLean et al. 1999). Climate change may also generate subtle shifts in terrestrial–aquatic linkages (e.g., changes in herbivory of riparian vegetation, with consequent effects on the quality and quantity of leaf litter fueling stream food webs) (Oswood et al. 1992). Although it is not possible at present to model the complex effects of climate change on Alaskan running water ecosystems quantitatively, a qualitative approach can consider some of the historical and current constraints on the lotic fish fauna of Alaska and the ways that climate change might influence these limitations.

Hydrothermal regimes in Alaskan streams and rivers can be categorized by their water sources. For example, glacially fed streams have a hydrothermal regime characterized by peak flows and cold temperatures in midsummer (low or no flow in winter), whereas spring-fed systems have near-constant year-round temperatures and serve as biological oases and winter refugia (Craig 1989; Milner et al. 1997; Reynolds 1997). Winter conditions are likely an important "filter" (Tonn 1990), influencing the current distribution and abundance of species. Juvenile fishes are probably the life stage most vulnerable to cold temperatures. An increase in winter minima will increase thermally suitable habitat for species currently excluded from areas by harsh physiological conditions that limit egg and juvenile survival (e.g., rainbow trout, *Oncorhynchus mykiss*, present in southern but not northern Alaska).

Such species could invade thermally suitable habitats provided there are no barriers to colonization.

It is likely that average air temperatures in northern Alaska are only slightly colder than the lower limits for some temperate coolwater species [e.g., rainbow trout and smallmouth bass (*Micropterus dolomieui*)]. Projections of climate warming for interior Alaska indicate that these lower limits will be reached within 100 years or less. As a consequence, Alaska will be less of a refugium for fish species specialized for extremely cold conditions. New, coolwater species will likely reach Alaska, either by long-distance migrations from marine sources or through introductions from temperate regions of North America. Present-day distributions of freshwater fishes in Alaska have been compared with redistributions expected to occur under a 4–8°C warming scenario. Western and northern coastal areas could host several Alaskan species not presently there, especially species with migratory marine life stages such as sturgeon, lamprey, and salmonids, with the latter group including brook trout (*Salvelinus fontinalis*), cutthroat trout (*O. clarki*), and rainbow trout.

Low temperatures at high latitudes limit population densities of humans (as well as other organisms), in turn limiting the ecological footprint of humans in the vast area of Alaska. Most of the landscape effects of humans that so plague the running waters of other regions (e.g., urbanization, agriculture, livestock grazing, mining, impoundments) are of very limited and localized importance in Alaska. Ecological constraints on economic development at high latitudes seem likely to continue to limit these impacts over the near future. Exploitation of natural resources (e.g., oil, timber, fisheries) as well as tourism fuels much of the Alaskan economy. Expansion of oil development has potential impacts on migratory waterfowl and mammals (both terrestrial and marine), but seems unlikely to substantially impact fresh waters. In contrast, timber harvest has had substantial effects on the streams and anadromous fishes of southeast Alaska (Murphy et al. 1997). There are plans to increase timber harvest in the boreal forest of interior Alaska. The extent to which the lessons learned from southeast Alaska (e.g., the value of riparian buffer strips, the central role of woody debris in streams, and the importance of logging roads in supply of sediments to streams) can be applied to the permafrost-riddled landscapes and low biotic productivity of the taiga is unclear.

Quantitative Examples

Although the magnitude of population and species losses cannot currently be quantified at the scale of a biome or climatic region, current relationships between diversity and habitat area can be used to estimate species loss under given scenarios of climate or land-use change that alter habitat area. We use

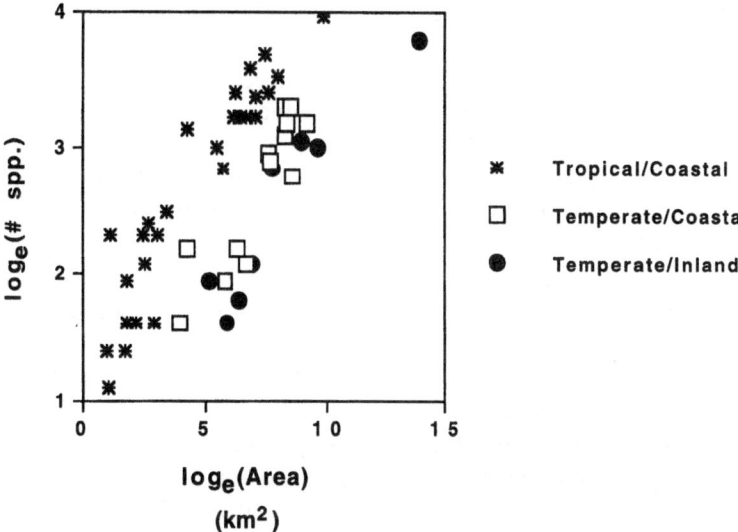

Figure 14.2. Relationship between catchment area (km²) and fish species in rivers in three climatic areas of Australia. Data were fit by Poisson regression [equations for (a) Temperate inland rivers [ln(#spp.) = 0.435 + 0.243 ln(area)] (b) Temperate coastal rivers [ln(#spp.) = 0.983 + 0.243 ln(area)], and (c) Tropical rivers [ln (#spp.) = 1.514 + 0.262 ln(area)]. (Data compiled from Cadwallader 1979, Allen and Hoese 1980, Department of Water Resources 1989, Cadwallader and Lawrence 1990, Hortle and Pearson 1990, Bishop and Forbes 1991, Pusey et al. 1995, Pusey and Kennard 1996.)

data for Australian fishes to illustrate this quantitative approach. Figure 14.2 shows how fish species diversity in three groups of rivers in Australia increase as the size of the catchment increases. One group of rivers is in tropical northern Australia (predominantly Queensland), whereas the other two groups are temperate rivers in southern mainland Australia, including coastal Victorian rivers and inland rivers that flow into the Murray-Darling River system. Despite the fact that many of these rivers have been extensively damaged by land-use practices in their catchments, by the construction of barriers, and by changes in the flow regime, the graphs reveal a strong relationship between richness and catchment size and, presumably, habitat quantity. Because count data for species richness against area usually follow a Poisson rather than a normal distribution, we used Poisson regression (e.g., Dobson 1990; kindly carried out by R. MacNally) to generate a best fit model for each of the three regions. The slopes of the Poisson regressions (Fig. 14.2) indicate the rate at which species are added to streams in the three regions as a function of catchment size. Tropical streams add species faster than temperate coastal streams, which add species faster than temperate inland streams. This pattern reveals an underlying gradient in annual precipitation, with tropical streams receiving 1200–1600 mm, temperate coastal areas 800–1000 mm, and temperate inland areas 300–600 mm.

Based on the Poisson models, we would expect that a change in climate that reduces annual precipitation would reduce species richness within a region by reducing effective habitat area. For Australia, changes in precipitation have been projected by the Commonwealth Scientific and Industrial Research Organization (CSIRO 1996) for the year 2070 using the CO_2 emission scenarios of Houghton et al. (1996). Two types of general circulation models were used to make the projections: "Slab" models assume a simple mixed-layer ocean formulation and "coupled" models incorporate a fully dynamic ocean formulation (e.g., Whetton et al. 1996). For the northern hemisphere both sets of GCMs agree, but for the southern hemisphere there are serious discrepancies between the two because slab models do not allow the Southern Ocean to absorb greenhouse-generated heat. For the tropical coastal, temperate coastal, and temperate inland regions, the coupled models predict annual precipitation declines of −10%, −10%, and −20%, respectively, whereas slab models predict +25%, −5%, and −10%, respectively. Thus, for five of six CSIRO projections, the models indicate that effective catchment size will decline in Australian streams and rivers, with attendant reductions in fish diversity likely being due to the loss of habitat. Projected increases in precipitation, however, would not necessarily lead to realized increases in fish richness in catchments with expanded habitat, unless assumptions are made about the accessibility of any new habitat to dispersing fish species.

It is worth noting that small streams in all three regions considered here appear to be "unsaturated" (i.e., they contain fewer species than expected from the Poisson regression). The most likely explanation for this "undersaturation" is that the small streams are generally shallow and have limited refugia in the face of the high flow variability commonly found in these small streams (Lake, personal observation). Thus, hydrologic variability probably superimposes an additional habitat constraint on species richness in small streams versus larger rivers in Australia, where a greater volume of water buffers the system against drying (thereby skewing the species–area relation). Hydrologic variability is known to influence many aspects of stream ecosystem structure and function (Poff and Allan 1995; see Poff et al. 1997); therefore, understanding how the timing and magnitude of river runoff will change in response to climate change are clearly important aspects of predicting potential species losses. This is especially true in regions or biomes that are prone to drying (e.g., deserts, savanna, grasslands).

Another example of quantifying changes in lotic species richness in response to climate change is shown in Figure 14.3. Fausch et al. (1984) derived relationships between maximum regional species richness and stream order, which is proportional to log catchment area (Leopold et al. 1964). Figure 14.3 shows such relationships for 10 streams and rivers in the Mississippi River Basin from which data have been assembled to date. Two groups of river systems can be identified based on the magnitude of species

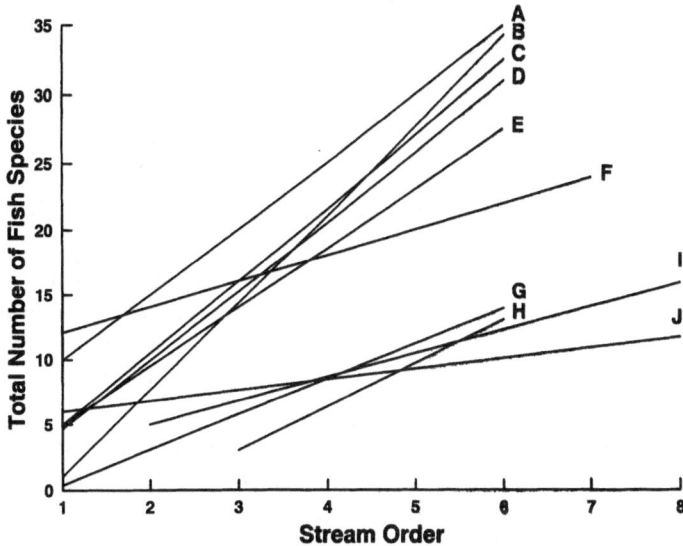

Figure 14.3. Maximum species richness lines for 10 fluvial systems in the Mississippi River basin (see Fausch et al. 1984, Schrader 1989, Fausch et al. 1990 for details). Basins are: (A) Raisin River, MI; (B) Red River, KY; (C) Embarras River, IL; (D) Wisconsin portion of the St. Croix River basin; (E) Chicago area rivers, IL; (F) Rock River region, IL and WI; (G) Salt Creek, NE; (H) James River, ND and SD; (I) South Platte River, CO; (J) Arkansas River, CO.

additions in the downstream direction as catchment size increases. The more speciose systems (upper cluster in Fig. 14.3) are in the eastern half of the Mississippi drainage (Temperate Deciduous Forest), where annual and seasonal precipitation is high. The less speciose systems are in the more arid, western half of the Mississippi basin (Great Plains). Because all these stream systems have been connected over geologic time, they share the same potential species pool. The contemporary differences in species–area relations demonstrated for these two groups of systems, therefore, reflect the response of the species pool to regional climatic differences over evolutionary time (Fausch et al. 1984; Cross et al. 1986). Thus, climate change in North America that shifts forests to plains with concomitant reductions in runoff would be predicted generally to reduce species diversity across all stream sizes. The same general pattern illustrated in Figure 14.3 exists for streams and rivers in other regions of the world (e.g., Australian tropical coastal streams are more speciose than are temperate inland streams of similar size—Fig. 14.2). The IMAGE predictions for altered precipitation in most of the Mississippi River Basin of North America range from −10 to +10 cm/year. These modest changes do not allow us to predict fish species responses across streams in the basin. In the future, more detailed projections of altered precipitation across this region could allow predictions of species losses, using an approach similar to that described earlier for Australian streams.

Prospectus

We are confident qualitatively that global climate and land-use changes will cause the extinctions of some fish and other riverine species. Changes in regional templates induced by climate and land-use changes will affect species in each of the climatic regions and biomes considered here. Modifications of thermal regimes, hydrologic regimes, and patterns of floodplain inundation will certainly reduce fluvial habitat and, hence, species diversity.

The relative contributions of hydrologic, temperature, and land-use changes to future declines in fish diversity appear to vary along the two major gradients we considered in this chapter: temperature and precipitation. When streams and rivers are arrayed from tropical to polar regions (thermal gradient), the importance of land use appears to decrease, primarily because most developing countries, with few land-use controls and high human population growth rates, are found in the tropics, and these countries have much greater potential for far-reaching land-use changes. By contrast, the relative importance of temperature change increases with increasing altitude or latitude (concordant with temperature-change projections). Of course, land-use changes can also be important in the temperate areas (e.g., logging in montane and boreal forests). The importance of hydrologic changes is evident along precipitation gradients, irrespective of latitude. In both tropical and temperate zones arid regions appear to be more sensitive to reductions in precipitation and runoff than do the wetter regions. The ultimate impact of altered hydrologic regimes will, of course, also depend on associated land-use change.

As climate and land-use changes progress over the coming years on local, regional, and global scales, aquatic habitat will become increasingly fragmented and isolated. Much of the reduction in aquatic species diversity is already occurring, as contemporary land- and water-use practices alter the riverine landscape by modifying physical, chemical, and biological conditions, and as water resources management creates barriers to species movement through river corridors. Species that are unable to move along river corridors to find suitable environmental conditions are under increasing threat of local extinction. We expect many regional extinctions of entire species to result from the cumulative losses of isolated and stressed populations. As indicated by our analyses, these losses will be particularly important to species that have small geographic range, ecological specialization, and migratory behavior—characteristics possessed by a significant fraction of existing species. Preventing further loss of sensitive fish (and other aquatic) species, therefore, requires managers to view streams and rivers as entire river basins, in and through which aquatic species are able to move to suitable local habitats (refugia) as regional environmental conditions change.

References

Allan JD, Flecker AS (1993) Biodiversity conservation in running waters. BioScience 43:32–43.

Allen GR, Hoese DF (1980) A collection of fishes from the Jardine River, Cape York Peninsula, Australia. Journal of the Royal Society of Western Australia 63:53–61.

Angermeier PL (1995) Ecological attributes of extinction-prone species: loss of freshwater fishes of Virginia. Conservation Biology 9:143–158.

Angermeier PL, Schlosser IJ (1995) Conserving aquatic biodiversity: beyond species and populations. American Fisheries Society Symposium 17:402–414.

Angermeier PL, Winston MR (1998) Local versus regional influences on local diversity in stream fish communities of Virginia. Ecology 79:911–927.

Arnell NW (1996) *Global Warming, River Flows, and Water Resources.* John Wiley and Sons, New York. 224 pp.

Balon EK (1975) Reproductive guilds of fishes: a proposal and definition. Journal of the Fisheries Research Board of Canada 32:821–864.

Baltz DM, Moyle PB, Knight NJ (1982) Competitive interactions between benthic stream fishes, riffle sculpin, *Cottus gulosus,* and speckled dace, *Rhinichthys osculus.* Canadian Journal of Fisheries and Aquatic Sciences 39:1502–1511.

Becker GC (1983) *Fishes of Wisconsin.* University of Wisconsin Press, Madison.

Behnke RJ (1992) *Native Trout of Western North America.* American Fisheries Society Monograph 6. Bethesda, MD.

Bell KNI, Brown JA (1995) Active salinity choice and enhanced swimming endurance in a 0 to 8-d-old larvae of diadromous gobies, including *Sicydium punctatum* (Pisces) in Dominica, West Indies. Marine Biology 121:409–417.

Benke AC (1990) A perspective on America's vanishing streams. Journal of the North American Benthological Society 9:77–88.

Bestgen KR, Platania SP (1990) Extirpation of *Notropis simus simus* (Cope) and *Notropis orca* Woolman (Pisces: Cyprinidae) from the Rio Grande in New Mexico, with notes on their life history. Occasional Paper, Museum of Southwestern Biology, University of New Mexico 6:1–8.

Bestgen KR, Platania SP (1991) Status and conservation of the Rio Grande silvery minnow, *Hybognathus amarus.* Southwestern Naturalist 36:225–232.

Bishop KA, Forbes MA (1991) The freshwater fishes of northern Australia. In: Haynes CD, Ridpath MG, Williams MAJ (eds) *Monsoonal Australia: Landscape, Ecology and Man in the Northern Lowlands,* pp. 79–107. A.A. Balkema, Rotterdam.

Briggs JC (1986) Introduction to the zoogeography of North American fishes. In: Hocutt CH, Wiley EO (eds) *The Zoogeography of North American Freshwater Fishes,* pp. 1–16. John Wiley and Sons, New York.

Brooks DR, McLennan DA (1993) Historical ecology: examining phylogenetic components of community evolution. In: Ricklefs RE, Schluter D (eds) *Species Diversity in Ecological Communities: Historical and Geographical Perspectives,* pp. 267–280. University of Chicago Press, Chicago.

Brown RS, Stanislawski SS, Mackay EC (1993) Effects of frazil ice on fish. In: Prowse TD (ed) *Proceedings of the Workshop on Environmental Aspects of River Ice,* pp. 261–278. National Hydrology Research Institute, Saskatoon, Saskatchewan, NHRI Symposium Series No. 12.

Bull JJ (1983) *Evolution of Sex-Determining Mechanisms.* Benjamin/Cummings, Menlo Park.

Burn DH, Arnell NW (1993) Synchronicity in global flood responses. Journal of Hydrology 144:381–404.

Cadwallader PL (1979) Distribution of native and introduced fish in the Seven Creeks River system, Victoria. Australian Journal of Ecology 4:361–385.

Cadwallader PL, Lawrence B (1990) Fish. In: Mackay N, Eastburn D (eds) *The Murray*, pp. 317–335. Murray Darling Basin Commission, Canberra.

Chapman DW (1966) Food and space as regulators of salmonid populations in streams. American Naturalist 100:345–357.

Cooper SD (1996) Rivers and streams. In: McClanahan TR, Young TP (eds) *East African ecosystems and their conservation*, pp. 133–170. Oxford, New York.

Cooper SD, Diehl S, Kratz K, Sarnelle O (1998) Implications of scale for patterns and processes in stream ecology. Australian Journal of Ecology 23:27–40.

Craig PC (1989) An introduction to anadromous fishes in the Alaskan Arctic. Biological Papers of the University of Alaska 24:27–54.

Cross FB, Moss RE (1987) Historic changes in fish communities and aquatic habitats in plains streams of Kansas. In: Matthews WJ, Heins DC (eds) *Community and Evolutionary Ecology of North American Stream Fishes*, pp. 155–165. University of Oklahoma Press, Norman.

Cross FB, Mayden RL, Stewart JD (1986) Fishes in the western Mississippi drainage. In: Hocutt CH, Wiley EO (eds) *The Zoogeography of North American Freshwater Fishes*, pp. 363–412. John Wiley and Sons, New York.

CSIRO Climate Impact Group (1996) Climate change scenarios for the Australian Region. CSIRO Division of Atmospheric Research, Aspendale.

Department of Water Resources, Victoria (1989) *Water Victoria. An Environmental Handbook*. Victorian Government Printing Office, Melbourne.

De Staso J III, Rahel FJ (1994) Influence of water temperature on interactions between juvenile Colorado River cutthroat trout and brook trout in a laboratory stream. Transactions of the American Fisheries Society 123:289–297.

Dobson AJ (1990) *An Introduction to Generalized Linear Models*. Chapman & Hall, London.

Dunne T (1979) Sediment yield and land use in tropical catchments. Journal of Hydrology 42:281–300.

Dunne T, Leopold LB (1978) *Water in Environmental Planning*. W.H. Freeman and Co., San Francisco.

Eaton JG, Scheller RM (1996) Effects of climate warming on fish thermal habitat in streams of the United States. Limnology and Oceanography 41:1109–1115.

Ely LL, Enzel Y, Baker VR, Cayan DR (1993) A 5000-year record of extreme floods and climate change in the southwestern U.S. Science 262:410–412.

Erdman DS (1984) Exotic fishes in Puerto Rico. In: Courtenay WR Jr., Stauffer JR Jr. (eds) *Distribution, Biology, and Management of Exotic Fishes*, pp. 162–176. The John Hopkins University Press, Baltimore.

Fausch KD (1989) Do gradient and temperature affect distributions of, and interactions between, brook charr (*Salvelinus fontinalis*) and other resident salmonids in streams? Physiol. Ecol. Japan, Spec. Volume 1:303–322.

Fausch KD, Bramblett RG (1991) Disturbance and fish communities in intermittent tributaries of a western Great Plains river. Copeia 1991:659–674.

Fausch KD, Bestgen KR (1997) Ecology of fishes indigenous to the central and southwestern Great Plains. In: Knopf FL, Samson FB (eds) *Ecology and Conservation of Great Plains vertebrates*, pp. 131–166. Springer-Verlag, New York.

Fausch KD, Karr JR, Yant PR (1984) Regional application of an index of biotic integrity based on stream fish communities. Transactions of the American Fisheries Society 113:39–55.

Fausch KD, Nakano S, Ishigaki K (1994) Distribution of two congeneric charrs in streams of Hokkaido Island, Japan: considering multiple factors across scales. Oecologia 100:1–12.

Firth P, Fisher SG (eds) (1992) *Global Climate Change and Freshwater Ecosystems.* Springer-Verlag, New York.

Fitzsimons JM, Zink RM, Nishimoto RT (1990) Genetic variation in the Hawaiian stream goby, *Lentipes concolor.* Biochemical Systematics and Ecology 18:81–83.

Fowler AM, Hennessy KJ (1995) Potential impacts of global warming on the frequency and magnitude of heavy precipitation. Natural Hazards 11:283–303.

Gaston KJ, Blackburn TM (1996) Conservation implications of geographic range size-body size relationships. Conservation Biology 10:638–646.

Goulding M, Carvalho ML, Ferreira EG (1988) *Río Negro: Rich Life in Poor Water.* SPB Academic, The Hague. 200 pp.

Granger O (1991) Climate change interactions in the greater Caribbean. Environmental Professional 13:43–58.

Greenwood PH (1958) *The Fishes of Uganda.* The Uganda Society, Kampala.

Grimm NB (1992) Implications of climate change for stream communities. In: Kingsolver JG, Kareiva PM, Huey RB (eds) *Biotic Interactions and Global Change*, pp. 293–314. Sinauer Assoc. Inc., Sunderland, MA.

Hanson K, Maul GA (1989) Analysis of the historical meteorological record at Key West, Florida (1851–1986) for evidence of trace gas-induced climate change. In: Maul GA (ed) Implications of climatic changes in the wider Caribbean region, pp. 63–71. UNEP/IOC Regional Task Team Report.

Harrison AD, Rankin JJ (1976) Hydrobiological studies of Eastern Lesser Antillean Islands. II. St Vincent: Freshwater fauna—its distribution, tropical river zonation, and biogeography. Archiv für Hydrobiologie/Supplement. 50(2/3):275–311.

Hayden BP (1988) Flood climates. In: Baker VR, Kochel RC, Patton PC (eds) *Flood Geomorphology*, pp. 13–26. John Wiley and Sons, New York.

Hinch SG, Healey MC, Diewert RE, Thomson KA, Hourston R, Henderson MA, et al. (1995) Potential effects of climate change on marine growth and survival of Fraser River sockeye salmon. Canadian Journal of Fisheries and Aquatic Sciences 52:2651–2659.

Hirschboeck KK (1988) Flood hydroclimatology. In: Baker VR, Kochel RC, Patton PC (eds) Flood geomorphology, pp. 27–49. John Wiley and Sons, New York.

Hortle KG, Pearson RG (1990) Fauna of the Annan river system, far north Queensland, with reference to the impact of tin mining. I. Fishes. Australian Journal of Marine and Freshwater Research 41:677–694.

Houghton JT, Meira Filho LG, Callander BA, Harris N, Kattenberg A, Maskell K (1996) *Climate Change 1995: The Science of Climate Change.* Cambridge University Press, Cambridge, UK.

Huguney B, Paugy D (1995) Unsaturated fish communities in African rivers. American Naturalist 146:162–169.

Hynes HBN (1971) Zonation of the invertebrate fauna in a West Indian stream. Hydrobiologia 38:1–8.

Irons JG III, Oswood MW (1997) Organic matter dynamics in 3 subarctic streams of interior Alaska, USA. Journal of the North American Benthological Society 16:23–28.

Jenkins RE, Burkhead NM (1994) The freshwater fishes of Virginia. American Fisheries Society, Bethesda, MD.

Johnson WC (1994) Woodland expansion in the Platte River, Nebraska: patterns and causes. Ecological Monographs 64:45–84.

Keleher CJ, Rahel FJ (1996) Thermal limits to salmonid distributions in the Rocky Mountain region and potential habitat loss due to global warming: a geographic information system (GIS) approach. Transactions of the American Fisheries Society 125:1–13.

Kenny JS (1995) Views from the bridge: a memoir of the freshwater fishes of Trinidad. Trinprint, Ltd., Barataria, Trinidad and Tobago. 98 pp.

Kinzie RA III (1988) Habitat utilization by Hawaiian stream fishes with reference to community structure in oceanic island streams. Environmental Biology of Fishes 22:179–192.

Knox JC (1985) Responses of floods to Holocene climatic change in the Upper Mississippi Valley. Quaternary Research 23:287–300.

Knox JC (1993) Large increases in flood magnitude in response to modest changes in climate. Nature 361:430–432.

Kramer DL, Lindsey CC, Moodie GEE, Stevens ED (1978) The fishes and the aquatic environment of the central Amazon basin, with particular reference to respiratory patterns. Canadian Journal of Zoology 56:717–729.

Kruse CG, Hubert WA, Rahel FJ (1997) Geomorphic influences on the distribution of Yellowstone cutthroat trout in the Absaroka Mountains, Wyoming. Transactions of the American Fisheries Society 126. (In press).

Labbe TR, Fausch KD (2000) Dynamics of intermittent stream habitat regulate persistence of a threatened fish at multiple scales. Ecological Applications 10:1774–1791.

Lake PS (1982) The relationships between freshwater fish distribution, stream drainage area and stream length of some streams in south-east Australia. Bulletin of the Australian Society for Limnology 8:31–37.

Lake PS, Palmer MA, Biro P, Cole J, Covich AP, Dahm C, Gibert J, Goedkoop W, Martens K, Verhoeven J (2000) Global change and the biodiversity of freshwater ecosystems: Impacts on linkages between above-sediment and sediment biota. BioScience 50:1099–1107.

Larson GL, Moore SE (1985) Encroachment of exotic rainbow trout into stream populations of native brook trout in the southern Appalachian Mountains. Transactions of the American Fisheries Society 114:195–203.

Lehtonen H (1996) Potential effects of global warming on northern European freshwater fish and fisheries. Fisheries Management and Ecology 3:59–71.

Leopold LB, Wolman MG, Miller JP (1964) Fluvial Processes in Geomorphology. W.H. Freeman and Co., San Francisco.

Lettenmaier DP, Gan TY (1990) Hydrologic sensitivities of the Sacramento-San Joaquin river basin, California, to global warming. Water Resources Research 26:69–86.

Lowe-McConnell RH (1987) Ecological Studies in Tropical Fish Communities. Cambridge University Press, Cambridge, UK.

Lowe-McConnell RH (1975) Fish Communities in Tropical Freshwaters. Longman, London.

Maciolek JA (1984) Exotic fishes in Hawaii and other islands of Oceania. In: Courtenay WR Jr., Stauffer JR Jr. (eds) Distribution, Biology, and Management of Exotic Fishes, pp. 131–161. The John Hopkins University Press, Baltimore.

Maciolek JA, Ford JI (1987) Macrofauna and environment of the Nanpil-Kiepw River, Ponape, Eastern Caroline Islands. Bulletin of Marine Science 41:623–632.

Matthews WJ (1987) Physicochemical tolerance and selectivity of stream fishes as related to their geographic ranges and local distributions. In: Matthews WJ, Heins DC (eds) Community and Evolutionary Ecology of North American Stream Fishes, pp. 111–120. Univeristy of Oklahoma Press, Norman.

Matthews WJ (1988) North American prairie streams as systems for ecological study. Journal of the North American Benthological Society 7:387–409.

Matthews WJ, Zimmerman EG (1990) Potential effects of global warming on native fishes of the southern Great Plains and the southwest. Fisheries 15(6):26–32.

McAllister DE, Platania SP, Schueler FW, Baldwin ME, Lee DS (1986) Ichthyofaunal patterns on a geographic grid. In: Hocutt CH, Wiley EO (eds) The Zoogeography of North American Freshwater Fishes, pp. 17–51. John Wiley and Sons, New York.

MacLean R, Oswood MW, Irons JG III, McDowell WH (1999) The effect of permafrost on stream biogeochemistry: a case study of two streams in the Alaskan (USA) taiga. Biogeochemistry 47:237–265.

McPhail JD, Lindsey CC (1970) Freshwater fishes of northwestern Canada and Alaska. Fisheries Research Board of Canada, Bulletin 173, Ottawa.

Meisner JD (1990) Effect of climate warming on the southern margins of the native range of brook trout, *Salvelinus fontinalis*. Canadian Journal of Fisheries and Aquatic Sciences 47:1065–1070.

Mertes LAK (1997) Documentation and significance of the perirheic zone on inundated floodplains. Water Resources Research 33:1749–1762.

Meyer JL, Sale MJ, Mulholland PJ, Poff NL (1999) Impacts of climate change on aquatic ecosystem functioning and health. Journal of the American Water Resources Association 35:1373–1386.

Milner AM, Irons JG III, Oswood MW (1997) The Alaskan landscape: An introduction for limnologists. In: Milner AM, Oswood MW (eds) *Freshwaters of Alaska: Ecological Ssyntheses*, pp. 1–44. Springer-Verlag, New York.

Minckley WL, Deacon JE (eds) (1991) *Battle Against Extinction: Native Fish Management in the American West*. University of Arizona Press, Tucson.

Minshall GW (1988) Stream ecosystem theory: a global perspective. Journal of the North American Benthological Society 7:263–288.

Mkanda FX (1996) Potential impacts of future climate change on Nyala *Tragelaphus angasi* in Lengwe National Park, Malawi. Climate Research 6:157–164.

Moore GA (1944) Notes on the early life history of *Notropis girardi*. Copeia 1944: 209–214.

Morrow JE (1980) *The Freshwater Fishes of Alaska*. Alaska Northwest Publishing Company, Anchorage.

Moyle PB (1986) Fish introductions into North America: patterns and ecological impact. In: Mooney HA, Drake JA (eds) *Ecology of Biological Invasions of North America and Hawaii*, pp. 27–43. Springer-Verlag, New York.

Moyle PB, Herbold B (1987) Life-history patterns and community structure in stream fishes of western North America: comparisons with eastern North America and Europe. In: Matthews WJ, Heins DC (eds) *Community and Evolutionary Ecology of North American Stream Fishes*, pp. 23–32. University of Oklahoma Press, Norman, OK.

Murphy ML, Milner AM (1997) Alaska timber harvest and fish habitat. In: Milner AM, Oswood MW (eds) *Freshwaters of Alaska: Ecological Syntheses*, pp. 229–263. Springer-Verlag, New York.

Myers N (1992) Synergisms: joint effects of climate change and othe forms of habitat destruction. In: Peters RL, Lovejoy TE (eds) *Global Warming and Biological Diversity*, pp. 344–354. Yale University Press, New Haven.

Myneni RB, Keeling CD, Tucker CJ, Asrar G, et al. (1997) Increased plant growth in the northern high latitudes from 1981 to 1991. Nature 386:698–702.

Nakano S, Kitano F, Maekawa K (1996) Potential fragmentation and loss of thermal habitats for charrs in the Japanese archipelago due to climatic warming. Freshwater Biology 36:711–722.

Nehlsen W, Williams JE, Lichatowich JA (1991) Pacific salmon at the crossroads: stocks at risk from California, Oregon, Idaho, and Washington. Fisheries 16(2):4–21.

Neilson RP, Marks D (1994) A global perspective of regional vegetation and hydrologic sensitivities from climatic change. Journal of Vegetation Science 5:715–730.

Oberdoff T, Guegan J, Hugueny B (1995) Global scale patterns of fish species richness in rivers. Ecography 18:345–352.

Oswood MW (1997) Streams and rivers of Alaska: a high latitude perspective on running waters. In: Milner AM, Oswood MW (eds) *Freshwaters of Alaska: Ecological Syntheses*, pp. 331–356. Springer-Verlag, New York.

Oswood MW, Milner AM, Irons JG III (1992) Climate change and Alaskan rivers and streams. In: Firth P, Fisher SG (eds) *Global Climate Change and Freshwater Ecosystems*, pp. 192–210. Springer-Verlag, New York.

Oswood MW, Irons JG III, Milner AM (1995) River and stream ecosystems of Alaska. In: Cushing CE, Cummins KW, Minshall GW (eds) *River and Stream Ecosystems. Volume 22, Ecosystems of the World*, pp. 9–32. Elsevier, Amsterdam.

Page LM, Burr BM (1991) *A Field Guide to Freshwater Fishes: North America North of Mexico.* Houghton Mifflin, Boston.

Peters RL (1992) Conservation of biological diversity in the face of climate change. In: Peters RL, Lovejoy TE (eds) *Global Warming and Biological Diversity*, pp. 15–30. Yale University Press, New Haven.

Pittock AB, Dix MR, Hennessy KJ, Jackett D, Katzfey JJ, McDougall TJ, et al. (1996) Progress towards climate change scenarios for the southwest Pacific. Weather and Climate 15:21–46.

Poff NL (1992) Regional hydrologic response to climate change: an ecological perspective. In: Firth P, Fisher SG (eds) *Global Climate Change and Freshwater Ecosystems*, pp. 88–115. Springer-Verlag, New York.

Poff NL (1996) A hydrogeography of unregulated streams in the United States and an examination of scale-dependence in some hydrological descriptors. Freshwater Biology 36:101–121.

Poff NL, Ward JV (1989) Implications of streamflow variability and predictability for lotic community structure: a regional analysis of streamflow patterns. Canadian Journal of Fisheries and Aquatic Sciences 46:1805–1818.

Poff NL, Ward JV (1990) The physical habitat template of lotic systems: recovery in the context of historical patterns of spatio-temporal heterogeneity. Environmental Management 14:629–646.

Poff NL, Allan JD (1995) Functional organization of stream fish assemblages in relation to hydrologic variability. Ecology 76:606–627.

Poff NL, Allan JD, Bain MB, Karr JR, Prestegaard KL, Richter BD, et al. (1997) The natural flow regime: a paradigm for river conservation and restoration. BioScience 47:769–784.

Power ME, Sun A, Parker M, Dietrich WE, Wootton JT (1995) Hydraulic food-chain models: an approach to the study of food-web dynamics in large rivers. BioScience 45:159–167.

Pusey BJ, Kennard MJ (1996) Species richness and geographical variation in assemblage structure of the freshwater fish fauna of the Wet Tropics Region of northern Queensland. Marine and Freshwater Research 47:563–573.

Pusey BJ, Arthington AH, Read MG (1995) Species richness and spatial variation in fish assemblage structure in two rivers of the Wet Tropics of northern Queensland, Australia. Environmental Biology of Fishes 42:181–199.

Rabeni CF (1996) Prairie legacies—fish and aquatic resources. In: Samson FB, Knopf FL (eds) *Prairie Conservation: Preserving North America's Most Endangered Ecosystem*, pp. 111–124. Island Press, Covallo, CA.

Radtke RL, Kinzie RA III (1996) Evidence of a marine larval stage in endemic Hawaiian stream gobies from isolated high-elevation locations. Transactions of the American Fisheries Society 125:613–621.

Radtke RL, Kinzie RA III, Folsom SD (1988) Age at recruitment of Hawaiian freshwater gobies. Environmental Biology of Fishes 23:205–213.

Rahel FJ, Keleher CJ, Anderson JL (1996) Potential habitat loss and population fragmentation for cold water fish in the North Platte River drainage of the Rocky Mountains: response to climate warming. Limnology and Oceanography 41:1116–1123.

Rapport DJ, Regier JA, Hutchinson TC (1985) Ecosystem behaviour under stress. American Naturalist 125:617–640.

Reiser DW, Wesche TA (1975) In situ freezing as a cause of mortality in brown trout eggs. The Progressive Fish Culturist 41:58–60.

Resh VH, De Szalay FA (1995) Streams and rivers of Oceania. In: Cushing CE, Cummins KW, Minshall GW (eds) *Ecosystems of the World 22: River and Stream Ecosystems*, pp. 717–736. Elsevier, Amsterdam.

Resh VH, Barnes JR, Craig DA (1990) Distribution and ecology of benthic macroinvertebrates in the Opunohu River catchment, Moorea, French Polynesia. Annals de Limnologie 26:195–214.

Resh VH, Barnes JR, Benis-Steger B, Craig DA (1992) Life history features of some macroinvertebrates in a French Polynesian stream. Studies on the Neotropical Fauna and Environment 27:145–153.

Reynolds JB (1997) Ecology of overwintering fishes in Alaskan freshwaters. In: Milner AM, Oswood MW (eds) *Freshwaters of Alaska: Ecological Syntheses*, pp. 281–302. Springer-Verlag, New York.

Richey J, Nobre C, Deser C (1989) Amazon River discharge and climate variability: 1903–1985. Science 246:101–103.

Rodríguez MA, Lewis WM Jr. (1994) Regulation and stability in fish assemblages of neotropical floodplain lakes. Oecologia 99:166–180.

Schlosser IJ (1985) Flow regime, juvenile abundance, and the assemblage structure of stream fishes. Ecology 66:1484–1490.

Schrader LH (1989) Use of the index of biotic integrity to evaluate fish communities in western Great Plains streams. Unpublished ms. thesis, Colorado State University.

Scott D, Poynter M (1991) Upper temperature limits for trout in New Zealand and climate change. Hydrobiologia 222:147–151.

Scott ML, Friedman JM, Auble GT (1996) Fluvial processes and the establishment of bottomland trees. Geomorphology 14:327–339.

Scott ML, Auble GT, Friedman JM (1997) Flood dependency of cottonwood establishment along the Missouri River, Montana, USA. Ecological Applications 7:677–690.

Sheldon AL (1968) Species diversity and longitudinal succession in stream fishes. Ecology 49:193–198.

Sioli H (1984) The Amazon and its main affluents: hydrography, morphology of the river courses, and river types. In: Sioli H (ed) *The Amazon, Monographiae Biologicae, Volume 56*, pp. 127–166. Dr. W. Junk, The Hague.

Skelton PH (1988) The distribution of African freshwater fishes. In: Leveque C, Bruton MN, Ssentongo GW (eds) *Biology and Ecology of African Freshwater Fishes*, pp. 65–91. Editions de l''ORSTOM, Travaux and Documents No. 216, Paris.

Smale MA, Rabeni CF (1995) Hypoxia and hyperthermia tolerances of headwater stream fishes. Transactions of the American Fisheries Society 124:698–710.

Smith RK, Fausch KD (1997) Thermal tolerance and vegetation preference of Arkansas darter and johnny darter from Colorado plains streams. Transactions of the American Fisheries Society 126:676–686.

Southwood TRE (1977) Habitat, the templet for ecological strategies? Journal of Animal Ecology 46:337–365.

Stone CP (1989) Hawai'i's wetlands, streams, fishponds, and pools. In: Stone CP, Stone DB (eds) *Conservation Biology in Hawai'i*, pp. 125–136. University of Hawaii Press, Honolulu.

Strange EM, Moyle PB, Foin TC (1992) Interactions between stochastic and deterministic processes in stream fish community assembly. Environmental Biology of Fishes 36:1–15.

Strange EM, Fausch KD, Covich AP (1999) Sustaining ecosystem services in human-dominated watersheds: Biohydrology and ecosystem processes in the South Platte River basin. Environmental Management 24:39–54.

Swanston DN (1991) Natural processes. In: Meehan WR (ed) *Influences of Forest and Rangeland Management on Salmonid Fishes and Their Habitats*, pp. 139–179. American Fisheries Society Special Publication 19, Bethesda, MD.

Taylor CM, Miller RJ (1990) Reproductive ecology and population structure of the plains minnow, *Hybognathus placitus* (Pisces: Cyprinidae), in central Oklahoma. American Midland Naturalist 123:32–39.

Thompson PD, Rahel FJ (1996) Evaluation of depletion-removal electrofishing of brook trout in small Rocky Mountain streams. North American Journal of Fisheries Management 16:332–339.

Tonn WM (1990) Climate change and fish communities: a conceptual framework. Transactions of the American Fisheries Society 119:337–352.

Tonn WM, Magnuson JJ, Rask M, Toivonen J (1990) Intercontinental comparison of small-lake fish assemblages: the balance between local and regional processes. American Naturalist 136:345–375.

Unganai LS (1996) Historic and future climatic change in Zimbabwe. Climate Research 6:137–145.

Vanden Bossche JP, Bernacsek GM (1990) Source book for the inland fishery resources of Africa, Volume 1. Food and Agriculture Organization, Committee on Inland Fisheries in Africa, Technical Paper 18/1, Rome.

van Someren VD (1952) *The Biology of Trout in Kenya Colony.* Govt. Printer, Nairobi. 114 pp.

van Someren VD (1962) The migration of fish in a small Kenya river. Revue de Zoologie et de Botanique Africaines 66:375–393.

Van Winkle W, Rose KA, Winemiller KO, DeAngelis DL, Christensen SW, Otto RG, et al. (1993) Linking life history theory, environmental setting, and individual-based modeling to compare responses of different fish species to environmental change. Transactions of the American Fisheries Society 122:459–466.

Ward JV (1994) Ecology of alpine streams. Freshwater Biology 32:277–294.

Welcomme RL (1969) The biology and ecology of the fishes of a small tropical stream. Journal of Zoology, London 158:485–529.

Welcomme RL (1976) Some general and theoretical considerations on the fish yield of African rivers. Journal of Fish Biology 8:351–364.

Welcomme RL (1979) Fisheries ecology of floodplain rivers. Longman, London.

Welcomme RL (1985) River fisheries. Food and Agriculture Organization, Fisheries Technical Paper 262, Rome.

Welcomme RL, de Merona B (1988) Fish communities of rivers. In: Leveque C, Bruton MN, Ssentongo GW (eds) *Biology and Ecology of African Freshwater Fishes*, pp. 251–276. Editions de l"ORSTOM, Travaux and Documents No. 216, Paris.

Weller G, Lynch A, Osterkamp T, Wendler G (1996) Climate change and its effects on the physical environment of Alaska. In: Anderson P, Weller G (eds) *Preparing for an Uncertain Future: Impacts of Short- and Long-term Climate Change on Alaska*, pp. 5–13. Center for Global Change and Arctic System Research, University of Alaska Fairbanks.

Whetton P, Adamson D, Williams M (1990) Rainfall and iver flow variability in Africa, Australia and East Asia linked to El Niño-Southern Oscillation events. In: Bishop P (ed) *Lessons from Human Survival: Nature's Record from the Quaternary*, pp. 71–82. Geological Society of Australia, Sydney.

Whetton PH, England MH, O'Farrell SPO, Watterson IG, Pittock AB (1996) Global comparison of the regional rainfall results of enhanced greenhouse coupled and mixed layer ocean experiments: implications for climate change scenario development. Climatic Change 33:497–519.

Williams JE, Johnson JE, Hendrickson DA, Contreras-Balderas S, Williams JD, Navarro-Mendoza M, et al. (1989) Fishes of North America endangered, threatened, or of special concern. Fisheries 14(6):2–20.

Williams MW, Losleben M, Caine N, Greenland D (1996) Changes in climate and hydrochemical responses in a high-elevation catchment in the Rocky Mountains, USA. Limnology and Oceanography 41:939–946.

Winemiller KO (1989) Patterns of variation in life history among South American fishes in seasonal environments. Oecologia 81:225–241.

Winemiller KO (1991) Ecomorphological diversification of freshwater fish assemblages from five biotic regions. Ecological Monographs 61:343–365.

Winemiller KO (1992) Life history strategies and the effectiveness of sexual selection. Oikos 62:318–327.

Winemiller KO, Rose KA (1992) Patterns of life-history diversification in North American fishes: implications for population regulation. Canadian Journal of Fisheries and Aquatic Sciences 49:2196–2218.

Winston MR, Taylor CM, Pigg J (1991) Upstream extirpation of four minnow species due to damming of a prairie stream. Transactions of the American Fisheries Society 120:98–105.

15. Potential Biodiversity Change: Global Patterns and Biome Comparisons

Osvaldo E. Sala, F. Stuart Chapin III, and
Elisabeth Huber-Sannwald

The purpose of the exercise reported in this book was to develop biodiversity scenarios for the year 2100. The scenarios focused on 10 terrestrial biomes and freshwater ecosystems, and were based on global scenarios of changes in the environment and current understanding about the specific biome sensitivity to global change. The first step was to identify the major drivers of biodiversity change at the global scale: changes in land use, climate, N deposition, biotic exchange (the deliberate or accidental introduction of species into an ecosystem), and atmospheric CO_2. Chapters 2 and 3 described these global patterns and the models used to predict their changes for the year 2100. Next, we estimated the magnitude of change in drivers for each biome. Finally, we estimated the sensitivity of each biome to a unit change in the drivers. The expected change in biodiversity due to each driver for each biome resulted from multiplying the expected change in each driver times the sensitivity to a unit change in driver. For each biome, Chapters 4 to 14 described the general patterns of biodiversity, the expected changes in drivers, the sensitivity to changes in drivers, and the expected patterns of biodiversity change. A first attempt at synthesizing this effort of developing global biodiversity scenarios has been published (Sala et al. 2000). This final chapter synthesizes the detailed information presented in each chapter, highlights similarities and differences among biomes, and develops the global biodiversity scenarios. First, we describe broadly the patterns of biodiversity and the location of high-diversity areas within each biome. Second, we develop global

scenarios of biodiversity change by combining the biome-specific information into a common framework.

Within-Biome Patterns of Biodiversity

The biological diversity of a biome is determined by the regional matrix
of biodiversity and by centers of high species diversity, which contribute
substantially to the overall biodiversity of a biome. The causes of diversity
patterns differ among biomes depending on the relative importance of long-
term geographic isolation, rapid evolutionary change, habitat diversity, and
human-induced and natural causes of extinctions. In some areas the elimination of dominant species may be compensated by species from the high-
biodiversity areas; however, loss of diversity will be substantial in situations
where human impact is focused in these areas. Where is biodiversity concentrated in each biome?

In *arctic and alpine tundra*, high-biodiversity areas occur in favorable
sites with relatively warm temperatures (e.g., steep slopes facing toward the
equator or at low altitudes and latitudes). The highly fragmented and isolated locations of alpine sites contribute to the significant proportion of rare
and endemic species. Endemism is especially high in the European Alps, as
well as in parts of the southern hemisphere and Himalayan alpine region.
Both global warming and land-use change will strongly affect the local
(mainly in the alpine) and regional (mainly in the arctic) diversity of the
tundra.

In the *boreal forest*, the prime diversity areas include early-successional
riparian floodplains and decaying logs in late-successional forests. The latter
support a rich beetle fauna associated with wood decay and a rich flora of
mosses and lichens. The floodplains support many migratory tropical birds.
In the Scandinavian countries, a high proportion of the beetle species is
threatened due to a long history of extensive forest harvest. Overall, there are
relatively few endemic species in the boreal zone.

The *savanna* biome is species-rich due to its long evolutionary history
without major glaciation events, allowing many taxa to co-evolve and co-exist
in space and time. The large interannual variability in climate contributes
to the diversity of life forms in savannas. Characteristic high-diversity areas
within this biome are wetlands and riparian habitats, which provide favorable
resource-rich conditions. In addition, there are rocky outcrops and ephemeral
hydromorphic vegetation types that harbor specialists with narrow ecological requirements.

The five *mediterranean-climate* regions of the globe all have high species
richness, due in part to their long complex evolutionary history without
glaciation events. In addition, early human intervention created a heterogeneous fine-grained pattern of many different land-use practices. Especially in
the Mediterranean Basin, a complex mainland and insular geography and a

high topographic variation resulted in unusually high landscape-scale diversity that explains the high level of floristic diversity and endemism in Europe.

The extreme environment of the *desert* has led to the evolution of high species diversity with a wealth of unique adaptations. The geomorphology and topographic diversity create diverse local moisture patterns and microhabitats that allow a multitude of animal and plant species to co-exist. In addition, mobile sand dunes and unique parent materials support areas of unusually high diversity of plants and animals.

Diversity varies enormously among *grassland* types, with many native grasslands such as the grasslands of the Pampas and the tallgrass prairie having levels of plant diversity as high as those typical of tropical forests, whereas others (e.g., the Patagonian grasslands) have less than 30 species of plants. Levels of plant-species diversity in grasslands are not associated with levels of diversity within other taxa. Grasslands that have high plant-species diversity may have low mammal diversity, but high bird diversity.

Temperate forests of the two hemispheres support a striking diversity in tree species, life form, structure, and function spanning a wide range of climate, geology, and evolutionary histories. Latitudinal and regional patterns in seasonality of temperature and rainfall determine the physiognomic diversity of these forests. The southern hemisphere has particularly high concentrations of endemism.

Tropical forests have particularly high species density. They support 14 of the 18 recognized global areas of highest endemism. Tropical forests also support a diversity of life forms and life histories.

Diversity in *lakes* develops during periods of geographic isolation from other water bodies, particularly in arid environments. In ancient lakes, speciation events are relatively frequent, contributing to high levels of endemism. The biodiversity of *streams* varies regionally, due to differences in history, temperature, hydrology, and geomorphology. The tropical freshwaters are a major reservoir of global fish diversity, whereas North American streams and rivers support the richest freshwater fish fauna of moist temperate regions.

Global Biodiversity Scenarios for the Year 2100

The Approach

The task of combining the effects of different drivers on the biodiversity change of different biomes required a common framework. Global models of environmental change express their output in different units [e.g., squared kilometers of land-use change or parts per million (ppm) of CO_2 in the atmosphere]. In addition, biomes differ substantially in current levels of species diversity. We used a business-as-usual scenario generated by global models of climate (Had CM2), potential vegetation (Biome 3) (Haxeltine and Prentice 1996), and land use (Scenario A from the Image 2 model) (Alcamo

1994) to estimate the change in the magnitude of the drivers of biodiversity change for each biome. We ranked the projected changes in drivers from small (value of 1) to large (value of 5) (Table 15.1A). The locations of the 10 biomes selected for this exercise were determined by aggregating the Bailey ecoregions (Bailey 1998).

The next step of the exercise was to estimate, for each biome, the impact that a unit change in each driver has on biodiversity independent of the expected change of the driver (Table 15.1B). As in the approach used to compare expected changes in drivers, we ranked the sensitivity of each biome to a unit change in driver from small (value of 1) to large (value of 5). We calculated the expected change in biodiversity due to the effect of one driver as the product of the expected change times the sensitivity.

Finally, to calculate the total change in biodiversity for each biome, we developed three different scenarios based on different assumptions regarding interactions among the drivers of biodiversity change. The first scenario was based on the assumption that there were no interactions among drivers; consequently, total biodiversity change, for each biome, was calculated as the sum of the effects of each driver. The second scenario was based on the assumption that there were antagonistic interactions among drivers; thus, the driver with the largest effect overshadowed the effects of the other drivers. In the antagonistic-interactions scenario, the biodiversity change of the biome was equal to the effect of the driver with maximum value. The third scenario assumed that there were synergistic interactions among drivers and that the effects on biodiversity of several drivers was larger than the sum of the effects of those same drivers acting independently. In the synergistic-interactions scenario, we calculated the biome change in biome diversity as the product of the effects of each driver.

Our current understanding does not allow us to predict which scenario will most closely represent biodiversity change by the year 2100. Evidence suggests that each scenario is plausible under particular circumstances. For example, the sum of the independent effects on biodiversity of elevated CO_2 and N deposition will be much smaller than the effect of enhanced CO_2 and N availability acting together (synergistic interaction). The effects of elevated CO_2 on several aspects of ecosystem functioning is amplified when combined with high N availability (Mooney et al. 1999). It is similarly very likely that the effect of biotic exchange on biodiversity will be enhanced if species introductions occur simultaneously with changes in land use or N availability. Other cases support the antagonistic-interactions scenarios. For example, it is unlikely that climate change or N deposition will further affect the biodiversity of tropical forest stands that have been cut, burned, and planted with a crop. We present the three scenarios as plausible alternatives for global biodiversity change because there is no clear evidence that any single scenario will best represent future patterns. Moreover, we expect that the shape of the interactions among drivers will differ among biomes, among drivers, and with the intensity of the change in drivers.

Table 15.1. (A) The expected changes for the year 2100 in the five major drivers of biodiversity change for the principal terrestrial biomes of the earth. (B) The impact of a large change in each driver on the biodiversity of each biome

	Arctic	Alpine	Boreal	Grassland	Savanna	Med.	Desert	N Temp.	S Temp.	Tropic
(A) Expected changes in drivers										
Land use	1.0	1.0	2.0	3.0	3.0	3.0	2.0	1.0	4.0	5.0
Climate	5.0	3.0	4.0	2.0	2.0	2.0	2.0	2.0	2.0	1.0
N deposition	1.0	3.0	3.0	3.0	2.0	3.0	2.0	5.0	1.0	2.0
Biotic exchange	1.0	1.0	2.0	3.0	3.0	5.0	3.0	3.0	2.0	2.0
Atmos CO_2	2.5	2.5	2.5	2.5	2.5	2.5	2.5	2.5	2.5	2.5
(B) Impact of a given change on diversity										
Land use	5.0	5.0	5.0	5.0	5.0	5.0	5.0	5.0	5.0	5.0
Climate	4.0	4.0	3.5	3.0	3.0	3.0	4.0	2.0	2.0	3.0
N deposition	3.0	3.0	3.0	2.0	2.0	2.0	1.0	3.0	3.0	1.0
Biotic exchange	1.0	1.0	1.0	2.0	2.0	3.0	2.0	1.5	3.0	1.5
Atmos CO_2	1.0	1.0	1.0	3.0	3.0	2.0	2.0	1.5	1.5	1.0

In this exercise, a unit change of the driver was defined for land use as conversion of 50% of land area to agriculture, for CO_2 as a 2.5-fold increase in elevated CO_2 as projected by 2100, for N deposition as 20 kg/ha/year, for climate as a 4°C, or 30% change, in precipitation, and for biotic exchange as the arrival of 200 new plant or animal species by 2100. Estimates vary from low (1) to high (5) and result from existing global scenarios of the physical environment and knowledge from experts in each biome (see text).

The Drivers

The IMAGE 2 model projects the largest changes in land use to occur in the tropical forest and southern temperate forest biomes (Table 15.1A). In contrast, biomes located in remote areas, such as the arctic and alpine tundra, which will continue to have low human density, are expected to show the least amount of land-use change. Grasslands, savannas, and mediterranean ecosystems will exhibit intermediate levels of land-use change for reasons that are specific to each biome.

The global circulation model (GCM) used in our exercise as well as GCMs included in the most current version of IPCC (Kattenberg et al. 1996) agreed with predictions of largest changes in temperature at high latitudes; consequently, we assigned the largest change in climate to the high-latitude biomes, arctic tundra and boreal forest (Table 15.1A). In contrast, tropical forest will experience the least climate change, and other biomes will show intermediate values.

Carbon dioxide mixes globally on an annual basis (Fung et al. 1987). We therefore assumed that all biomes will experience the same change in atmospheric CO_2 (Table 15.1A). Patterns of nitrogen deposition vary significantly among regions, with the highest levels occurring in Eastern North America, Western Europe, and Eastern Asia associated with the intensity of industrial and urban activities (Holland et al. 1999). We assigned the highest values for N deposition to northern temperate forests and the lowest to biomes located in regions distant from industrial areas such as the arctic tundra and southern temperate forests (Table 15.1A). Biotic exchange is driven by activities such as trade and agriculture and is therefore related to the pattern of human activity. Remote areas receive fewer exotic species than areas with intense human activity (Drake et al. 1989).

The Biome Sensitivity

The sensitivity of biodiversity in a particular biome to changes in each driver is generally poorly documented in carefully controlled experiments. These sensitivities, however, can be estimated from general principles, from ecological patterns of species distribution along gradients, and from changes in diversity that have occurred in response to a variety of human-induced environmental changes. In assigning values of biome sensitivity to each driver, authors of each chapter reviewed the available literature for their biome and consulted widely with other ecologists familiar with that biome. We then used these literature reviews and the experience of chapter authors to develop a set of "sensitivity rankings" that were consistent across biomes.

Land-use change is the driver with the largest impact on biodiversity (Sala 1995). The impact is so large and equally negative for all biomes that we assigned land-use change the maximum value in all biomes (Table 15.1B). Land-use change affects biodiversity primarily by reducing habitat availability. For example, when an area of tropical forest is logged, burned, plowed, and seeded with a soybean crop, most native plant species disappear, and the below-ground biota are drastically modified (Anderson 1995).

We expect that a given change in climate will have a larger effect in biomes characteristic of extreme environments (Table 15.1B). Climate change likely will affect biodiversity in all ecosystems but the rate of change per unit temperature change will be larger for the arctic, alpine tundra, and desert biomes that possess species narrowly adapted to extreme climatic conditions.

Nitrogen deposition will have the largest effect on ecosystems that are most limited by N (e.g., temperate and boreal forest) and the least effect on biomes that are most frequently limited by other factors [e.g., water availability (deserts) or phosphorus (tropical forests)] (Table 15.1B). N deposition affects biodiversity by changing N availability in the soil. Numerous studies reported a negative relationship between N additions and species diversity (Berendse and Elberse 1990; Huenneke et al. 1990; Tilman 1993). For example, fertilization with $27\,gN/m^2/year$ in a grassland, characteristic of the North American tallgrass prairie, resulted, after 11 years, in a 50% reduction in species richness (Tilman 1993). Changes in soil N will alter first the competitive balance of plant species by favoring species with high relative growth rate that can take advantage of this resource. If changes persist in time or space, they will result in local extinctions.

The vulnerability of different ecosystems to invasions is an issue of current debate and one that is attracting a significant research effort. The severity of climate is one of the factors that has been suggested as an important determinant of vulnerability to invasions, with more mesic environments being more vulnerable than xeric ecosystems (Rejmánek 1989). Experimental studies have reinforced the idea that high initial biodiversity may reduce vulnerability to invasions (Levine 2000). The same study, however, highlighted the role of other factors that may overshadow the effect of the original biodiversity level. We assigned the lowest sensitivity to biotic exchange to arctic, alpine tundra, boreal forest, and tropical forest, and the highest to mediterranean ecosystems (Table 15.1B).

The sensitivity of different biomes to elevated CO_2 is associated with the degree of water limitation (Mooney et al. 1991). One of the most consistent effects observed in elevated CO_2 experiments has been a reduction in stomatal conductance and a consequent increase in water-use efficiency (Jackson et al. 1994). We therefore assigned the highest sensitivity values to grasslands and savannas because they are water-limited ecosystems with a combination of functional groups with different rooting patterns, photosynthetic pathways, phenology, and woodiness (Table 15.1B). In contrast, we assigned the lowest sensitivity values to arctic, alpine, boreal forest, tropical forest, and freshwater ecosystems.

Ranking of Drivers

The exercise of developing biodiversity scenarios yielded a ranking of drivers according to their expected global effect on biodiversity for the year 2100

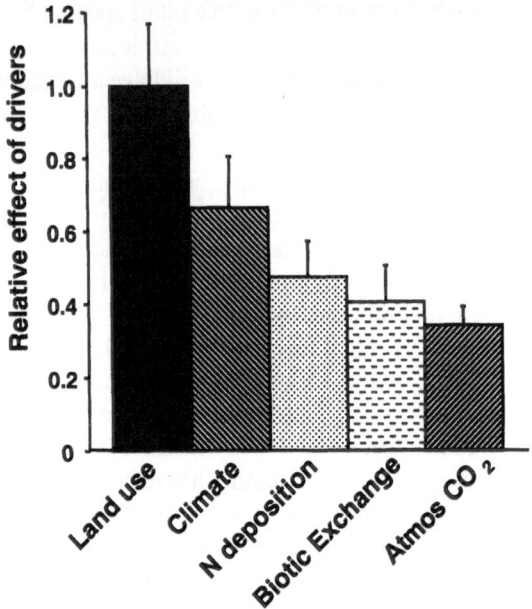

Figure 15.1. Relative effects of the major drivers of changes on biodiversity. The expected biodiversity change for each biome for year 2100 was calculated as the product of the expected change in drivers times the effect of each driver on biodiversity for each biome. Values represent the average across biomes and they are made relative to the maximum change, which resulted from change in land use. Thin bars are standard errors and represent variability among biomes. Redrawn with permission from Sala et al. (2000). Copyright 2000 American Association for the Advancement of Science.

(Fig. 15.1). Land-use change is expected to be the driver with the largest effect on biodiversity as indicated by the average effect across biomes. Land-use change will affect biodiversity by changing habitat availability that will result in local and global species extinctions. Climate change is the second most important driver primarily due to the strong effect of warming at high latitudes. N deposition, biotic exchange, and atmospheric CO_2 follow land-use and climate change in the ranking of global effects on biodiversity. Variability among biomes is maximal for land use (Fig. 15.1) due to the large variability among biomes in expected land-use change and the uniformly high sensitivity of all biomes to changes in land use. In contrast, the effect of elevated CO_2 shows small variability because CO_2 is well mixed in the atmosphere and because differences in sensitivity to CO_2 among biomes are relatively narrow (Table 15.1B).

Variation Among Biomes

Biomes differ strikingly in the expected effect that different drivers of biodiversity change will have by the year 2100 (Fig. 15.2). Tropical forests and

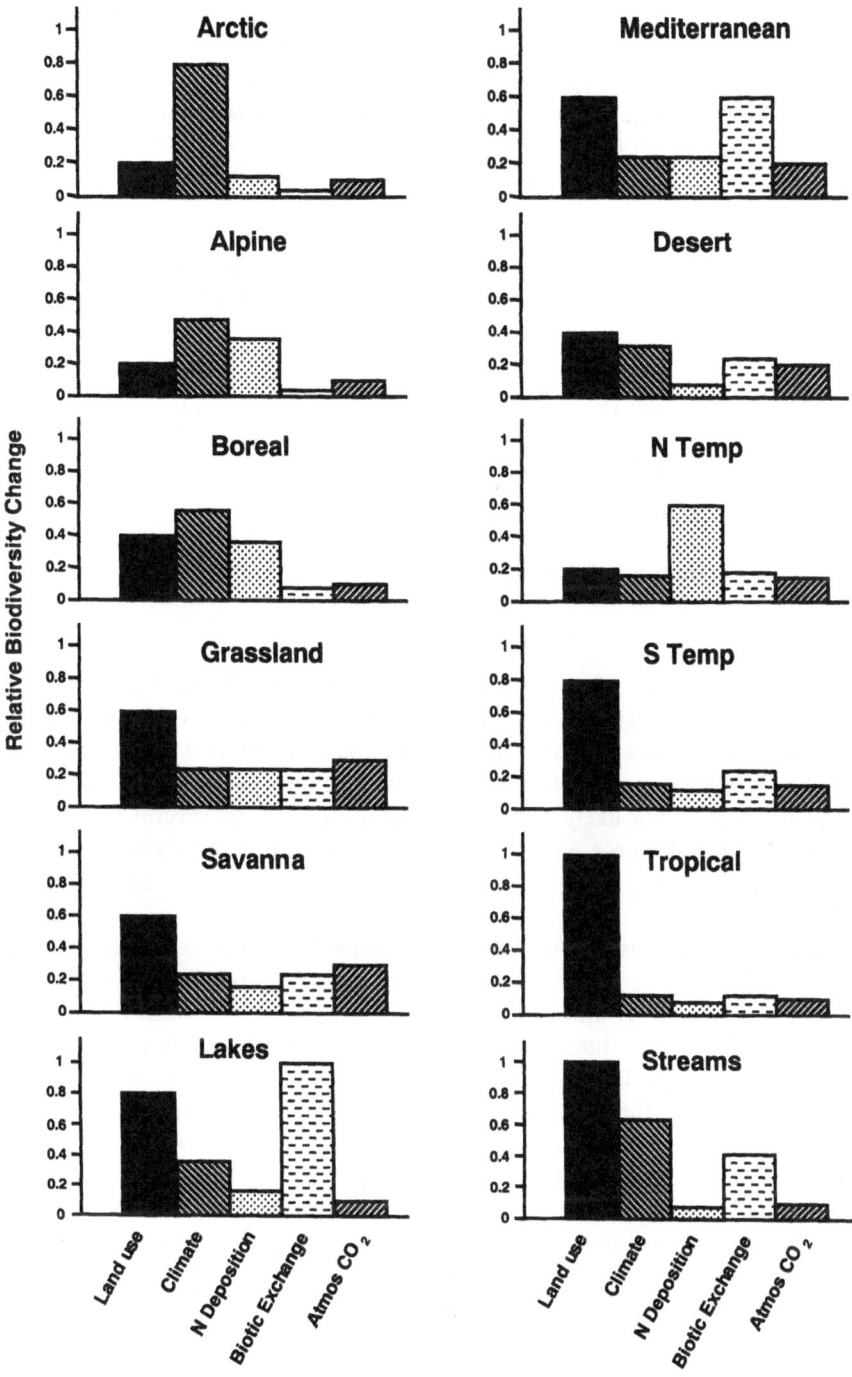

Figure 15.2. The effect of each driver on biodiversity change for each terrestrial biome and freshwater ecosystem type calculated as the product of the expected change of each driver times its effect for each terrestrial biome or freshwater ecosystem. Expected changes and impacts are specific to each biome or ecosystem type and are presented in Tables 15.1 and 15.2. Values are relative to the maximum possible value. Redrawn with permission from Sala et al. (2000). Copyright 2000 American Association for the Advancement of Science.

Table 15.2. (A) Expected changes for the year 2100, in the major drivers of biodiversity change for lakes and streams. (B) The impact of a large change in each driver on the biodiversity of each major freshwater-ecosystem type

	(A) Expected changes in drivers		(B) Impact on diversity	
	Lakes	Streams	Lakes	Streams
Land use	4.0	5.0	5.0	5.0
Atmos CO_2	2.5	2.5	1.0	1.0
	2.0	2.0	2.0	1.0
N deposition				
Climate	3.0	4.0	3.0	4.0
Biotic exchange	5.0	3.5	5.0	3.0

Methods and assumptions are as in Table 15.1 (see text).

southern temperate forests will likely be affected mostly by a single factor (i.e., land-use change) whereas the other drivers will have a relatively small effect. Arctic tundra will also likely be affected mostly by a single driver, in this case climate change. In contrast, Mediterranean ecosystems, savannas, and grasslands will likely be affected simultaneously by several factors, all with moderate-to-large effects. Finally, biomes such as northern temperate forests and deserts will likely experience low-to-moderate impacts of all drivers.

Freshwater ecosystems will likely experience large changes in biodiversity that result from changes in land use, biotic exchange, and climate (Table 15.2 and Fig. 15.2). Lakes and streams will both be affected significantly by land-use change because human activities are disproportionately concentrated around waterways. Urban areas and agriculture tend to be located on riparian zones or near them. Human activity results in increase input of nutrients, sediments, and pollutants. Biotic exchange, which results from both intentional human actions and unintentional consequences of these actions, is also relatively larger in freshwater ecosystems than in terrestrial biomes (Lodge et al. 1998). For example, fish stocking in lakes and streams has driven many native fish to extinction, and the unintentional exchange of biota in ballast water that has had large negative effects on the biota of several lakes. N deposition and elevated CO_2 will likely have smaller effects in freshwater ecosystems than in terrestrial ecosystems (Tables 15.1 and 15.2). The combined effect of all these factors currently has resulted in a larger decline of biodiversity in freshwater ecosystems than in the most strongly impacted terrestrial ecosystems (Ricciardi and Rasmussen 1999).

Streams in tropical regions will likely be affected most strongly by land-use change, whereas climate change and biotic exchange will have relatively smaller effects. Temperate streams will likely be affected equally by land-use change and biotic exchange (Richter et al. 1997; Harding et al. 1998). Finally, high-latitude streams will likely be affected the most by climate change with land-use change and biotic exchange playing a small role (Oswood et al. 1992). Streams are more sensitive than lakes to changes in climate because of the large effect of climate on run-off and its large effects on stream biodiversity (Poff et al. 1997).

We developed three scenarios of biodiversity change taking into account the effect of all drivers for the ten terrestrial biomes. The three scenarios were based on assumptions of no-interactions, antagonistic interactions, and synergistic interactions among drivers of biodiversity change (Fig. 15.3; see color insert). In the first scenario, which is based on the assumption of no interactions among drivers, mediterranean ecosystems and grasslands appear as the biomes that will experience the largest proportional change in biodiversity, mostly because of the additive effects of most drivers that all have moderate-to-high values (Fig. 15.3A). In contrast, arctic, alpine, and desert ecosystems will experience the least proportional change, mostly as a result of the low-to-moderate effect of most drivers. The range of change from the biome that will change the most to the one that will change the least is relatively narrow in this scenario with the minimum change being 60% of the maximum.

In the second scenario, which was based on the assumption of antagonistic interactions among drivers, the ranking of biomes changed drastically (Fig. 15.3B). Tropical and temperate forests and arctic ecosystems will be the biomes with the largest proportional change in biodiversity, whereas they were among the biomes with lowest change in the previous scenario. Biomes with large changes in this scenario respond to the effect of a single driver, which will be land-use change for tropical and temperate forests and climate for arctic tundra.

In the third scenario, which was based on the assumption of synergistic interactions among drivers, the ranking of biomes is similar to the ranking of the first scenario with the largest proportional change in mediterranean and grasslands biomes and the least proportional change in tropical forest and arctic tundra (Fig. 15.3C). Biomes that will be affected by multiple drivers show larger changes in this scenario than biomes that will be affected by a single factor, even when the expected change of the driver will be very large. The assumption of synergistic interaction among drivers amplifies the differences among biomes. The change expected for the biome with the least change (arctic) will be just 2% of the biome with the largest change (mediterranean ecosystems). The other two scenarios showed narrower differences between the biomes with the highest and lowest expected change.

Despite the differences that result from the three different assumptions about interactions among drivers, common patterns emerge in the comparison of all three scenarios. Mediterranean ecosystems and grasslands appear, in all three scenarios, among the biomes that will experience the greatest proportional change in biodiversity by the year 2100. Savannas, independent of the scenario chosen, appear as a biome that will experience moderate change. In all three scenarios, deserts and northern temperate forests will also experience moderate-to-low proportional change in diversity. In contrast, the expected change in tropical and southern temperate forests differs dramatically among scenarios. These two biomes range from being the biomes that will show the greatest proportional change in biodiversity in the antagonistic-interactions scenario to being among the biomes that will change the least in the no-interactions and the synergistic-interactions scenarios. All of these

Biodiversity Scenarios

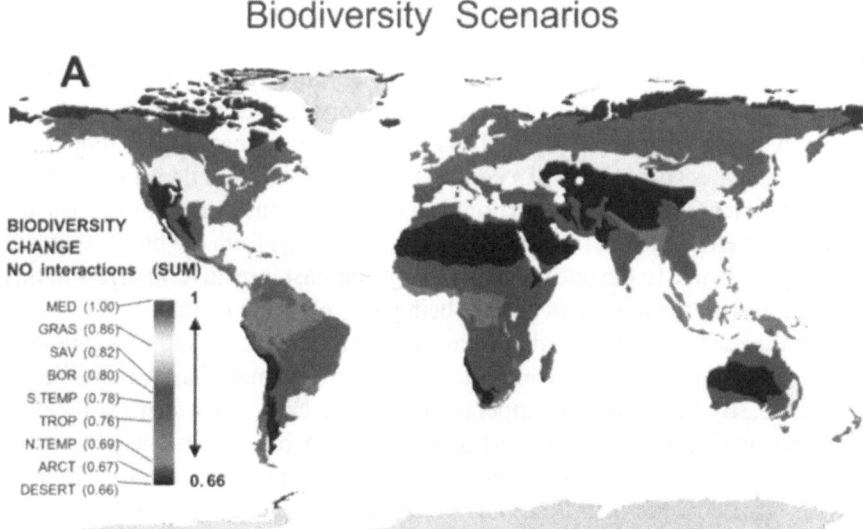

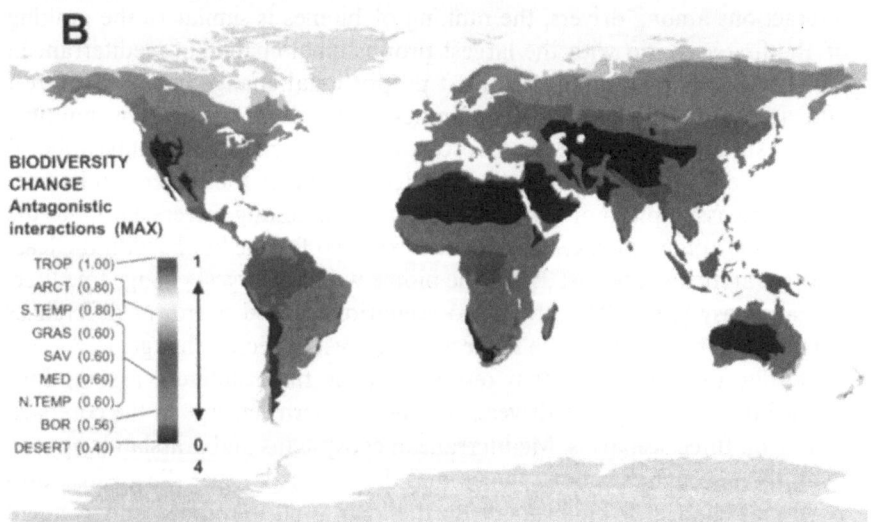

scenarios project *proportional* changes in biodiversity. Because the tropical forest has many more species than the arctic, for example, the changes in absolute number of species lost would be greater in those biomes with greatest current biodiversity.

Conclusions and Future Research Needs

Biodiversity is quite sensitive in all biomes to drivers of global change. Land-use change appears as the driver with the largest global effect on biodiversity by the year 2100; however, the importance of the different drivers varies enor-

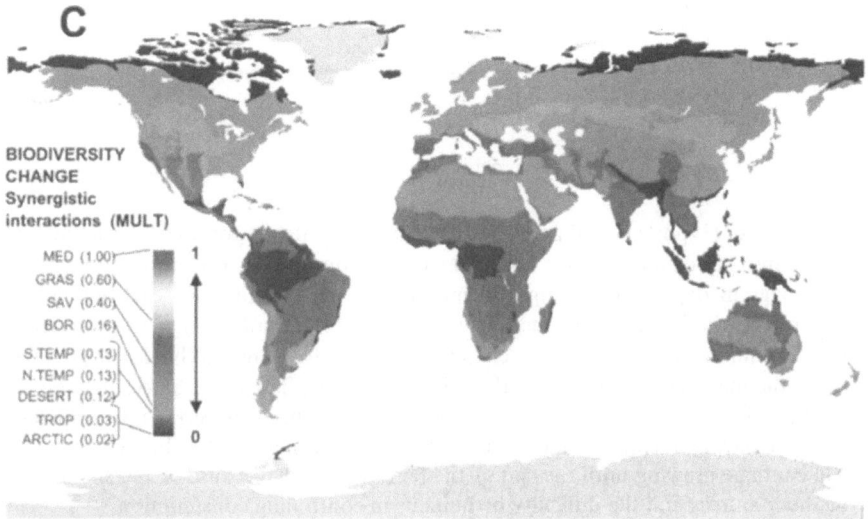

Figure 15.3. Maps of three scenarios of the expected change in biodiversity for the year 2100. Scenario A assumes that there are no interactions among drivers of biodiversity change; consequently, total change is calculated as the sum of the effects of each driver, which in turn results from multiplying the expected change in the driver for a particular biome (Table 15.1A) times the effect of the driver that is also a biome-specific characteristic (Table 15.1B). Scenario B assumes that total biodiversity change equals the change resulting from the driver that is expected to have the largest effect and is calculated as the maximum of the effects of all the drivers. Scenario C assumes synergistic interactions among the drivers; consequently, the total change is calculated as the product of the changes that result from the action of each driver. The different colors represent the expected change in biodiversity from moderate to maximum for the different biomes of the world ranked according to the total expected change. The numbers in parentheses represent the total change in biodiversity relative to the maximum value projected for each scenario. The biomes are MED (mediterranean ecosystems), GRAS (grasslands), SAV (savannas), BOR (boreal forest), S. TEMP (southern temperate forest), TROP (tropical forest), N. TEMP (northern temperate forest), ARCT (arctic ecosystems), DESERT (desert). Values for alpine, stream, and lake ecosystems are not shown. (Redrawn with permission from Sala et al. 2000. Copyright 2000 American Association for the Advancement of Science.)

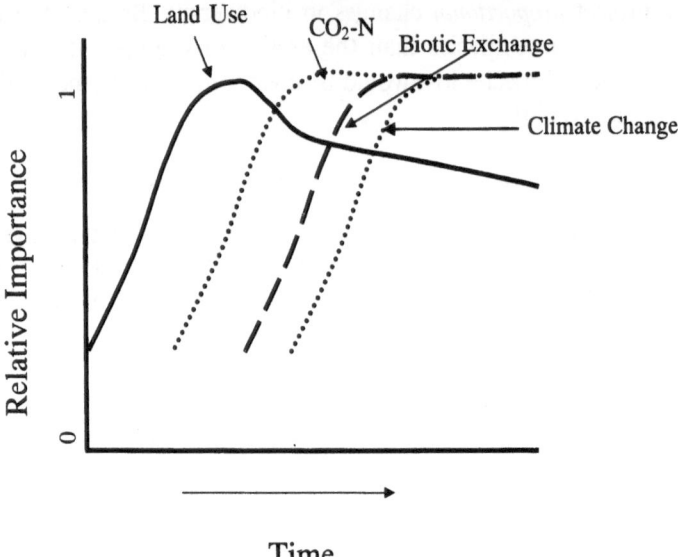

Time

Figure 15.4. Hypothetical diagram of the changes in the importance of each driver of biodiversity change relative to its maximum, from present time to the year 2100. The diagram depicts the changes in importance through time for each individual driver from the lowest (0) to the highest (1), but it does not attempt to make comparisons among drivers. Land-use change is expected to change at the fastest rate and reach the maximum sooner than the other drivers do. The importance of land-use change will likely decline when most of arable land has been converted into cropland. On the contrary, the importance of the other drivers will grow at a slower rate, but it will continue growing until the end of the period of study because of the abundance of the resources and the difficulty of humans in controlling consumption.

mously among biomes ranging from those affected by a single factor, land use or climate, to those affected by most drivers.

In addition to the idiosyncratic geographical patterns of drivers and their differential effects on biodiversity, their relative importance will likely vary with time (Sala et al. 1999). We expect that land-use change will be the driver that will have the steepest rate of change and will achieve a maximum value the soonest (Fig. 15.4). The rate of change in land use has been documented extensively. Since the beginning of the twentieth century a very large fraction of native ecosystems have been transformed into croplands and urban areas (Richards 1993). The rate of change of land use, however, will be reduced in the medium-to-short term when most of the arable land is converted into agricultural land. Changes in the composition of the atmosphere have also been clearly documented (Keeling 1986). The rate of change of greenhouse gases in the atmosphere does not seem to be limited in the medium term due to the abundance of fossil fuels (Schimel et al. 1996) and the difficulties in implementing a global policy that constrains energy consumption. Changes in climate will result from changes in the composition of the atmosphere and consequently will lag behind changes in the concentration of CO_2 (Kattenberg et al. 1996). The impact of humans on the nitrogen cycle has also been documented (Vitousek 1994), and it is unlikely to decrease in the medium term. The importance of biotic exchange will be magnified by changes in land use, CO_2, and N deposition and will consequently lag behind them. During the time period explored by this scenario, we expect that the importance of land-use change will grow quickly, but that its relative importance will decline, whereas the other drivers will continue increasing their effects on biodiversity.

This exercise highlights the sensitivity of biodiversity change to the assumptions about interactions among drivers of biodiversity change. We suggest that this is one of the most important sources of uncertainty and that decreasing the level of uncertainty will require a major interdisciplinary research effort. We hypothesize that the shape of the interactions among drivers will vary among biomes and among sets of drivers. The shape of the interaction may also change with the intensity of drivers. At low levels of change, synergistic interactions may prevail, but the antagonistic scenario may be the most realistic at high levels of any driver. Another source of uncertainty in this exercise is the future state of the drivers. Any improvement in the scenarios of change in climate, land use, and CO_2, will result in a reduction of the uncertainty associated with the biodiversity scenarios.

The scale at which the global scenarios were constructed influences the error of the exercise and limits its applicability. Scenarios were developed for ten terrestrial biomes and two types of freshwater ecosystems. Each of the biome chapters highlighted major differences within biomes that were overshadowed by the scale at which results were synthesized. Most management decisions occur at a finer scale than the one used in this study. Humans manage primarily paddocks, watersheds, and regions and struggle to manage larger units that encompass a variety of ecological, political, and social conditions.

Actions tending to mitigate biodiversity change include those actions that decrease the rate of change of global change drivers. For example, reductions in the rate of change of climate and land use would reduce the rate of change of biodiversity. Those changes should be complemented with specific actions at a finer scale and tailored for the biological, social, and economic conditions of each region. Different management plans will be required for different regions and must be based on a thorough understanding of the ecological and social characteristics of each region. The fine-scale understanding of the determinants of biodiversity change is as important as the understanding of the global patterns and will be an important challenge for the future.

References

Alcamo J (1994) *Image 2: Integrated Modeling of Global Climate Change*. Kluwer Academic Publishers, Dordrecht.

Anderson JM (1995) The soil system. In: Mooney HA, Lubchenco J, Dirzo R, Sala OE (eds) *Global Biodiversity Assessment: Section 6*, pp. 406–412. Cambridge University Press, Cambridge, UK.

Bailey RG (1998) *Ecoregions: The Ecosystem Geography of the Oceans and Continents.* Springer, New York.

Berendse F, Elberse WT (1990) Competition and nutrient availability in heathland and grassland ecosystems. In: Grace JB, Tilman D (eds) *Perspectives on Plant Competition*, pp. 93–116. Academic Press, San Diego.

Drake JA, Mooney HA, di Castri F, Groves RH, Kruger FJ, Rejmanek M, et al. (1989) *Biological Invasions: A Global Perspective.* John Wiley and Sons, Chichester.

Fung IY, Tucker CJ, Prentice KC (1987) Application of advanced very high resolution radiometer vegetation index to study atmosphere-biosphere exchange of CO_2. Journal of Geophysical Research 92D:2999–3015.

Harding JS, Benfield EF, Bolstad PV, Helfman GS, Jones EBD (1998) Stream biodiversity: the ghost of land use past. Proceedings of the National Academy of Sciences of the United States of America 95:14843–14847.

Haxeltine A, Prentice IC (1996) BIOME3: an equilibrium terrestrial biosphere model based on ecophysiological constraints, resource availability, and competition among plant functional types. Global Biogeochemical Cycles 10:693–709.

Holland EA, Dentener FJ, Braswell BH, Sulzman JM (1999) Contemporary and pre-industrial global reactive nitrogen budgets. Biogeochemistry 46:7–43.

Huenneke LF, Hamburg SP, Koide R, Mooney HA, Vitousek PM (1990) Effects of soil resources on plant invasion and community structure in California serpentine grassland. Ecology 71:478–491.

Jackson RB, Sala OE, Field CB, Mooney HA (1994) CO_2 alters water use, carbon gain, and yield for the dominant species in a natural grassland. Oecologia 98:257–262.

Kattenberg A, Giorgi F, Grassl H, Meehl GA, Mitchell JFB, Stouffer RJ, et al. (1996) Climate models—Projections of future climate. In: *Climate Change: The IPCC Scientific Assessment*, pp. 285–358. Cambridge University Press, Cambridge, UK.

Keeling CD (1986) *Atmospheric CO_2 Concentrations. Mauna Loa Observatory, Hawaii 1958–1986.* Carbon Dioxide Information Analysis Center, Oak Ridge.

Levine J (2000) Species diversity and biological invasions: relating local process to community pattern. Science 288:852–854.

Lodge DM, Stein RA, Brown K, Covich WAP, Bronmark C, Garvey JE, et al. (1998) Predicting impact of freshwater exotic species on native biodiversity. Australian Journal of Ecology 23:53–67.

Mooney HA, Drake BG, Luxmoore RJ, Oechel WC, Pitelka LF (1991) Predicting ecosystem responses to elevated CO_2 concentrations. BioScience 41:96–104.

Mooney HA, Canadell J, Chapin FS, Ehleringer J, Körner C, McMurtrie R, et al. (1999) Ecosystem Physiology Responses to Global Change. In: Walker BH, Steffen WL, Canadell J, Ingram JSI (eds) *The Terrestrial Biosphere and Global Change: Implications for Natural and Managed Ecosystems: A Synthesis of GCTE and Related Research*, pp. 141–189. Cambridge University Press, Cambridge, UK.

Oswood MW, Milner AM, Irons JG (1992) Climate change and Alaskan rivers and streams. In: Firth P, Fischer SG (eds) *Global Climate Change and Freshwater Ecosystems*, pp. 192–210. Springer-Verlag, New York.

Poff NL, Allan JD, Bain MB, Karr JR, Prestegaard KL, Richter BD, et al. (1997) The natural flow regime: a paradigm for river conservation and restoration. BioScience 47:769–784.

Rejmánek M (1989) Invasibility of plant communities. In: Drake JA, Mooney HA, di Castri F, Groves RH, Kruger FJ, Rejmánek M, et al. (eds) *Biological Invasions: A Global Perspective.* John Wiley and Sons, New York.

Ricciardi A, Rasmussen J (1999) Extinction rates of North American freshwater fauna. Conservation Biology 13:1220–1222.

Richards JF (1993) Land transformation. In: Turner BL II, Clark WC, Kates RW, Richards JF, Mathews JT, Meyer WB (eds) *The Earth as Transformed by the Human Action*, pp. 163–178. Cambridge University Press, Cambridge, UK.

Richter BD, Braun DP, Mendelson MA, Master LL (1997) Threats to imperiled freshwater fauna. Conservation Biology 11:1081–1093.

Sala OE (1995) Human-induced perturbations on biodiversity. In: Mooney HA, Lubchenco J, Dirzo R, Sala OE (eds) *Global Biodiversity Assessment, Section 5*, pp. 318–323. Cambridge University Press, Cambridge, UK.

Sala OE, Chapin FS III, Gardner RH, Lauenroth WK, Mooney HA, Ramakrishnan PS (1999) Global change, biodiversity and ecological complexity. In: Walker BH, Steffen WL, Canadell J, Ingram JSI (eds) *The Terrestrial Biosphere and Global Change: Implications for Natural and Managed Ecosystems*, pp. 304–328. Cambridge University Press, Cambridge, UK.

Sala OE, Chapin FS III, Armesto JJ, Berlow E, Bloomfield J, Dirzo R, Huber-Sanwald E, et al. (2000) Global biodiversity scenarios for the year 2100. Science 287:1770–1774.

Schimel D, Alves D, Enting I, Heimann M, Joos F, Raynaud D, et al. (1996) Radiative forcing of climate change. In: Houghton J, Meira Filho L, Callander B, Harris N, Kattenberg A, Maskell K (eds) *Climate Change 1995: The Science of Climate Change*, pp. 76–86. Cambridge University Press, Cambridge, UK.

Tilman D (1993) Species richness of experimental productivity gradients: how important is colonization limitation? Ecology 74:2179–2191.

Vitousek PM (1994) Beyond global warming: ecology and global change. Ecology 75:1861–1876.

Index

Ecological Studies

Volumes published since 1995